T0239160

Machine Learning Contests: A Guidebook

Wang He • Peng Liu • Qian Qian

Machine Learning Contests: A Guidebook

 Springer

Wang He
Commercial Algorithm Dept.
Xiaomi (China)
Beijing, China

Peng Liu
Huawei Technologies (China)
Dongguan, Guangdong, China

Qian Qian
ShuCheng Technology
Shanghai, China

ISBN 978-981-99-3722-6   ISBN 978-981-99-3723-3   (eBook)
https://doi.org/10.1007/978-981-99-3723-3

This Springer imprint is published by the registered company Springer Nature Singapore Pte Ltd.
The registered company address is: 152 Beach Road, #21-01/04 Gateway East, Singapore 189721, Singapore

Preface

Algorithm Competition Era

In 2010, Kaggle, the world's leading algorithm contest platform, introduced its first Forecast Eurovision Voting competition with a prize of $1000. In 2015, the first algorithm competition in China was held on Tianchi platform, with the title Ali Mobile Recommendation Algorithm, and the reward was 300,000 yuan, attracting more than 7000 competitors. In spite of the fact that algorithm competitions started later in China than in other countries, China has held 400 of the 1000 worldwide events in this field from 2015 onwards, witnessing an average annual growth rate of 108.8%, a total of more than 1.2 million participants, and 280-million-yuan bonus. With such a high increase rate in the number of contests organized, the technical value, business value, and the inspiration for innovations accumulated from these algorithm competitions are incredible.

Why Write

Regarding the reason for writing this book, I would like to go back to a message sent to me by Xinglu Chen, the planning editor of the Posts and Telecom Press, on Zhihu (a Chinese question-and-answer website) on April 19, 2019. She said in the message that she had read many of my articles about algorithm competitions, and she knew that I had won many algorithm contests. Therefore, she was expecting me to publish a book about algorithm competitions. From the beginning of 2018, I already have created a column and started to share articles related to such contests. Along the way, I have been continuing to produce. These pieces of writing currently have seen a total of one million views. So, when I received the invitation to write a book on algorithm competitions, I deemed it a great recognition of my effort for sharing competition knowledge as well as to my achievement and accomplishment. I, thus, readily accepted the request and decided to title it *Machine Learning Contests: A Guidebook*.

In order to complete the work, I invited Peng Liu, the champion and runner-up of many competitions in China, who is also my former teammate in a contest, to co-author. And Xinglu Chen recommended Qian to me, who was the grandmaster of the competition platform Kaggle, and was among the first batch of contestants in China. Considering that we excel in different parts, each of us is responsible for different chapters to ensure the quality of the work.

Features of the Book

The book is organized based on suggestions from many top competitors and our careful discussion. In fact, algorithm competitions cover a great variety of different aspects. However, what we try to do is to explore the most essential part of them, and then to explain actual cases by using modules of multiple domains, which is also a major feature of this work. The book is divided into the following five parts.

Part I—Half the Work, Twice the Effect. This part focuses on the general process of algorithm competitions and introduces the core content and specific work of each part of a contest. Each chapter involved in this part is with a specific practical case for better illustration.

Part II—Birds of a Feather Flock Together. This part mainly explains the problems related to user profiles. There is no denying that constructing a sound labeling system is the basis for user profiles and the key to solving relevant competition questions, such as customized recommendation and financial risk control, which all need to be based on user profiles. To enable readers to speed up their learning and grasp these competition questions, we will present a particular competition case, namely Elo Merchant Category Recommendation on the Kaggle platform.

Part III—Learn from History to Create a Bright Future. This part concentrates on time series forecasting problems, first describing the common problem-solving ideas and techniques for such issues, and then analyzing two specific cases—Global Urban Computing AI Challenge on the Tianchi platform and Corporación Favorita Grocery Sales Forecasting on Kaggle.

Part IV—Precise Delivery, Optimized Experience. Most of the businesses related to Computational Advertising are good competition topics, and this part mainly demonstrates the core technologies and businesses of Computational Advertising, including ad recall, ad ranking, and ad bidding. The real-world cases involved in this part are 2018 Tencent Advertising Algorithm Competition—Audience Lookalike Expansion, and TalkingData AdTracking Fraud Detection Challenge on Kaggle.

Part V—Listen to What You Say and Understand What You Write. This part presents common tasks and technologies related to natural language processing; the case shown is the well-known competition Quora Question Pairs on the Kaggle platform.

This book is a systematic introduction to contests in the field of algorithms, not only explaining the theory behind the practice, but also elaborating in detail the guide to scoring and necessary skills needed from various angles, using different cases.

Target Readers

The potential readers of this book will be divided into three categories.

- *Those who are interested in algorithmic contests.* Interest is the biggest driving force. In order to make algorithm competitions more interesting and diverse, this book adds a lot of expanded and exploratory content, introducing and carrying out practice from multiple directions and in many fields.
- *Those who want to study machine learning and explore practice on algorithms in depth.* Taking part in an algorithm competition is the best way for practice, which could enhance understanding of theory. This is also what the book emphasizes.
- *Those who major in Computer Science.* Machine learning or the deep learning algorithm, as a hot career in the computer industry, is worth further study. This book provides a very good explanation to real situations to help readers know how and why.

Welcome to Contact with Us

In view of our limited time and level, there are inevitably errors and mistakes; if you find any problems while reading, please feel free to contact us by sending emails to fish_ml@foxmail.com.

Beijing, China   Wang He
January 2023

Acknowledgments

Writing this book is not an easy job. I mainly used the time after work in the evening to write. I also had regular online meetings with Peng Liu and Qian to discuss the writing progress and review the content. Here I am very grateful to Peng Liu and Qian for their great contributions. Their rich competition experience is an important factor for this book being completed in such good quality.

In addition, the work cannot be done without the help of many other people. They contributed significantly to this book though they did not co-author it. I would like to express my gratitude here to them all.

I would also like to thank Xinglu Chen, Junhua Wang, and Yan Wang, editors of Turing Company of Posts and Telecom Press. They gave us sufficient time to finish the writing, and also put forward many valuable suggestions. They played a part in the successful publication of this book.

Finally, I would like to thank my wife, who gave me a lot of support and care in my process of writing my book.

Thank you!

Contents

Part V   Listen to What You Say and Understand What You Write

Part I
Half the Work, Twice the Effect

Chapter 1
Guide to the Competitions

With the advent of the Internet era and the improvement made on computer hardware capabilities, artificial intelligence (AI) has witnessed immerse growth in recent years. The Internet era has brought a large amount of information, which truly is the big data. In addition, the excellent performance of hardware has also greatly enhanced the computing power of computers. The combination of the two makes it very natural for artificial intelligence to flourish. As a traditional and highly interpretable algorithm, machine learning also plays an important role in the troika of artificial intelligence. After several discussions, the book was finally named *Machine Learning Contests: A Guidebook*, which was intended to help machine learning beginners break away from various formulas and theories that are seemingly beautiful but slightly boring through practical methods, so they can appreciate the mysteries of machine learning in practical applications, among which contests are in fact some of the most special exercises they can get.

The reason why a competition is highly recommended as a fundamental way to put machine learning into practice is that it is really an excellent approach for people to get started quickly with machine learning. For beginners in this field, they are not equipped with enough skills yet to directly work in enterprises and cope with real application scenarios, and the knowledge they gained from reading books is somewhat shallow after all. When it comes to competitions, people always think of various math, physics, and chemistry contests in high school. These events have a very high participation threshold and are available at home and abroad. In fact, a good ranking in such contests guarantees that contestants get admission to well-known colleges and universities both within and outside the country. Therefore, the mentioning of competition always brings a daunting feeling. But in recent years, the rise of artificial intelligence has spawned various algorithmic competitions that are relatively friendly and much more interesting. The current trend of the times is that people from all walks of life are seeking a way to survive the fierce competitions. Taking advantage of advanced technology is certainly a good way to complete a transformation; thus, some enterprises begin to seek the help of AI, and to solicit excellent algorithm solutions from the society; in addition, researchers in the

© The Author(s), under exclusive license to Springer Nature Singapore Pte Ltd. 2023
W. He et al., *Machine Learning Contests: A Guidebook*,
https://doi.org/10.1007/978-981-99-3723-3_1

academic field are also eager to obtain corporate scenarios and data for algorithm research, which has brought forth a variety of competition platforms. This book mainly introduces readers to the competition experience in algorithms related to machine learning.

For beginners who are interested in entering the machine learning field to engage in research or relevant work, a competition is a practical choice with great cost-effectiveness. It can be said to have no threshold at all since anyone is qualified to participate. Of course, the organizers' own employees are not allowed to take part in the competitions, and even if they take part in, they cannot compete for the ranking list. The various competitions can cover typical application scenarios in many industries, so that participants can not only get training in practice, but also experience the magic brought by the implementation of machine learning in a variety of industries, and even meet many experts in different areas during the competitions and make some like-minded friends.

This chapter mainly introduces competitions from three aspects—competition platforms, competition procedures, and competition types. Section 1.1 aims to demonstrate well-known algorithm competition platforms both at home and abroad to help readers quickly understand the competition channels. Section 1.2 illustrates the general process of completing a machine learning algorithm competition, as well as the functions and contribution of each module. More detailed content will be given in Chaps. 2–6. Section 1.3 will show readers common competition types in order to enable them to be thoroughly familiar with the applicable scenarios and industry requirements for machine learning algorithm competitions.

1.1   Competition Platforms

The various contests we participate in are published by numerous competition platforms, either large or small, such as the international Kaggle and the competition platform in China—Alibaba Cloud Tianchi. You can find the competition events you excel at or like on these platforms.

1.1.1   Kaggle

Mr. Andrew Ng, a master in the field of machine learning, once said that machine learning was mostly just mathematical statistics, the data-related feature engineering directly determined the upper limit of the model, and the algorithm just kept approaching this upper limit. In the field of machine learning, there is a very vivid analogy: the modeling process is like cooking, with the data being the ingredients, the algorithm representing the cooking procedure, and the final taste of the dish standing for the effect of the model. Watching many food videos, such as the famous food documentaries *A Bite of China* and *Once Upon a Bite* broadcast by CCTV

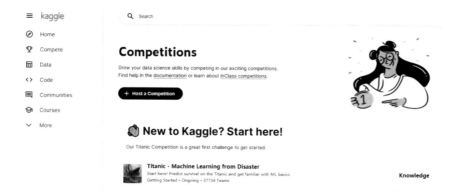

Fig. 1.1 Kaggle Page

(China Central Television), you will find that a large part of these films is about how to obtain fresh ingredients. The old saying also says that you cannot make bricks without straw, which implies the importance of ingredients. Analogous to the machine learning algorithm competition, the importance of data is self-evident. This is what the international competition platform Kaggle, which will be introduced below, describes itself as: the home of data science.

Open the home page of Kaggle's website and click Compete at the top. The interface that appears is shown in Fig. 1.1. The left sidebar has five main parts besides Home and More, namely Compete (competition unit), Data (data set), Code (code notes), Communities (community discussion), and Courses (online courses). As the world's most popular data science competition website, its front page also shows that here you can get access to all the data and code you need to complete data science work. As of October 5, 2020, there are more than 50,000 data sets and more than 400,000 public code notes. This book will focus on Kaggle's competition units. Click Compete; what is presented is a list of all the competitions in history from top to bottom. There will always be various contests in progress existing at the top. Click on one of them casually, and you will also see the relevant information of the competition, roughly Overview (general information), Data (data set), Code (code notes), Discussion (community discussion), Leaderboard (ranking list), Rules (competition rules), etc. Next, we will take the contest Microsoft Malware Prediction as an example to elaborate the main elements of a competition. The home page of the competition is shown in Fig. 1.2.

- **Overview**, which is the general information. Here is a brief overview of the competition, which consists of four parts: Description, Evaluation, Prizes, and Timeline.

  - **Description**: It is the background introduction of the competition and information about the organizers. The competition Microsoft Malware Prediction wrote that malicious software was committed to circumventing traditional security measures; Once a computer was infected by malicious software,

Fig. 1.2 Home Page of Microsoft Malware Prediction Competition

criminals would hurt consumers and businesses in many ways. Microsoft has more than 1 billion clients ranging from companies to customers, so it takes this issue very seriously and has invested a lot of money to improve safety. As part of its overall security strategy, Microsoft is working on the development technology of data science community, so as to predict whether a machine will soon be attacked by software intended to cause damage. As with the previous malicious software challenge (Malware Challenge, 2015), Microsoft is providing Kaggle with an unprecedented set of malicious software data to encourage it to make open-source progress in effective technology for predicting malicious software. Can you help protect more than 1 billion machines from damage?

–   **Evaluation**: The criteria for judging the contest and the format of the submissions will be listed here. Microsoft Malware Prediction competition uses the area under the ROC curve (area under curve, i.e. AUC) of the predicted probability and the real label as the model score, so this competition is a binary classification problem.

–   **Prizes**: It is shown here that the total bonus of this competition is 25,000 US dollars, in which 12,000 US dollars for the champion accounts for almost half. Generally speaking, 25,000 US dollars is a common amount in Kaggle prize-winning competitions, and sometimes it can reach up to 100,000 US dollars. It should be noted that this competition will have certain requirements for the winners: after the competition is over, the winners need to submit the modeling scheme doc within the specified time period; Microsoft internal employees are not allowed to take part in the contest. These are basically what most competitions will require.

–   **Timeline**: It mainly introduces the timeline of the competition. The deadline for forming teams and submission DDL are quite important. Generally, a competition lasts for 2–3 months, so it is very necessary to arrange the time reasonably.

•   **Data**. After understanding the background and tasks of the competition, participants can begin to familiarize themselves with the data. The usual data format will be in the form of a CSV wide table. There is a separate Data Description in the Data section, where the data information of all sheets is usually given, including the collection source, task description, and detailed meaning of each field. Take the competition Microsoft Malware Prediction as an example; its Data Description is as follows.

The goal of the competition is to predict the probability of a Windows machine being infected by a variety of malicious software families based on its different characteristics. The organizers combine heartbeat and threat reports collected by Windows Defender, Microsoft's endpoint protection solution, to generate a telemetry dataset containing relevant attributes and machine infections. Each row in this dataset corresponds to a machine and is uniquely identified by MachineIdentifier. HasDetections are tags for machines that indicate whether malicious software has been detected on the machine. Participants need to use the information and tag training set in the train.csv to predict the HasDetections value for each computer in the test.csv. The sampling method used to create this data set is designed to meet certain business restrictions, including user privacy and the time period during which the machine is running. Detecting malicious software is essentially a time series forecasting problem, but it has become more complicated due to the introduction of new machines, online and offline machines, machines receiving patches, machines receiving new operating systems, etc. The data set provided here has been roughly divided by time. The complexity and sampling requirements mentioned above mean that you may see that your cross-validation, public leaderboard score, and private leaderboard score are not exactly the same! In addition, this data set does not represent that in Microsoft customers' machines, because it is sampled and contains a larger proportion of malicious software.What participants should do first when getting involved in the competition is to familiarize themselves with the questions and data, which often contain many important details. Take the above topic as an example. The data looks very simple. The organizers have clearly classified the training set and the testing set, as well as the standard feature field and the label field. The task of the event is also very clear; that is, to predict whether the machine will be infected by malicious software. However, one point that cannot be ignored is mentioned in the introduction: this question is essentially a time series forecasting problem. The training set and testing set are only roughly divided according to time, and in order to highlight the malicious software, the machine has carried out a certain up-sampling of the positive sample. Therefore, this complexity and sampling will bring great uncertainty to the modeling, resulting in slight difference in the ups and downs of cross-validation, and in public leaderboard score and private leaderboard score, which is reflected in the private leaderboard after the competition. Compared with the public leaderboard, its ranking fluctuates extremely sharply.

- **Code.** This part is where the open-source community of this competition is located. Kaggle has become one of the world's largest data competition platforms, thanks to its open-source features and discussion atmosphere. Here, you can see a variety of exploratory data analysis (EDA), feature engineering, modeling methods, and completely different code styles and personal preferences. Some code even shows the score of this code on the list below its title. Here, participants can learn all kinds of tools and code writing to their heart's content. In order to achieve the same goal, players will also find more concise, elegant, and fast implementations here. At the same time, they can integrate various modeling methods and learn from others. Even in the competition circle, it is said that as long as open source is well integrated, it is not a problem to win a silver medal.

- **Discussion.** Unlike Code, which undertakes the function of code notes, Discussion is a real place for contestants to exchange ideas and discuss. There are few codes here, but there are various QA (Q & A) and different understandings and discovery to the competition. Here participants can freely discuss the relevant experience of the competition with data science enthusiasts around the world, which is even to explore and verify corresponding theories. You can see various masters and even grandmasters, and the interaction between them is also very exciting.
- **Leaderboard.** It is used to display the ranking list, where all participants who have successfully submitted the result document can find their place. The ranking list is refreshed in real time, which can be said to be very stimulating for participants who are head-scratching in a race against time. Kaggle's competitions are usually divided into Public Leaderboard and Private Leaderboard, which are often called A list and B list in the competition circle. This also shows a very important concept in the field of machine learning, i.e., model generalization. Although the existence of the real-time list facilitates the contestants, so that they can constantly verify their ideas and compare the scores of different schemes, this is only the results on the public leaderboard. One point that machine learning modeling pays great attention to is the generalization performance of the model, which can also be said to be its robustness. Only a model with strong robustness can always maintain good results in future predictions, which is very important for practical applications, so there is such a division as A list and B list. In a general sense, the same batch of data is divided into two parts: one is used to evaluate the A list score, and the other is used to assess the B list score. Participants usually need to continuously revise and improve the modeling scheme based on the A list score in the first stage of the competition. Finally, they have two opportunities to choose the result file used to calculate the B list score. The final ranking is based on the B list score. The generalization of machine learning modeling is often a difficult and painful point. For some competitions, the ranking on the A list and the B list would undergo earth-shaking changes. There are often models of contestants' overfitting the A list, which means the model will perform very poorly on data other than that on the A list. This is also the reason why people sometimes call artificial intelligence modeling such as machine learning alchemy or metaphysics.
- **Rules.** This section gives the relevant rules of the competition. These are more detailed supplement to those in the Overview section. It is usually necessary to pay attention to several important time points such as the opening time of the A list evaluation at the beginning of the competition, the deadline for team merging, and the time to switch to the B list. In addition, there are restrictions on the

number of team members, the total times of team submissions, the judgement on what competition cheating is, and other prohibited behaviors. It is suggested that participants should not only be familiar with the competition background and content, but also know the competition rules well, so as not to accidentally violate the competition rules and lead to wasted efforts.

1.1.2 Tianchi

Tianchi is a large big data intelligence platform in China. It opens high-quality desensitized data sets (Ali data and third-party authorized data) and computing resources to the society, attracts high-level talents from all over the world to create excellent solutions, effectively helps the industry and government to solve business pain points, and recruits and transports talents for enterprises. As the leader of China's AI industry, Tianchi provides data intelligence solutions integrating brand, ecology, talent, and computing power to create value for the industry. Since 2014, Tianchi has successfully operated more than 400 data competitions of high specification, covering 600,000 data developers in 98 countries and regions around the world. The competition topics on the Tianchi platform are mainly to solve business pain points in actual scenes, with strong practicality and applicability. In addition to the machine learning algorithm competition involved in this book, there are also innovative application competitions and program design competitions, and the rewards are also very generous. See Fig. 1.3 for the Tianchi big data competition platform.

1.1.2.1 Registration

Like most competition websites, in order to prevent cheating, such as creating a ghost account, and ensure fair and just competition, platforms such as Tianchi need users to register through their email or mobile phone number and upload personal certificates for authority identification.

Fig. 1.3 Tianchi Big Data Competition Platform

1.1.2.2 Competition System

Tianchi is also composed of introduction to competition questions, open source community, and other sections, and also has A list and B list. Unlike Kaggle, Tianchi usually has preliminaries and semi-finals, and each has A list and B list respectively. Tianchi's B list is usually a data change test, and it will last for a few days. Therefore, compared with the A list, it is only shortened in time, while Kaggle's test data is given in advance; only the part in A list is calculated when scoring, and finally two are selected as the result file of the B list calculation. In addition, Tianchi is evaluated at a fixed point in time during the preliminary round, that is, the offline round stage; in the semi-round round, i.e., the platform round stage, the players debug the algorithm locally and complete the model training, and submit the Docker image of the inferential process, and then the prediction result generated by the image will be evaluated in real time. Tianchi will also limit the number of submissions each day. For one thing, the limit is to shorten the gap between the resource allocation of different contestants and prevent some participants from gaining improper advantages by virtue of their powerful computing resources, and for another, it is to prevent contestants from relying too much on test results for modeling, which will lead to the model falling into the mire of overfitting, thus making the model less generalized and causing a lot of useless work.

1.1.2.3 Points

The design of Tianchi has points rules. In accordance with points or conditions, five levels of titles are designed for participants. From low to high, they are data novice, data geek, data god, data scientist, and data master. Tianchi will display the top 100 players according to their points, which is also a special practice of Tianchi. Other competition platforms basically only display the top 100 on the leaderboard.

1.1.3 DF

DF (DataFountain) is a professional big data and artificial intelligence competition platform designated by CCF (China Computer Federation), which has close ties with academia. The DF platform categorized competitions according to technologies (such as data mining, natural language processing, computer vision, etc.) and industries (such as finance, medical care, Internet, security, electricity, entertainment, transportation, smart cities, communications, industry, retail, society, automobiles, education, logistics, real estate, big data, etc.), closely connecting academia and industry. Although the bonus amount may not be as large as that on other large platforms, its understanding of segmentation of industries and diversification of application scenarios are still very attractive.

1.1.4 DC

The full name of the DC competition platform is DataCastle, which is a company located in Chengdu, China. Its website structure and competition methods are similar to those of Kaggle and Tianchi. What is unique of it is that it has a special part for government-enterprise competitions. Usually, participants can see many related entrepreneurial competitions supported by the government, state-owned enterprises, and central enterprises on the DC platform. In addition to the algorithm competitions that this book mentions, there are also creative contests.

1.1.5 Kesci

Kesci, whose Chinese name is Hejing Community, is a strategic cooperation platform for the Big Data Challenge, an annual Chinese university and college computer science competition. Compared with the DF platform and DC platform, Kesci can also provide an online notebook training environment, which is more friendly to some participants who do not have sufficient hardware resources.

1.1.6 JDATA

JDATA Zhihui platform is a competition platform under JD.com. Its section settings are mostly similar to those of Tianchi and Kaggle, but the details are different. Every spring is the peak time for JD.com's own competition events. It is interesting to note that JD.com's competitions mainly involve e-commerce and logistics. They usually customize some evaluation indicators. Participants need to consider their own modeling solutions after receiving the wide table data, including building training sets and testing sets, and selecting sample labels. The data quality and difficulty of the competition questions are extremely high. Of course, there are many bonuses, and college students also have the opportunity to get an "Express Entry" in campus recruitment.

1.1.7 Corporate Websites

In addition to the domestic and foreign mainstream competition platforms listed above, some companies will hold their own competitions; they do not cooperate with the platforms, but build a simple website by themselves, such as Tencent's social advertising algorithm competition; although there is only a website, this competition is still very popular. In addition, there are FlyAI, AI Challenger, etc. Participants do not need to know everything, but can follow some official accounts to learn the latest relevant information of the competition, such as Coggle Data Science, Kaggle Competition Book, Mapo Tofu AI, etc.

1.2   Competition Procedures

How many steps does it take to successfully complete a competition? The answer is three. First download the data, then run the results with code, and finally submit the results. Of course, this is just kidding. Machine learning algorithm competition also cannot escape the so-called routine, and after we sum up a lot of actual experience, we roughly divide the whole process of completing a competition into five parts, namely problem modeling, data exploration, feature engineering, model training, and model integration, as shown in Fig. 1.4. Of course, there are still some preparations to be done before the competition, such as registering accounts, perfecting personal information, and even real-name authentication; then click on the competition you want to participate in to sign up. This section only briefly introduces the competition procedures, and the details will be discussed in Chaps. 2–6 of this book.

1.2.1   Problem Modeling

We believe that everyone still remembers that in the early days of the college entrance examination, the teachers stressed the importance of examining the questions, and understanding the questions is always the first and foremost step. Accurately making clear the meaning of questions can avoid many detours. In the problem modeling of machine learning, not all data is in the form of feature tags, which can already be directly added to the model training. In many cases, it is necessary to analyze the data to abstract the modeling goals and solutions. Although the goal of the competition is usually evident, not all competition data is in the form that can be directly added to the training. Some competitions, such as those on JDATA Zhihui platform, often have some evaluation methods different from general classification and regression evaluation indicators. Participants often need to use the data provided by the organizers to construct training sets and testing sets based on their understanding of the competition questions. This kind of competition greatly tests the participants' problem modeling ability, which is also the difficulty of this kind of contest. Now, the choice of problem modeling methods greatly affects the performance of participants.

Fig. 1.4 Competition Procedures

1.2.2   Data Exploration

Data exploration is one of the most important concepts in the field of machine learning, customarily known as EDA (Exploratory Data Analysis). After understanding the game prompts and roughly knowing the way the problem is modeled, participants need to combine the comprehension to the background business of the game questions to see what the data looks like, whether the data matches the description, what information the data contains, and the data quality. First of all, contestants need to have a clear idea of the data, mainly the meaning, scope, and data structure of the values of each field in the wide table. Then a deeper level is to combine the labels to analyze the distribution state of features, the identical distribution of the training set and the testing set, the business association between the features, and the representation of implicit information, etc. In general, data exploration is a connecting step, which can help participants better learn problem modeling and prepare for the feature engineering to be carried out next.

1.2.3   Feature Engineering

Like EDA, Feature Engineering is also an important concept in the field of machine learning. Its name suggests that this is a module that can be called engineering. Mr. Andrew Ng, a leading machine learning expert, once said in his famous CS229 machine learning course at Stanford University that machine learning was mostly feature engineering; features determined the upper limit of the prediction effect of machine learning, and the algorithm just kept approaching this upper limit; this shows the importance of feature engineering. In fact, whether in competition or in practical application, feature engineering is the module that takes the most time and takes up most of the energy of modelers.

1.2.4   Model Training

After the model scheme is established according to the questions, relevant data exploration is carried out according to the business understanding, and then the feature engineering is gradually improved, the standard training set and testing set structure can be obtained, and subsequently how to carry out model training can be considered. In general machine learning algorithm competitions, most participants prefer GDBT-like tree models. Of course, this is also because their effects are really excellent. The commonly used tree models are mainly XGBoost and LightGBM. Both models have scikit-learn interface functions, which are very easy to use. In addition, sometimes the contestants need to use algorithms such as LR, SVM, and RF, while other times they need to use deep learning models such as DNN, CNN,

RNN, and their derivative models, as well as the popular FFM in the advertising field, etc. If the previous step takes the contestant's own time and energy, then this step, by comparison, mainly depends on the contestant's computing resources. Of course, if it is not a particularly large amount of data, model training will generally be fast. In addition to selecting the right model, this module of model training also needs to spend time on parameter tuning. Although the effect is not very unusual or distinctive as long as the parameters are not set outrageously, for many contestants, even a little improvement in performance may mean an improvement in ranking.

1.2.5 Model Integration

After all kinds of tedious and arduous attempts in the early stage, contestants can finally come to the popular model integration, that is, the stage of finding teammates. Each algorithm has its own advantages and limitations; therefore, by fostering strengths and avoiding weaknesses, and integrating the advantages of each algorithm, participants can make their model more effective. There are many ways to model fusion, such as Stacking, weighted voting, etc., which will be described in detail in Chap. 6. The reason why model fusion is called finding teammates is that in the competition, every single contestant has great unique personal characteristics, including different preference for problem modeling, feature engineering, model training, and other processes, which leads to great variations in the schemes between different contestants; in spite of this, the performance of model integration brought by the differences is excellent, and the greater the difference, the better the effect improvement. It is also recommended that if you do not have a particularly familiar teammate, you can do it yourself first and go through the whole process alone. This is also an exercise for yourself. In the later period, if you really have no idea and badly need help, you can consider finding participants with similar results to form a team. The importance of team strength in the competition is self-evident. Meanwhile, setting up a team in the later stage means the results are submitted independently by each team member in the early stage, which implies more opportunities for verifying ideas. Reasonably making good use of the rules of the competitions is allowed and advocated.

1.3 Competition Types

All kinds of dazzling competitions are exciting to try. Data competitions with many categories can meet the different needs of distinct participants. At the same time, they also promote the development of the AI + industry, which could enable all sectors of society to actively explore artificial intelligence. Therefore, it is necessary to introduce the common types of data competitions today. The following part will focus on three aspects: data types, task types, and application scenarios.

1.3.1   Data Types

The field of artificial intelligence can be roughly divided into three main directions: computer vision (CV), natural language processing (NLP), and data mining (DM). From the perspective of data types, the three can be simply distinguished. The field of computer vision is mostly processing image data, which, of course, also includes videos; natural language processing is mostly text data, involving word segmentation in various languages, etc. Both of them have received common attention from academia and industry in recent years with the continuous improvement of computer hardware performance and the rapid development of broadband networks. The competition on Kaggle will give data types under the title, such as picture data, audio data, text data, wide table data, etc. This book will concentrate on the related competitions of traditional wide table data types. In traditional wide table data, there are usually unique id indexes and feature columns that match samples. According to the meaning, features can be divided into category features (such as user gender) and numerical features (such as age, height, weight, etc.). The above features are in the form of single-value features, in addition to multi-value features, such as the column with user interests and hobbies, which can include fitness, running, and photography at the same time. For this special multi-value feature, we have special processing skills, which will also be explained in detail in Chap. 13.

1.3.2   Task Types

Competitions related to machine learning are mainly based on algorithms, and occasionally there are scheme innovation and design competitions, etc. This book will mainly explain machine learning algorithm related competitions, primarily associated with supervised learning - that is, to carry out modeling according to task requirements and through existing labeled training set data, so as to predict the testing set data and give the results of the corresponding labels, and then to conduct score evaluation. Task types can be roughly divided into classification and regression in accordance with the type of problem, and Chap. 2 will specifically give the evaluation indicators of corresponding tasks.

1.3.3   Application Scenarios

When we mention application scenarios, we naturally think of the application of machine learning in various industries, and the needs and pain points of the industry. Throughout the major competition platforms, the main industries involved are medicine, manufacturing product lines, finance, e-commerce, the Internet, etc. Among them, the richness and diversity of user data in the Internet industry, as well as the less encountered challenges such as medical ethics, have resulted in many application scenarios, such as advertising, search, and recommendation, which are the situations in which artificial intelligence is involved more today.

1.4   Thinking Exercises

1. Please register your account and browse on Kaggle, Tianchi, DF, DC, Kesci, JDATA, and other websites to get familiar with the content introduced in this chapter.
2. What are the main parts of the complete competition procedures? What is the role of each part in the procedure?
3. Take the scenes that come into contact with daily life as an example to list 5 applications that may use machine learning algorithms.

Chapter 2
Problem Modeling

After the contestants get the competition topic, the first thing they should consider is problem modeling, and at the same time complete the pipeline construction of the baseline model, so that they can get feedback on the results in the first place to facilitate the follow-up work. In addition, the existence of the competition depends on real business scenarios and complex data. Participants usually have many ideas about this, but there is always limitation on the times of online submissions for verification of results. Therefore, it becomes very important to reasonably segment the training set and verification set, and build a credible offline verification, which is also the basis for ensuring the generalization of the model. The problem modeling in the competition can be mainly divided into three parts: question understanding, sample selection, and offline evaluation strategy. This chapter first introduces the corresponding work of problem modeling from these three parts, and in the meantime explains the corresponding skills of using them as well as application codes. Then readers will be led to conduct a practice by using a real case, so that they could understand and apply what is learnt in this chapter.

2.1 Understanding the Competition Question

In fact, the aim of examining the competition question is to sort out the problem intuitively and analyze the method that the problem can be solved, the background of the game question, and the main pain points of the contest. To clarify what a contest is for, we should start from figuring out the competition task triggered by the contest background, understanding the business logic and what external data may be meaningful to the event, and having a preliminary comprehension of the game question data, such as knowing what data are related to the mission now, and what is the correlation logic between the data. Usually, the competition task will show the question background, game data, and evaluation indicators. This part of work for getting to know the game question will become an important part and a prerequisite

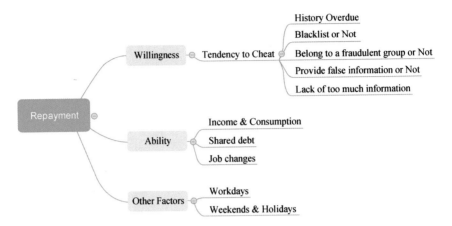

Fig. 2.1 User Repayment Forecast Business

of the competition; through the comprehension of the prompts and the analysis of the real business, we can use our own prior knowledge for preliminary analysis, well paving the way for the next part.

2.1.1   Business Background

2.1.1.1   Go Deep into the Business

The competition itself exists for specific scenarios, and many operations will vary greatly depending on a specific scenario. The scenario mentioned here refers to the business. So, how can we analyze the business? For example, to analyze the purchase behavior of users, here you need to know the purpose of the user's purchase, how the purchased products attract the user, the goods that the company can provide, whether the product is consistent with the user request, target user positioning, the user re-purchase rate, the user's purchasing power, and the method of payment. In short, it is to consider and sort out this process either as a merchant or a user.

Next, let's take a more intuitive business comprehension to show the analysis process in the actual competition, as shown in Fig. 2.1. This is a competition question with the background of Internet financial credit business. The goal is to predict the repayment amount and date from the user side, that is, to predict the repayment situation of the user. If you consider it from the perspective of the merchant, the user's willingness and ability to repay will become the key factors affecting the repayment situation. Next, let's string the business lines together. First, when a user goes to borrow money, the merchant will consider the amount of money the user wants to borrow and whether the user has a tendency to cheat and historical borrowing; then, when the user successfully borrows money, the merchant will consider the user's current debt situation, time left for repayment, and the salary day; finally, when the user successfully repays money, the merchant will analyze the user's overdue situation, remaining arrears, and current installments, etc. This simulates the basic business line.

What has been described above is how to carry out business understanding. In the following part, we will demonstrate how to closely combine the objectives of the competition with the business to bring benefits to the competition results.

2.1.1.2 Be Clear About the Goals

The real business covers a wider range of content than the competition, so the challenge goal is only part of it, and the real business also includes the data provided by the organizer. In the above example, the contest goal is to predict the user's repayment amount and date; the contestants, therefore, can first analyze the relevant business in accordance with this goal, that is, the factors that affect the user's repayment, etc., then feed the information in the business back to the competition objective, namely, the salary day, the total number of borrowings, etc. What closely connects the goal of the competition question with the real business is the data. With specific data, the characteristics can be extracted in line with the business to explicitly represent the repayment of users. Therefore, in order to further learn the competition, it is necessary to have a preliminary understanding of the data.

2.1.2 Understanding Data

We can divide data comprehension into two parts, the data base layer and the data description layer. Of course, in the problem modeling stage, you do not need to gain particularly deep mastery over the data; rather, you just need to do basic analysis. In the later data exploration stage, you can further understand the data and discover key information from the data.

- **Data base layer.** The quality of the original data sources provided by various competition organizers is mixed, and the data forms such as data types, storage formats, etc. are also varied. In order to further analyze and model, it is often necessary to clean, process, and calculate the original data sources. The data base layer focuses on the source, production process, retrieval logic, calculation logic, etc. of each data field. Only by comprehending these can each original field be correctly understood, selected, and used, thus processing and calculating more derived fields which are required. The final presentation of data is usually a data sheet.
- **Data description layer.** The data description layer mainly carries out statistical analysis and general description on the processed data base layer. The focus of this level is to summarize the overall data status through some simple statistics

(such as mean value, maximum—minimum value, distribution, growing rate, trend, etc.) as much as possible, so that participants can clearly have a picture of the basic situation of the data. However, there is no uniform standard for which statistics are used, which depends on the specific scene presented by the data. For example, for time series problems, you can count their growth rate, trend, and period; for conventional numerical features, you can observe statistics such as their means, maximum and minimum values, and variances; for sample sets with multiple categories, you can use distributions, quantiles, etc. for description.

Based on the above two levels of data exploration, participants can obtain a basic comprehension over the data, and these understandings will play a key role in the subsequent data preprocessing and feature extraction.

2.1.3   Evaluation Indicators

2.1.3.1   Classification Indicators

The classification problem is not only a core concern that often appears in competitions, but also a common machine learning issue in practical applications. It is much more difficult to evaluate the effect of classification problems than to assess the effect of regression problems. These two types of issues contain a variety of evaluation indicators. This book will put aside the traditional introduction method and combine practical applications to summarize the characteristics, advantages, and disadvantages of evaluation indicators to help participants obtain certain benefits in the competition.

Common classification indicators in contests include error rate, accuracy, precision, (also known as precision rate), recall (also known as recall rate), F1-score, ROC curve, AUC, and logarithmic loss (logloss), etc. In fact, these indicators measure the effectiveness of the model, and they are related to each other, but their respective emphases are different. After we understand the definition of each indicator, we can figure out their differences and connections. The following part will briefly introduce the above indicators and give an example to illustrate these indicators.

Error Rate and Accuracy

In the classification problem, the error rate is the proportion of the number of samples with wrong classification results to the total number of samples, and the accuracy is the proportion of the number of samples with correct classification results to the total number of samples. That is, the error rate $= 1 - $ accuracy.

Precision and Recall

Taking the simplest binary classification as an example, Fig. 2.2 shows the source of the definition of the confusion matrix, where TP, FN, FP, and TN represent the number of samples in their respective populations.

The basic logic in it is to give a threshold value to the probability value predicted by the model. If the probability value exceeds the threshold, the sample is predicted to be 1 (Positive, positive class); otherwise the prediction is 0 (Negative, negative class).

- **True Positive (TP)**: The prediction category is 1, the real category is 1, and the prediction is correct.
- **False Positive (FP)**: The prediction category is 1, the real category is 0, and the prediction is wrong.
- **True Negative (TN)**: The prediction category is 0, the real category is 0, and the prediction is correct.
- **False Negative (FN)**: The prediction category is 0, the real category is 1, and the prediction is wrong.

The precision rate P refers to the proportion of real positive samples in the samples judged by the classifier to be positive—that is, how many of all samples judged by the classifier to be positive are real positive samples, and the formula is defined in formula (2.1):

$$P = \frac{TP}{TP + FP} \tag{2.1}$$

It is easy to know that if only a single positive sample prediction is made and the prediction category is correct, 100% precision can be obtained through this formula. However, this is meaningless, which will make the classifier ignore data other than positive samples, so another indicator needs to be considered, that is, the recall rate R.

Recall rate refers to the proportion of positive samples correctly judged by the classifier to the total positive samples, that is, how many of all positive samples are judged by the classifier as positive samples, defined as formula (2.2):

		Real Category	
		1	0
Prediction Category	1	True Positive(TP)	False Positive(FP)
	0	False Negative(FN)	True Negative(TN)

Fig. 2.2 Confusion Matrix

$$R = \frac{TP}{TP + FN} \tag{2.2}$$

Precision and recall rate reflect two aspects of classifier performance. Relying on one of them alone cannot comprehensively evaluate the performance of a classifier. Generally speaking, you cannot have your cake and eat it too. The higher your precision rate, the lower the recall rate; on the contrary, the higher the recall rate, the lower the precision rate. Then, in order to balance the impact of precision and recall rate, a more comprehensive evaluation of a classifier will need the F1-score, the indicator that combines the two.

F1-score

Many machine learning classification problems are in need of both a high precision rate and a high recall rate at the same time, so you can consider using the harmonic average formula to weigh up these two indicators, so as to avoid the phenomenon of false high means because one is higher, and the other is lower while making use of arithmetic average. The F1-score can play such a role, and its definition is as formula (2.3):

$$\text{F1-score} = 2 \times \frac{P \times R}{P + R} \tag{2.3}$$

We can easily observe that its maximum value is 1 and its minimum value is 0. It is also very simple to build an evaluation code for calculating precision rate, recall rate, and F1-score. The specific implementation code is as follows:

```
from sklearn.metrics import precision_score, recall_scroe, f1_score
precision = precision_score(y_train, y_pred)
recall = recall_scroe(y_train, y_pred)
f1 = f1_score(y_train, y_pred)
```

ROC Curve

In addition to the above evaluation indicators, there is also a tool commonly used to measure the imbalance in classification, namely the ROC curve (receiver operating characteristic curve). The ROC curve is used to draw the TP rate and FP rate when different classification thresholds are used. Lowering the classification threshold will cause more samples to be classified as positive categories, thereby increasing the number of false positive cases and true positive cases. Figure 2.3 is a more typical ROC curve. In addition, the ROC curve and AUC are often used to evaluate the pros and cons of a binary classifier, so here is a question: now that there are so many evaluation indicators, why the ROC curve is still used?

In actual data sets, there are often uneven positive and negative samples; that is to say, there are much more negative samples than positive samples (or the opposite is

Fig. 2.3 TP Rate and FP
Rate under Different
Classification Thresholds

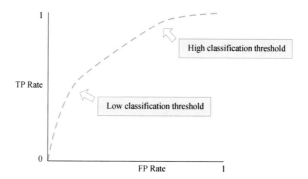

true), and the distribution of positive and negative samples in the testing set may also change over time. The ROC curve has a good characteristic; that is, it can still remain unchanged in this case. However, the ROC curve is not common in competitions. On the contrary, AUC can be said to be our old friend and often appears in classification problems.

AUC

In the ranking business for Internet search, recommendation, and advertising, AUC is an extremely typical evaluation index. It is defined as the area under the ROC curve. Because the ROC curve is generally above the straight line $y = x$, the value range is between 0.5 and 1. The reason why AUC is used as an evaluation indicator is that the ROC curve cannot clearly explain which classifier has the better effect in many cases; for AUC, however, as a numerical value, the larger its value, the better the effect of the classifier will be.

What is worth mentioning is the ranking characteristics of AUC. Compared with indicators such as precision rate and recall rate, the AUC index itself has nothing to do with the absolute value of the probability predicted by the model. It only focuses on the ranking effect between samples, so it is especially suitable to be used as an evaluation index for modeling ranking-related problems. AUC is a probability value. If we randomly select a positive sample and a negative sample, then the probability that the current classification algorithm will rank this positive sample in front of the negative sample according to the calculated score is the AUC value. Therefore, the larger the AUC value, the more likely the current classification algorithm will list the positive sample in front of the negative sample value, i.e., it can be better classified.

In-depth Thought
Now that the AUC has nothing to do with the score value predicted by the
model, why is this a good feature? Suppose you use indicators such as
precision and F1-score, and the score predicted by the model is a probability
value, then you must choose a threshold to determine which samples are
predicted to be 1 and which predictions are 0. Different threshold selection
will lead to different precision rate, recall rate, and F1-score values, and AUC
can directly use the model to predict the score itself by referring to the relative
order, so that it is more useful. In the competition task, the contestants can
even save the trouble of testing the threshold.

Logarithmic Loss

This indicator can be used to evaluate the probability output of the classifier.
Logarithmic loss quantifies the accuracy of the classifier by punishing the wrong
classification. Minimizing the logarithmic loss is basically equivalent to maximizing
the accuracy of the classifier. In order to calculate the logarithmic loss, the classifier
must provide a probability result, which means after the input samples are fed into
the model, the probability value of each category (between 0 and 1) will be
predicted, not just the most likely category. The standard form of the logarithmic
loss function[1] is shown in formula (2.4):

$$logloss = - \log P(Y|X) \tag{2.4}$$

For the sample point (x, y), y is the real label, while in the binary classification
problem, its value can only be 0 or 1. Suppose the real label of a sample point is y_i,
and the probability of taking $y_i = 1$ for the sample point is y_p, then the loss function
of the sample point is as formula (2.5):

$$logloss = - \frac{1}{N} \sum_{i=1}^{} (y_i \log p_i + (1 - y_i) \log(1 - p_i)) \tag{2.5}$$

Let's think about this: AUC also only needs to give the probability value predicted by
the model to calculate and measure the effect of the model, so what is the difference
between logarithmic loss and it?

Logarithmic loss mainly judges whether the probability predicted by the model is
accurate enough. It pays more attention to the degree of coincidence with the
observed data, while AUC evaluates the model's ability to rank positive samples in
front. Since the two indicators have different emphases in evaluation, the

---

[1] Unless otherwise specified, the log function in this book represents the logarithm with e as
the base.

participants will consider different issues and choose different evaluation indicators. For the problem of advertising CTR estimation, if you consider the effect of advertising ranking, you can choose AUC, which will not be affected by extreme values. In addition, the logarithmic loss reflects the average deviation, and it is more inclined to classify accurately the category with a large number of samples.

In-depth Thought
Among the classification problems of various data competitions, AUC and logarithmic loss are basically the most common model evaluation indicators. In general, AUC and logarithmic loss are more commonly used than error rate, accuracy, precision rate, recall rate, and F1-score. Why is this? Because the prediction results of many machine learning models for classification problems are probability values. If you want to calculate the above indicators, you need to convert the probability into categories first, which requires setting a threshold by human. If the prediction probability of a sample is higher than this threshold, the sample will be classified into the corresponding category; if it is lower than this threshold, it will be put into another category. Therefore, the selection of the threshold greatly affects the calculation of the score, which is not conducive to the accurate evaluation of the effect of models from the contestants. By comparison, the use of AUC or logarithmic loss can avoid the trouble of converting the prediction probability into categories.

2.1.3.2   Indicators of Regression

Mean Absolute Error

First of all, please consider the question of how to measure the effect of a regression model. What may occur to you naturally is using the mean of the residual (the difference between the real value and the predicted value), that is, the formula (2.6):

$$\text{residual}(y, y') = \frac{1}{m} \sum_{i=1}^{n} \left(y_i - y_i'\right) \tag{2.6}$$

However, there is a problem here. When the real value is distributed on both sides of the fitting curve, the residual can be positive or negative for different samples, and the direct addition will result in set-off against each other. Therefore, the method of using the distance between the real value and the predicted value is considered to measure the effect of the model, that is, the mean absolute error (MAE, Mean Absolute Error), also known as L1 norm loss, which is defined as in formula (2.7):

$$\text{MAE}(y, y') = \frac{1}{m} \sum_{i=1}^{n} |y_i - y_i'| \tag{2.7}$$

Although the mean absolute error solves the problem of positive and negative cancellation of residual sum, which can better measure the quality of regression model, the existence of absolute value leads to unsmooth function and cannot carry out derivation at certain points, which indicates the mean absolute error is not continuously differentiable in the second order, and the secondary derivative has always been 0.

Extended Learning
In XGBoost, the mean absolute error can be used as a loss function for model training, but due to its limitations, users usually choose Huber loss to replace it. So why do they choose Huber loss? Since the mean absolute error is not continuously derivable (not at 0), it is necessary to use the derivable objective function to approximate the mean absolute error. For the mean squared error (MSE) that will be discussed below, the gradient will decline as the loss decreases, making the prediction result more accurate. In this case, Huber loss is very useful. It will fall near the minimum due to the reduction of the layer. Compared to the mean square error, Huber loss is more robust to outliers. Therefore, Huber loss has combined the advantages of mean absolute error and mean square error. However, the problem of Huber loss may require us constantly adjust the hyper-parameter delta.

Mean Squared Error

What corresponds to the mean absolute error is the mean squared error (MSE, Mean Squared Error), also known as L2 norm loss, which is defined as formula (2.8):

$$\text{MSE}(y, y') = \frac{1}{m} \sum_{i=1}^{n} (y_i - y_i')^2 \tag{2.8}$$

Because the mean squared error is inconsistent with the dimension of the data label, in order to ensure the dimensional consistency, it is usually necessary to extract the square root of the mean squared error, which leads to the root mean squared error (RMSE).

> **In-depth Thought**
> So, what is the difference between the mean absolute error and the mean squared error? The mean squared error squares the error (the true value—the predicted value), so if the error is >1, the mean squared error will further increase the error. At this time, if there are outliers in the data, the error value will be large, and the square of the error will be much larger than the absolute value of the error. Therefore, compared with using the mean absolute error to calculate the loss, the model using the mean squared error will give more weight to the outliers. In other words, the mean squared error is sensitive to outliers, and the mean absolute error is insensitive to outliers.

Root Mean Squared Error

The root mean squared error is used to evaluate the quality of the regression model and will extract the square root of the mean squared error, which reduces the value of the error. Its definition is as formula (2.9):

$$\text{RMSE}(y, y') = \sqrt{\frac{1}{m} \sum_{i=1}^{n} (y_i - y_i')^2} \tag{2.9}$$

The values of the above-mentioned measures are all related to specific application scenarios, so it is difficult to define uniform rules to measure the quality of the model. Similarly, the root mean squared error also has certain flaws. For example, in the application scenario of Computational Advertising, when it is necessary to predict the traffic of advertisements, some outlier points may cause the root mean squared error index to become very poor, so even in 95 percent of the data samples, the model predicts well. If we do not choose to filter out outliers, we need to find a more appropriate indicator to evaluate the prediction effect of advertising traffic. The following will present the average absolute hundred percentage error (MAPE), which is a more robust evaluation indicator than the root mean squared error. This is equivalent to normalizing the error of each point, reducing the impact of individual outliers on absolute error.

Average Absolute Percentage Error

The average absolute percentage error (MAPE) and the second-order derivative of the average absolute error do not exist. But unlike the average absolute error, the average absolute percentage error not only considers the error between the predicted value and the real value, but also takes the ratio between the error and the real value into account. For instance, in the 2019 Tencent advertising algorithm competition, although the difference between the predicted value and the real value is the same, due to the use of the average absolute percentage error to evaluate, the result would be: the larger the real value, the smaller the error. The definition of the average absolute percentage error is as formula (2.10):

$$\text{MAPE}(y, y') = \frac{1}{m} \sum_{i=1}^{n} \frac{|y_i - y'_i|}{y'_i} \tag{2.10}$$

2.2   Sample Selection

Even in real-world competitions, the data provided by the organizer may have quality problems that make participants very headache. This will undoubtedly have a great impact on the final prediction result, so it is necessary to consider how to select the appropriate sample data for training. Then how can we select the appropriate sample? Before answering this question, let's look at the specific reasons that influence the results. Here, four main reasons are summarized: the performance of the model is affected by a too large data set, noise and abnormal data lead to insufficient accuracy, redundant or irrelevant sample data do not bring benefits to the model, and uneven distribution of positive and negative samples leads to data skew.

2.2.1   Main Reasons

2.2.1.1   Too Large Data Set

Machine learning algorithms related competitions involve application scenarios in all walks of life, and the amount of data is also different. The data magnitude of related competitions such as search recommendations and advertisements reaches tens of millions or even hundreds of millions, and excessive data sets seriously affect the rapid verification of various feature engineering and modeling methods. In most cases, our computing resources are limited, so we need to consider data sampling processing, and then model and analyze on smaller data sets. In addition, in specific business scenarios, it may be possible to filter out some data that is not meaningful for modeling, which can help improve model performance.

2.2.1.2   Data Noise

There are two main sources of data noise. One is that improper operation during data collection leads to errors in information representation, and the other is that the characteristics of the data itself have a reasonable range of jitter leading to noise and anomalies. The existence of data noise has both negative and positive sides. For one thing, the existence of noise will cause lower data quality and impact the effect of the model; for another, we can also make the model more robust by introducing noisy data into the training set. In addition, if the source of the noise data is the first, it is necessary to see whether the correct data can be decoded accordingly, which sometimes greatly enhances the modeling effect. Therefore, when it is necessary to deal with noise data, please first consider whether it is caused by acquisition errors, and then weigh the generalization of the model against the current effect of the model. Sometimes denoising will lead to poor generalization performance of the model, and the effect of the model cannot be well guaranteed after changing the data set.

To remove noise, we must first identify the noise, and then take a variety of methods such as direct filtering or modifying the noise data. The noise data may be incorrect eigenvalues, such as missing eigenvalues, exceeding the range of eigenvalues, etc.; it may also be incorrectly labeled; an instance of this situation is that the positive sample of the binary classification problem is labeled as a negative sample. Many data denoising is an example of detecting and removing noise from training data. Section 3.1 will show specific approaches for removing noise or abnormal data.

2.2.1.3 Data Redundancy

In a general sense, data redundancy is a different concept than a too large data set. When you mention a data set, you will naturally think of a set of samples. Its size usually indicates the number of samples in the vertical direction, while data redundancy focuses on describing the redundancy of data characteristics. The redundancy existing in the data will not only make a difference to the performance of the model, but also introduce noise and anomalies, which may have the adverse effect on the model. A typical solution to data redundancy is feature selection, which will be explained in Sect. 4.4.

2.2.1.4 Uneven Distribution of Positive and Negative Samples

In the machine learning scenario where the positive and negative samples of binary classification are not balanced, the data set is often relatively large. In order for the model to better learn the features in the data and make the model more effective, data sampling is sometimes required, which also avoids the trouble of insufficient computing resources due to the large data set. This is a relatively shallow understanding. More essentially, data sampling is to simulate random phenomena and simulate a random event according to a given probability distribution. In addition, there is a saying that a small number of sample points are used to approximate an overall distribution and characterize the uncertainty in the overall distribution. Most of the data provided by the competition is part of the real and complete data extracted by the organizer, and will ensure the consistency of the data distribution, reduce the difficulty of the competition, and guarantee efficiency. Furthermore, a subset (training set) can also be extracted from the overall sample data to approximate the population distribution, and then the purpose of training model is to minimize the loss function on the training set. After training is completed, another data set (testing set) is needed to evaluate the model. Data sampling also has some advanced usage, such as performing oversampling or undersampling on samples, or constantly changing the distribution of samples to adapt to model training and learning when the target information remains unchanged, which is often used to solve the problem of unbalanced samples.

2.2.2   Accurate Methods

In the competition, after getting the data, if it is found that the data has two situations—the data set is too large, and the positive and negative samples are unbalanced—it is necessary to give targeted solutions at the beginning; that is, how to deal with the following two problems: how to improve the model training speed in order to reduce costs when the amount of data is very large, and how to solve such problems through data sampling for scenarios where the positive and negative samples are unevenly distributed.

First of all, for the first problem, the following two solutions are mainly recommended.

- **Simple random sampling.** The solution here is divided into non-replacement and replacement. The practice is relatively simple, so no specific further introduction will be made.
- **Stratified sampling.** This method samples each category separately. This is a method of randomly selecting samples from different categories according to a specified proportion from a data set that can be divided into different subsets (or called layers or categories). Its advantages are the sample is more representative and the sampling error is relatively small; the disadvantage is the sampling process is more complicated than simple random sampling.

For the second problem, there are mainly the following three solutions.

- **Score weighting processing.** The problem of uneven distribution occurs from time to time, including fraudulent transaction identification and email spam identification, etc., and the number of positive and negative samples varies greatly. Figure 2.4 illustrates the distribution of positive and negative samples of a certain competition, and the proportion of positive samples is only about 2%. Considering that the importance of positive samples is higher than that of negative samples, the corresponding score weights can be designed during model training and evaluation, so that the model can learn the parts that need attention. The scoring weighting method is a quite common one. Of course, different weighting methods can be selected in different application scenarios, such as the Micro Fscore index in the multi-classification problem and the Weighted Fscore index used in the KDD CUP 2019 competition. The two

Fig. 2.4 Distribution of
Target Variables

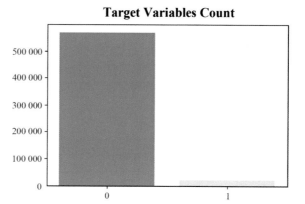

evaluation indicators have different weights for different categories, and the prediction effect of the model can be improved by weighting different categories.

The specific steps of this method are to first examine all samples and give them weights according to whether the samples meet a certain requirement. For example, in an unbalanced binary classification, if the sample label is 1, we will set its weight to w_1 (custom); if the sample label is $\mathbf{0}$, we will set its weight to w_2 (custom). Then the sample weights are substituted into the model for training and testing.

The intuitive meaning of weighting is to assume that the value of a positive sample is greater than that of multiple negative samples from a business perspective. Therefore, it is hoped that the model can learn more key information from the positive samples during training. When it does not learn well, it should be punished more severely.

- **Undersampling.** It is to randomly select a portion of a sample category with larger samples and eliminate it, so that the target category of the final sample is not too unbalanced. Commonly used methods include random undersampling and Tomek Links, in which Tomek Links first finds two contrast class samples with very close indicators, and then removes the one with higher proportion of labels in such samples. This kind of algorithm can provide a very good decision boundary for the classifier.
- **Oversampling.** It mainly recombines the categories with fewer samples to construct new samples. Commonly used methods include random oversampling and SMOTE algorithm. SMOTE algorithm does not simply copy existing data, but generates new data based on the original data by using the algorithm. The schematic diagram of undersampling and oversampling is shown in Fig. 2.5.

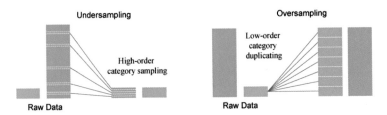

Fig. 2.5 Undersampling and Oversampling

2.2.3 Application Scenarios

So, in what scenarios do you need to deal with the imbalance of the sample? The following gives some specific scenarios to help participants better deal with such problems.

- If the competition task has a particularly large demand for recall, or the prediction of each positive sample is far more important than the prediction of negative samples, then it is difficult to obtain better modeling results if no measures are taken at this time.
- If the evaluation index of the competition is AUC, then the participants will find in the actual contest process that the difference between processing and not processing the sample imbalance problem is not a big concern. But it is also like a fluctuation of a parameter. After combining the processed results with the unprocessed results, the evaluation index is generally slightly improved.
- If positive samples and negative samples are of the same significance or value in the competition task—that is, it is equally important to predict a positive sample correctly and predict a negative sample correctly—then it does not really matter if you do not do anything else at all.

2.3 Offline Evaluation Strategy

Usually in data competitions, participants cannot use all the data for training models, because doing so will result in no data set for offline verification of the effect of the model, and then will make it impossible to evaluate the prediction effect of the model. In order to solve this problem, it is necessary to consider how to slice the data and build a suitable offline verification set. For different types of problems, different offline verification methods are needed. This book roughly divides these problems into two types - strong time sequence and weak time sequence, and then determines the offline verification methods.

Fig. 2.6 Time Series Segmentation Verification

2.3.1   Strong Time Sequence Problems

For the contest questions with obvious time series factors, it can be regarded as a strong timing problem, that is, the time of online data is after that of the offline data set. In this case, the data closest to the testing set in time can be used as the verification set, and the time range of the verification set is after the training set. Figure 2.6 is the time series segmentation verification method.

For example, in the "Passenger Car Retail Volume Forecast" competition on the Tianchi platform, the preliminary competition provides Yancheng sales configuration data of different types of cars from January 2012 to October 2017, requiring participants to predict Yancheng sales data for November 2017 for each type of car. This is a very obvious problem with time series factors, so we can choose the last month of the data set as the verification set, that is, data of 2017 October.

2.3.2   Weak Time Sequence Problems

The verification method of this kind of problem is mainly K-fold Cross Validation. According to the different values of K, different cross-validation methods will be derived, as follows.

1. When $K = 2$, this is the simplest K-fold cross validation, that is, 2-fold cross-validation. At this time, the data set is divided into two parts: D1 and D2. First, D1 is the training set and D2 is the verification set; then, D2 is the training set and D1 is the verification set. There are obvious disadvantages in the 2-fold cross validation, that is, the selection of the final model and parameters will greatly depend on the prior division method of the training set and the verification set. For different division methods, the results vary dramatically.
2. When $K = N$, that is N-fold cross validation, which is called "Leave-one-out Cross Validation". The specific method is to use only one data as the testing set, and all other data as the training set, and repeat N times (N is the amount of data in the data set). The advantages and disadvantages of this method are obvious. Its positive sides are: first, it is not affected by the testing set and training set division

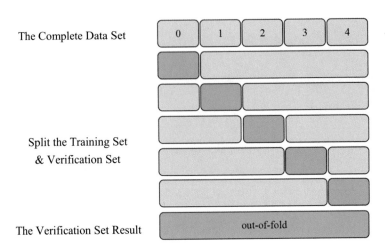

Fig. 2.7 Five-fold Cross Validation

method, because each data has been tested separately; second, it uses *N*-1 data to train the model, and almost all the data is used to ensure that the deviation of the model is smaller. At the same time, its negative side is also obvious: that is, the computation is tremendous. If the data set is ten million levels, then it needs to be trained ten million times.

3. In order to solve the defects of (1) and (2), we generally take $K = 5$ or 10 as a compromise, which is also the most common offline verification method used. For example, when $K = 5$, as shown in Fig. 2.7, we divide the complete training data into five parts, and use four parts of the data to train the model, and the last one part to evaluate the quality of the model. Then loop this process over five pieces of data in turn, and combine the five evaluation results obtained, such as averaging or voting. The following is a common cross-validation code, in which the parameter NFOLDS is used to control the number of folds. The specific code is as follows:

```
from sklearn.model_selection import KFold
NFOLDS = 5
folds = KFold(n_splits= NFOLDS, shuffle=True, random_state=2021)
for trn_idx, val_idx in folds.split(X_train, y_train):
    train_df, train_label = X_train.iloc[trn_idx, :], y_train[trn_idx]
    valid_df, valid_label = X_train.iloc[val_idx, :], y_train[val_idx]
```

2.4   Cases in Practice

As a summary of this chapter, the following will cause readers to apply what they have learned in this chapter and conduct a classic Kaggle introductory competition actual practice—housing price forecast. The home page of the competition is shown in Fig. 2.8. This section includes the understanding of the competition questions and

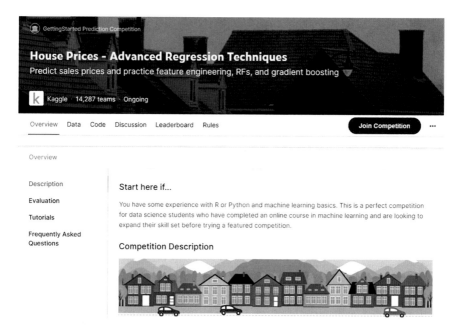

Fig. 2.8   Home Page of the Competition

offline verification. I hope that readers can quickly build a baseline (baseline) and get satisfactory results after understanding the content of this chapter, carefully analyzing the business, and improving the offline evaluation work.

2.4.1   Understanding the Competition Question

First of all, let's get familiar with the business background of the next question. This contest demands predicting the final cost of housing. There are 79 variables in its data set, covering almost all aspects of Ames, Iowa housing. It can be seen that there are many factors that affect the cost of housing, as shown in Fig. 2.9.

Among the above factors affecting housing prices, the followings are the key value factors.

- **Location.** Location is the key to high valuation. For example, being close to large business districts and famous schools or close to the city center may contribute to a relatively high housing price.
- **Shape & Area.** The more space and rooms a house has, and the more land a house occupies, the higher the valuation of it.
- **Internal Structure.** The latest utilities and add-ons (such as garages) are highly desirable factors influencing value.

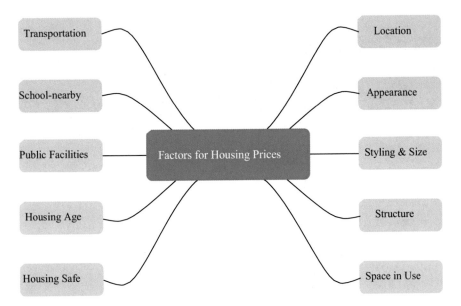

Fig. 2.9 Factors Affecting Housing Prices

These preliminary understanding and business analysis are of great help to Chap. 4. Next, we will import the data and observe the basic information of the data to get a basic understanding, which will play a key role in the data preprocessing, feature extraction and so on afterwards.

First, import the basic modules:

```
import numpy as np
import pandas as pd
from sklearn.model_selection import KFold
from sklearn.metrics import mean_squared_error
from sklearn.preprocessing import OneHotEncoder
import lightgbm as lgb
```

Next, load the data:

```
train = pd.read_csv('train.csv')
test = pd.read_csv('test.csv')
```

Then, look at the basic information of the data:

```
train.describe()
```

Finally, implement the basic processing of the data:

```
all_data = pd.concat((train,test))
all_data = pd.get_dummies(all_data)
# Fill in missing values
all_data = all_data.fillna(all_data.mean())
# Data splitting
X_train = all_data[:train.shape[0]]
X_test = all_data[train.shape[0]:]
y = train.SalePrice
```

This question uses the mean squared error as the evaluation index, and its calculation method is as formula (2.11):

$$\mathrm{MSE}(y, y') = \frac{1}{m} \sum_{i=1}^{n} (y_i - y_i')^2 \qquad (2.11)$$

2.4.2   Offline Verification

The verification code is as follows:

```
# K-fold cross validation
from sklearn.model_selection import KFold
folds = KFold(n_splits= 5, shuffle=True, random_state=2021)
# model parameters
params = {'num_leaves': 63, 'min_child_samples': 50,
'objective': 'regression',
    'learning_rate': 0.01, 'boosting_type': 'gbdt', 'metric': 'rmse'}
for trn_idx, val_idx in folds.split(X_train, y):
    trn_df, trn_label = X_train.iloc[trn_idx, :], y[trn_idx]
    val_df, val_label = X_train.iloc[val_idx, :], y[val_idx]
    dtrn = lgb.Dataset(trn_df, label = trn_label)
    dval = lgb.Dataset(val_df, label = val_label)
    bst = lgb.train(params,dtrn, num_boost_round=1000,
valid_sets=[dtrn, dval],
        early_stopping_rounds=100, verbose_eval=100)
```

At this point, we have completed the basic problem modeling. We not only have a preliminary understanding of the problem, but also set up a baseline, which can quickly feed the prediction results back. Continuous optimization of the model based on the preliminary results is the important work afterwards. In the following chapters, we will take you to explore the data, discover the characteristics of the data, and acquire more useful information from them.

2.5   Thinking Exercises

1. Can the multi-classification problem be treated as a regression problem? What are the requirements for category labeling?
2. At present, all the existing evaluation indexes are given. How can the loss function of these evaluation indexes (classification indexes and regression indexes) be realized?
3. When solving the problem of unbalanced distribution of samples, try to use codes to implement sample weighting, category weighting, and sampling algorithm, and compare the score changes before and after the use of weights.
4. When sampling unbalanced data sets, will it affect the independent identical distribution relationship between training sets and testing sets?
5. When performing K-fold cross validation, is the larger the K value, the better?
6. In most cases, we will choose to use K-fold cross validation, so why can K-fold cross validation help improve the effect?

Chapter 3
Data Exploration

Data exploration is one of the core modules of the competition. It runs through the competition and is also the key to the victory of many competitions. So, what is data exploration? What problems can be solved? First of all, three points should be made clear: that is, how to ensure that you are ready to use the algorithm model for the competition, how to choose the most appropriate algorithm for the data set, and how to define the characteristic variables that can be used for the algorithm model.

Data exploration can help answer the above three questions and guarantee the best results of the competition. It is a way to summarize, visualize, and familiarize yourself with the important features of the data set. In general, data exploration can be divided into three parts: the first is pre-competition data exploration (i.e. data preliminary exploration), which helps us obtain a holistic understanding of the data and discover the problems existing in the data, such as missing values, outliers, and data redundancy, etc.; the second is data exploration during the competition, which discovers the characteristics of variables through analysis of data and helps extract valuable features, when it is possible to analyze from univariate, multivariate, and variable distribution; the last part is the analysis of the model, which can be divided into feature importance analysis and result error analysis, enabling us to find problems from the results and further optimize them.

Data exploration is useful for us to find some features of the data, the correlation between the data, and is helpful for subsequent feature construction. This chapter will explain data exploration in combination with real competition cases.

3.1 Preliminary Data Exploration

Data preliminary exploration can be regarded as pre-competition data exploration, which mainly includes analysis ideas, analysis methods, and clear purposes. Through systematic exploration, we can deepen our comprehension of data.

3.1.1 Analytical Thinking

In a real contest, it is best to use a variety of exploration ideas and methods to explore each variable and compare the results. After fully understanding the data set, you can enter the data preprocessing stage and feature extraction stage to convert the data set according to the desired business results. The goal of this step is to make sure that the data set is ready to be applied to machine learning algorithms.

3.1.2 Analysis Methods

The analysis of data exploration mainly adopts the following methods.

- **Univariate visual analysis**: providing summary statistics for each field in the raw data set.
- **Multivariate visual analysis**: used to understand the interaction between different variables.
- **Reduced-dimension analysis**: making it easier to find the fields with the largest variance between characteristic variables in the data and meanwhile reducing the data dimension while retaining the maximum amount of information.

By applying these methods, we can verify our assumptions in the competition and determine the direction of attempt in order to understand the problem and select the model and verify whether the data is generated as expected. Therefore, the distribution of each variable can be checked, some missing values can be defined, and finally possible ways to replace them can be found.

3.1.3 Purpose Clarification

It would be an irrational decision to skip the data exploration stage in a competition. Due to the rush to enter the algorithm model stage, many players often either omit the data exploration process completely or only do a very superficial analysis work. This is a very serious and common mistake for most contestants. This kind of inconsiderate behavior may lead to skewed data, outliers, and too many missing values. For the competition, this will produce some bad results as follows.

- Generate inaccurate models.
- Generate accurate models on wrong data.
- Choose the wrong variable for the model.
- Make use of resources inefficiently, including reconstruction of models.

Being familiar with the possible negative effects helps us to clarify the main purpose of data exploration. First, data exploration is used to answer questions, test business assumptions, and generate assumptions for further analysis. Second, you can also use data exploration to prepare the modeling data. The two have one thing in common, which is to give you a good understanding of your data, either getting the answers you need, or developing an intuition to explain the results of future modeling.

The purpose of data exploration is further visualized. The following are seven things that must be made clear in the data exploration phase.

1. **Basic information of data set**: for example, how big the data is and what type each field is.
2. **Duplicate values, missing values, and outliers**: whether removing duplicate values and missing values is serious, whether missing values have special meanings, and how outliers are to be discovered.
3. **Whether there is redundancy between features**: For example, if the unit of height is expressed in both cm and m, this is redundancy. We can find redundant features through similarity analysis between features.
4. **Whether there is time information**: When there is time information, analysis of correlation, trend, periodicity, and outliers is usually carried out, and potential data penetration problems may also be involved.
5. **Label distribution**: For classification problems, make sure whether there is an uneven distribution of categories. For regression problems, make clear whether there are outliers, how the overall distribution is, and whether target transformation is needed.
6. **Distribution of training set and testing set:** whether there are many feature fields in the testing set that do not exist in the training set.
7. **Univariate/Multivariate distribution:** to be familiar with the distribution of features and the relationship between features and labels.

When we have known why we need to conduct data exploration and understood the things that must be clear in data exploration, data exploration will become more purposeful.

To start exploring the data, you first need to import the basic library, and then load the given data set. You may already know how to do this, but do not know where to start. Thanks to the pandas library for making it become a very simple task. First, import the package as pd, then use the read_csv () function, and pass the path and parameters where the data is to the function. The parameters involved can be used to ensure that the function can read the data correctly, and the first row of the data will not be interpreted as the column name of the data.

One of the most basic steps in data exploration is to obtain a basic description of the data, and to acquire a basic sense of the data by getting a basic description of the data. The following methods are used to help us understand the data.

- **DataFrame.describe()**: View the basic distribution of data. Specifically, statistics is carried out on each column of data; statistical values include frequency, mean, variance, minimum, percentiles, maximum, etc. It helps us quickly understand the data distribution and find outliers and other information.

	Feature	Unique_values	Percentage of missing values	Percentage of values in the biggest category	type
72	PoolQC	3	99.520548	99.520548	object
74	MiscFeature	4	96.301370	96.301370	object
6	Alley	2	93.767123	93.767123	object
73	Fence	4	80.753425	80.753425	object
57	FireplaceQu	5	47.260274	47.260274	object
3	LotFrontage	110	17.739726	17.739726	float64
59	GarageYrBlt	97	5.547945	5.547945	float64
64	GarageCond	5	5.547945	90.821918	object
58	GarageType	6	5.547945	59.589041	object
60	GarageFinish	3	5.547945	41.438356	object

Fig. 3.1 Basic Distribution of Data

- **DataFrame.head** (): You can load the first five rows of the dataset directly.
- **DataFrame.shape:** Get the row and column situation of the data set.
- **DataFrane.info** (): You can quickly get a simple description of the dataset, such as the type of each variable, the size of the dataset, and missing values.

The methods listed above can help us understand the basic information of the data. Next, we will show the powerful functions of these methods through specific operations. First, the situation of nunique and missing values is shown in a piece of code here:

```
stats = []
for col in train.columns:
    stats.append((col, train[col].nunique(),
        train[col].isnull().sum() * 100 / train.shape[0],
        train[col].value_counts(normalize=True,
        dropna=False).values[0] * 100, train[col].dtype))
    stats_df = pd.DataFrame(stats, columns=['Feature', 'Unique_values',
        'Percentage of missing values',
        'Percentage of values in the biggest category', 'type'])
stats_df.sort_values('Percentage of missing values', ascending=
False)[:10]
```

Figure 3.1 shows the basic information of the data generated by the above code. We find special variables from it for detailed analysis. Here we select variables with low nunique values and more missing values for observation. Generally, if nunique is 1, it does not have any meaning. It means that all values are the same, and there is no distinction, so it needs to be deleted. It can be found that some variables have many missing values, such as the missing proportion reaching more than 95% when we can consider deleting them.

As shown in Fig. 3.2, the histogram can be used to demonstrate the distribution of missing values of variables more intuitively. The following is the specific generation code for the visualization map of missing values of variables:

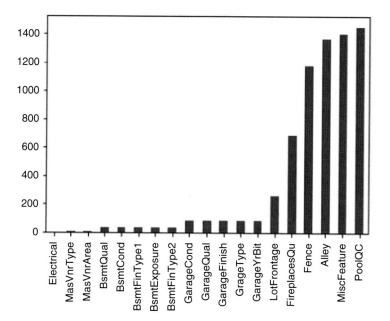

Fig. 3.2 Distribution of Missing Values of Variables

```
missing = train.isnull().sum()
missing = missing[missing > 0]
missing.sort_values(inplace=True)
missing.plot.bar()
```

3.2 Variable Analysis

Next, we will conduct a more detailed analysis, not only for each variable, but also to analyze the relationship between variables and the correlation between variables and labels, as well as to conduct hypothesis tests to help us extract useful features.

3.2.1 Univariate Analysis

Univariate can be divided into label, continuous type, and category type.

3.2.1.1   Labels

There is no doubt that labels are the most important variable and the goal pursued by a competition. We should first observe the distribution of labels. For housing price forecast, the label SalePrice is a continuous variable. The basic description of this label is shown in Fig. 3.3.

The basic information generation code is as follows:

```
train['SalePrice'].describe()
```

In Fig. 3.3, SalePrice looks pretty normal. Next, let's take a more detailed look at the distribution of SalePrice in a visual way. The relevant code is as follows:

```
plt.figure(figsize=(9, 8))
sns.distplot(train['SalePrice'], color='g', bins=100, hist_kws=
{'alpha': 0.4})
```

The obtained results are shown in Fig. 3.4.

As you can see in Fig. 3.4, SalePrice has a deviation from the normal distribution, belongs to the right skewed type, and has a peak state, and some outliers of it are more than 500,000. We will eventually find a way to remove these outliers and come up with variables that can make the algorithm model learn well and conform to the normal distribution. The following code will perform a logarithmic conversion of SalePrice and generate a visual diagram. The conversion result is shown in Fig. 3.5:

```
plt.figure(figsize=(9, 8))
sns.distplot(np.log(train['SalePrice']), color='b', bins=100,
hist_kws={'alpha': 0.4})
```

3.2.1.2   Continuous Type

Here we can first observe the basic distribution of continuous variables, as shown in Fig. 3.6.

Fig. 3.3 Basic Description of the Label SalePrice

count	1460.000000
mean	180921.195890
std	79442.502883
min	34900.000000
25%	129975.000000
50%	163000.000000
75%	214000.000000
mas	755000.000000

Name: SalePrice, dtype: float64

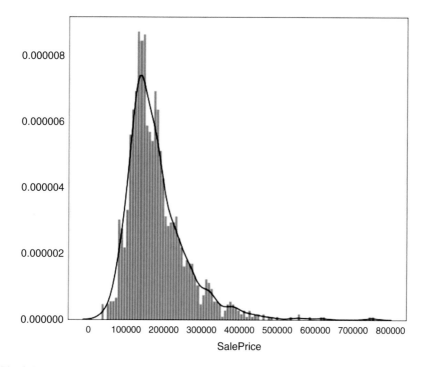

Fig. 3.4 Distribution of SalePrice

Similar to the way labels are viewed, histograms are mainly used here to observe the distribution of values, the frequency of occurrence of each value, etc. The following is the generation code for the distribution visualization diagram of continuous variables:

```
df_num = train.select_dtypes(include = ['float64', 'int64'])
df_num = df_num[df_num.columns.tolist()[1:5]]
df_num.hist(figsize=(16, 20), bins=50, xlabelsize=8, ylabelsize=8)
```

Next, a more scientific analysis is brought off, starting with correlation analysis. It is worth noticing that correlation analysis can only compare numerical features, so for letter or string features, you need to encode them first and convert them into numerical values before you can see what correlation there is between the features. In actual competitions, correlation analysis can well filter out features that are not directly related to labels, and this method can help achieve good results in many competitions.

When we look at a visual chart of correlation analysis, we need to understand what the diagram represents and what information can be obtained from it. Please first learn the most basic concepts: positively correlated to and negatively correlated to.

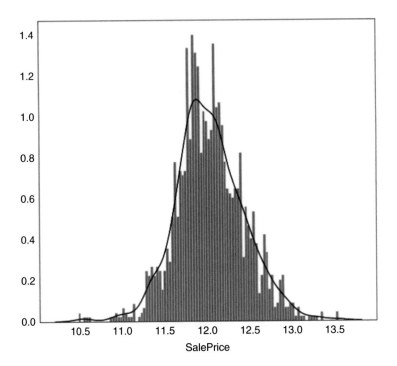

Fig. 3.5 Logarithmic Conversion of SalePrice

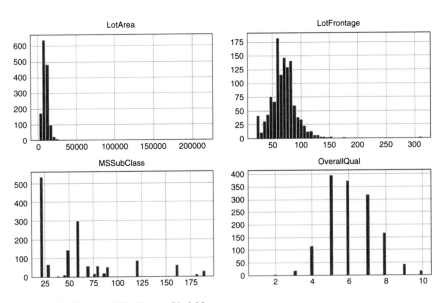

Fig. 3.6 Distribution of Continuous Variables

- **Positively correlated to**: If an increase in one feature causes an increase in another, they are positively correlated. The value 1 indicates perfect positive correlation.
- **Negative correlated to**: If an increase in one feature causes a decrease in another, they are negatively correlated. The value - 1 implies a perfect negative correlation.

Now suppose that feature A and feature B are completely positively correlated to each other, which means that the two features contain highly similar information with little or no difference in the information. This is called multiple linear for the two features contain almost the same information.

When building or training models, if you use these two features at the same time, then one of them may be redundant. We should try to eliminate redundant features as much as possible, because it will make the training time longer and some other advantages will disappear. The following code is used to generate a similarity matrix diagram related to SalePrice:

```
corrmat = train.corr()
f, ax = plt.subplots(figsize=(20, 9))
sns.heatmap(corrmat, vmax=0.8, square=True)
```

The similarity matrix diagram generated is shown in Fig. 3.7, from which variables with strong correlation to housing prices can be found. Among them, OverallQual (the general evaluation), GarageCars (garages), TotalBsmtSF (basement area), GrLivArea (living area), and other characteristics are positively correlated to SalePrice, which is also very consistent with our business intuition. From the similarity matrix, not only the relationship between housing price and variables, but also the relationship between variables can be found, so how to use the similarity matrix for analysis becomes the key point.

3.2.1.3 Category Type

As known to all, the purpose of data exploration is to help us understand data and build effective features. For example, if we find a feature that has a strong correlation with a label, then we can make a series of extensions around this strong correlation feature, specifically, cross-combinations such as strong correlation plus weak correlation, strong correlation plus strong correlation, etc., to find out potential information in higher dimensions.

First, please observe the basic distribution of category variables, that is, the frequency of observing each attribute. According to the frequency, we can not only quickly find hot attributes and attributes that appear rarely, but also further analyze the reasons for this situation. For example, Taobao, often referred to as "Amazon of China", has more female users than male users, which is mainly due to the platform's strong influence in the clothing and beauty business. This is from a

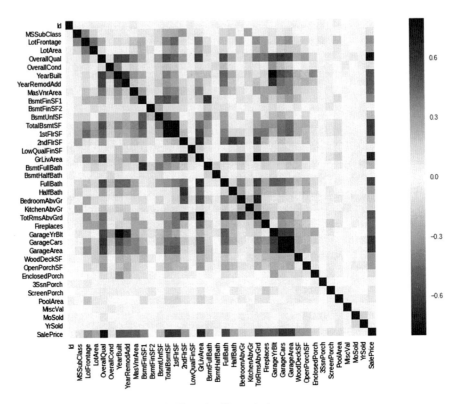

Fig. 3.7 Similarity Matrix (see also the color illustration)

business perspective, and of course it may also be the reason for data sampling. The distribution of some category-based variables is visualized as shown in Fig. 3.8.

3.2.2 Multivariate Analysis

Univariate analysis is too simple to dig out the internal connections between variables and obtain more fine-grained information, so multivariate analysis has become a must. Analyzing the relationship between characteristic variables helps to build better features while reducing the probability of building redundant features. Here we will select the characteristic variables that need special attention in this competition question for analysis.

From the above similarity matrix, we have learnt that housing evaluation is positively correlated to SalePrice. Now please further expand the analysis to consider whether there is a certain relationship between housing evaluation and housing location. Next, we will show the connection between the two visually. The specific implementation code is as follows:

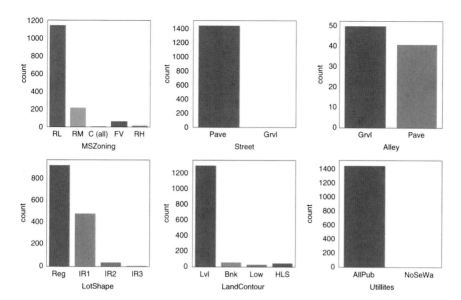

Fig. 3.8 Basic Distribution of Category Variables

```
plt.style.use('seaborn-white')
type_cluster = train.groupby(['Neighborhood','OverallQual']).size()
type_cluster.unstack().plot(kind='bar',stacked=True, colormap=
'PuBu', figsize=(13,11),
grid=False)
plt.xlabel('OverallQual', fontsize=16)
plt.show()
```

Figure 3.9 shows a bar chart of the evaluation distribution of different housing locations. We can find that the darker the color, the higher the evaluation. NoRidge, NridgHt, and StoneBr all have good evaluation.

By looking at Fig. 3.10, we can further see what the SalePrice of houses in different locations is.

Completely in line with our intuition, high-rated locations (NoRidge, NridgHt, and StoneBr) correspond to high SalePrice, which also indicates that housing location evaluation has a relatively strong correlation with housing price. In addition to proving that the original feature is strongly related to SalePrice through such analysis, how can new features be constructed through analysis?

Since the combination of housing location and housing evaluation can lead to houses with higher selling prices, we can construct the cross-combination features of these two categories of features to make a more detailed description, and we can also construct the average housing price under this combination feature and so on.

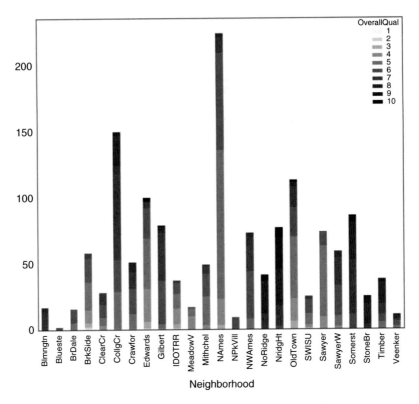

Fig. 3.9 Bar Chart of Evaluation Distribution of Different Housing Locations

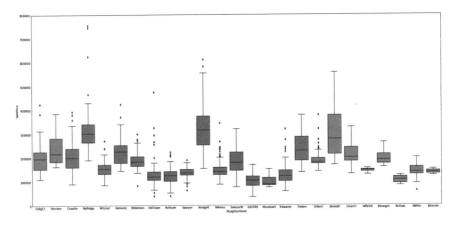

Fig. 3.10 SalePrice Box Diagram Corresponding to Different Housing Locations

3.3 Model Analysis

3.3.1 Learning Curve

The learning curve is a widely used effect evaluation tool in machine learning. It can reflect the score changes of the training set and verification set in the training iteration and can help us quickly comprehend the learning effect of the model. We can observe whether the model is overfitting through the learning curve, so as to determine how to improve the model by judging the degree of fitting.

A learning curve is widely applied in model evaluation in machine learning, and the model will gradually learn (optimize its internal parameters) along with the training iteration, such as neural network models. At this time, the indicators used to evaluate learning may be maximized (classification precision) or minimized (regression error), which also means that higher scores indicate more information learned, and lower scores mean less information acquired. Next, please look at some common shapes observed in the learning curve diagram.

3.3.1.1 Underfitting Learning Curve

Underfitting means that the model cannot learn the information presented by the data in the training set. Here, the learning curve of the training loss can be used to determine whether underfitting occurs. Under normal circumstances, the underfitting learning curve may be a flat line or has a relatively high loss, which indicates that the model cannot learn the training set at all.

Figure 3.11 shows two common types of underfitting learning curves. The left figure shows that the fitting ability of the model is insufficient, and the figure on the right shows that it needs to reduce loss through further training.

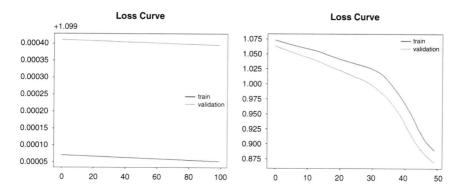

Fig. 3.11 Underfitting Learning Curve (see also the color illustration)

3.3.1.2   Overfitting Learning Curve

Overfitting means that the model learns well from the training set, including statistical noise or random fluctuations in the training set. The problem with overfitting lies in the more specialized the model is to the training data, the worse its generalization ability to the new data will be, which will lead to an increase in generalization error. The increase in generalization error can be measured by the performance of the model on the verification set. This often happens if the capacity of the model exceeds the capacity required by the problem and there is too much flexibility. This also happens if the model takes too long to train.

As shown in Fig. 3.12, the left figure shows the overfitting learning curve. It can be seen that the verification set loss curve starts to increase after it decreases to a point, while the training set loss keeps decreasing. The right figure is a normal learning curve; there is neither underfitting nor overfitting. The loss of both the training set and the verification set can be reduced to a stable point, and the difference between the two final loss values is very small, from which it can be determined that the degree of fitting is good.

3.3.2   Feature Importance Analysis

Feature importance can be obtained through model training. For tree models (such as LightGBM and XGBoost), the importance score of the feature is obtained by calculating the information gain or the number of splits of the feature. For model LR and SVM, the feature coefficient is used as the feature importance score. Take LR as an example, each feature corresponds to a feature coefficient w; the larger the w, the greater the influence of the feature on the model prediction results, and the more significant the feature can be considered. We assume that the feature importance score and feature coefficient w are both used to measure the importance of features in the model and can play their role in feature selection.

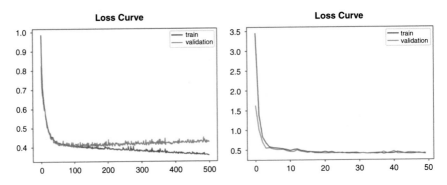

Fig. 3.12 Comparison of Learning Curves (see also the color illustration)

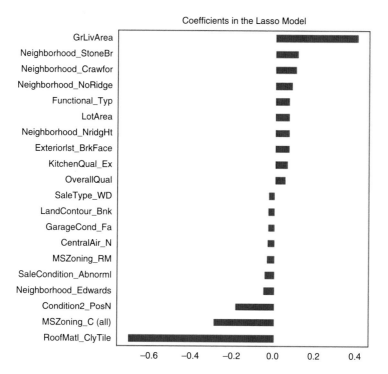

Fig. 3.13 Feature Coefficients in Lasso Model Training

The analysis of the importance of features can be used for business understanding. Some strange features play a key role in the model and can help us better learn business operations. At the same time, if some features are unconventional, then we can also see that they may be overfitting features, etc. Figure 3.13 is the coefficient of each feature in the Lasso model training. It can be seen that there are both features with high coefficient and features with positive and negative correlation.

In Fig. 3.13, the feature that has the highest positive correlation to SalePrice is the GrLivArea (housing area), which is very intuitive to us. The larger the housing area, the higher the selling price. There are also some location features, such as Street StoneBr, NridgHt and NoRidge, which also play a positive role. Of course, there are also many negative features. These negative features are meaningless and can usually be eliminated directly.

3.3.3 Error Analysis

Error analysis is the key to pinpointing problems through model prediction results. Broadly speaking, in regression problems, we look at the distribution of prediction results, while in classification problems, we look at confusion matrices, etc. This can help us find out which samples or types of samples the model does not predict well enough to cause inaccurate results, then we can analyze the possible factors that cause the result errors, and finally correct the training data and models.

In real problems, error analysis becomes more detailed. For example, in a binary classification task of user default estimation, there are 200 wrong classification samples in the verification set result. Further analysis shows that 70% of the wrong classification samples are misjudgments caused by a large number of missing features. At this time, adjustments are needed. Not only can the prediction ability of the model be enhanced by creating more feature information that can describe these misjudgment samples, but also these misjudgment samples can be given higher weights in model training.

3.4   Thinking Exercises

1. This chapter only draws visual maps of some variables. Please try to draw distribution maps of other variables and observe them.
2. There are many types of charts for data visualization, so it becomes difficult to choose charts that are suitable for certain data features or analysis objectives. This includes different analysis objectives such as trend analysis, multi-class comparison, data connection, data distribution, etc. You can try to summarize the data characteristics and goals displayed by different charts.
3. How can we draw the image of a confusion matrix? How can we locate the category with high error?

Chapter 4
Feature Engineering

In this chapter, we will introduce to you the key part in the algorithm competition that has the heaviest workload and that determines whether the contestants can get a good ranking—feature engineering. Andrew Ng once mentioned in his CS229 machine learning course at Stanford University that machine learning was mostly feature engineering; features determined the upper limit of the prediction effect of machine learning, and the algorithm just kept approaching this upper limit. This shows the significance of feature engineering. Basically, participants will spend 80% of their time and energy on building feature engineering. During the application of machine learning, feature engineering is what between data and algorithms. Feature engineering is to transform the original data source into features, which in turn enable us to characterize samples from a variety of new dimensions. Features can better describe potential problems to the prediction model, thereby improving the accuracy of the model's predictive analysis of unseen data. High-quality features help to improve the generalization performance of the model as a whole, and features are largely associated with basic problems. Feature engineering is both a science and a fun art, which is why data scientists spend a lot of their time on data preparation before modeling while enjoying it.

Feature engineering is mainly divided into four parts: data preprocessing, feature transformation, feature extraction, and feature selection. In this chapter, we will also introduce the corresponding work of feature engineering from these four parts, and at the same time give the using skills and application codes.

4.1 Data Preprocessing

In the algorithm competition, the data set we get may contain a large number of errors and omissions, either because of manual entry errors resulting in the existence of abnormal points that make the data "dirty", or some sample information not able to be collected in practice. These errors and omissions information is very

© The Author(s), under exclusive license to Springer Nature Singapore Pte Ltd. 2023
W. He et al., *Machine Learning Contests: A Guidebook*,
https://doi.org/10.1007/978-981-99-3723-3_4

unfavorable to model training, and thus will make it impossible for models to learn more accurate rules from the data set. So, in most cases, the initial data is basically not directly used for model training, or even if it is used, you can only get a relatively bad result. If the data provided by the organizer is good enough, then the contestants are really very happy. The quality of the data directly determines the accuracy and generalization ability of the model, and at the same time affects its smoothness during feature construction. Therefore, in the case of the low quality of the data provided by the competition, it is necessary to preprocess the data and process all kinds of dirty data in a corresponding way, so as to obtain standard, clean, continuous data for data statistics, data mining, etc. At the same time, we should also try to deal with the missing values depending on the situation, such as whether it needs to be filled, and if so, whether the mean or median should be filled, and so on. In addition, the data provided by some competitions and the corresponding storage methods may require more memory than the participants' own hardware conditions, so it is necessary to carry out certain memory optimization, which is also helpful for operation on larger data sets with limited memory space.

4.1.1   Processing Missing Values

No matter in competitions or in dealing with practical problems, it is often encountered that data sets have data missing. For example, information cannot be collected, the system fails, or users refuse to share their information, resulting in data missing. In the face of data missing problems, in addition to algorithms such as XGBoost and LightGBM that can directly handle missing values during training, many other algorithms (such as LR, DNN, CNN, RNN, etc.) cannot directly approach the problem of missing values. In the data preparation stage, it takes more time to cope with the trouble than in the algorithm construction stage, because operations such as filling in missing values need to be handled carefully to avoid errors in the processing process and prevent the model training effect from being influenced.

4.1.1.1   Distinguishing Missing Values

First, participants need to find the manifestation of missing values. In addition to None, NA, and NaN, missing values also include other special values used to represent missing numerical values, such as missing values filled with -1 or -999. There is also a kind of business that looks like a missing value, but has practical significance, which requires special way to handle. For example, users who do not fill in the "marital status" item may be more sensitive to their privacy and should be set to a separate category, such as using the value 1 to indicate married, the value 0 to indicate unmarried, and the value -1 to indicate unfilled; users who do not fill in the "driving experience" item may not have a car, so it is reasonable to fill it with 0. When the missing values are found, they need to be filled reasonably according to the information that the missing values may contain in different application scenarios.

4.1.1.2 Processing Method

Data missing can be divided into loss of category features and loss of numerical value features, and their filling methods are quite different. For the case of missing category features, a new category is usually filled, which can be 0, −1, negative infinity, etc. For the loss of numerical value features, the most basic method is filling means, but this method is more sensitive to outliers, so you can choose to fill in medians, because this method is not sensitive to outliers. In addition, when filling data, be sure to consider whether the selected filling method will affect the accuracy of the data. A summary of the filling method is as follows.

- **For category features**: you can choose the most common type of filling method, namely filling the modal number; or directly fill in a new category, such as 0, −1, negative infinity.
- **For numerical value features**: you can fill in the average, median, modal number, maximum, minimum, etc. Which statistical value to choose requires specific analysis of specific problems.
- **For ordered data (such as time series)**: you can populate adjacent values next or previous.
- **Model predictive filling**: Ordinary filling is only the normal state of a result and does not consider the influence of the interaction between other features. The column containing the missing value can be modeled and the result of the missing value can be predicted. Although this method is more complex, the final result is intuitively better than what could be achieved through direct filling, but the effect in actual competitions needs to be specifically tested.

4.1.2 Dealing with Outliers

In real data, it is often found that after one or some fields (features) are sorted according to a certain variable (such as the time in the time series problem); it could be observed that some values are much higher or lower than other values within a certain range. There are also some unnormal situations, such as the age of being 0 or more than 100 among ad clickers. We can regard all these as outliers, and their existence may have a negative effect on the performance of the algorithm.

4.1.2.1   Looking for Outliers

Before dealing with outliers, we first need to find outliers. Here we summarize two commonly used methods for outliers of numerical features.

The first method is to find outliers through visual analysis. Simply using scatter plots, we can clearly observe the existence of outliers. Points that deviate seriously from dense areas can be treated as outliers, as shown in Fig. 4.1.

The second is to find outliers through simple statistical analysis, that is, to judge whether there are abnormalities in the data according to basic statistical methods. For example, quartile interval, extreme difference, average deviation, standard deviation, etc. This method is suitable for mining numerical data of single variables, as shown in Fig. 4.2.

Fig. 4.1 Data Scatter Plot Visualization

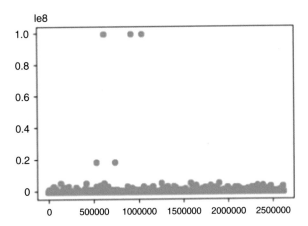

Fig. 4.2 Quartile Interval Box Diagram

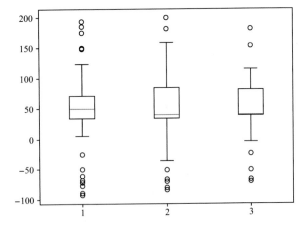

Extended Thinking

Discrete outliers (all values outside the definition range of discrete attributes are outliers), knowledge-based outliers (e.g. height 10 m), etc. can be treated as category missing values.

4.1.2.2 Coping with Outliers

- **Delete records containing outliers**. This approach has the advantage of eliminating the uncertainty caused by samples containing outliers but has the disadvantage of reducing the sample size.
- **Regarding outliers as missing values.** Treat outliers as missing values and use the method of missing value processing to process them. This method has the advantage of centralizing outliers into one category and increasing the availability of data; the disadvantage is that confusing outliers with missing values will affect the accuracy of data.
- **Average values (medians) correction.** The outlier can be corrected using the average value corresponding to the same category, with the same advantages and disadvantages as "regarding outliers as missing values".
- **Do not process.** Data mining is performed directly on data sets with outliers. The effectiveness of this method depends on the source of the outlier. If the outlier is caused by an input error, it will have a negative impact on the effectiveness of data mining. If the outlier is only a record of the real situation, direct data mining can retain the most authentic information.

4.1.3 Optimizing Memory

In competitions related to machine learning, the data involved in the competition questions is often large, and the participants' own computer hardware conditions are limited, so there are often memory errors in the code due to insufficient memory, which brings troubles to the participants. Therefore, it is necessary to introduce some methods that help optimize memory and run the code to the maximum. Here we will introduce two common methods of Python—memory recycling mechanism and numerical type optimization.

- **Memory recycling mechanism.** In Python's memory recycling mechanism, the gc module mainly uses "referencing counting" to track and recycle garbage. On the basis of referencing counting, it can also solve the problem of circular references that may be generated by container objects through "mark sweep", and further improve the efficiency of garbage recycling by "intergenerational recycling" to exchange space for time. In general, when we delete some variables, we use gc.collect () to free up memory.

- **Numerical type optimization.** The data storage formats commonly used in competitions are csv and txt, which need to be read as tabular data, namely DataFrame format, when processing. This requires the use of pandas toolkit for operations; pandas can represent numerical data as NumPy arrays at the bottom and make them continuously stored in memory. This storage method not only consumes less space, but also allows us to quickly access data. Because pandas use the same number of bytes to represent each value of the same type, and the NumPy array stores the number of these values, pandas can quickly and accurately return the number of bytes consumed by columns of numeric value types.

Many data types in pandas have multiple subtypes, and they can use fewer bytes to represent different data. For example, float types have subtypes such as float16, float32, and float64. The numeric part of these type names indicates how many bits this type uses to represent data. An int8 data type uses 1B (8bit) to store a value, which can represent 256 (2^8) binary values, which means that we can use this subtype to act as values between -128 and 127 (including 0).

We can use np.iinfo class to confirm the minimum and maximum values of each int subtype. The code is as follows:

```
import numpy as np
np.iinfo(np.int8).min
np.iinfo(np.int8).max
```

Then, we can judge the subtype to which the feature belongs by selecting the minimum value and maximum value of the features in a column. The code is as follows:

```
c_min = df[col].min()
c_max = df[col].max()
if c_min > np.iinfo(np.int8).min and c_max < np.iinfo(np.int8).max:
    df[col] = df[col].astype(np.int8)
```

In addition, without affecting the generalization performance of the model, for category variables, if the number of encoded IDs is large, extremely discontinuous, and there are relatively fewer types, they can be encoded from 0 again (natural number coding), which can also reduce the memory occupancy of variables. For numerical variables, memory occupancy is often excessive due to floating-point numbers. It can be considered to normalize the minimum and maximum values first, then multiply them by 100, 1000, etc., and then round them up. This not only can retain the size relationship between the same variables, but also greatly reduce memory occupancy.

4.2 Feature Transformation

After the data preprocessing is completed, sometimes the contestants need to make some numerical transformations on the features, and in the real world competition, many original features cannot be directly used, so some adjustments need to be made to help the contestants better construct the features.

4.2.1 Non-dimensionalization Processing of Continuous Variables

Non-dimensionalization refers to the conversion of data of different specifications to the same specification. Common non-dimensional processing methods include standardization and interval scaling method. The premise of standardization is that the eigenvalues obey the normal distribution. After standardization, the eigenvalues obey the standard normal distribution. The interval scaling method uses boundary value information to scale the value interval of features to a specific range, such as [0, 1].

Single feature transformation is the key to building some models (such as linear regression, KNN, and neural networks), and has no effect on models related to decision trees. This is also one of the reasons why the decision tree and all its derived algorithms (random forest, gradient boost) are becoming increasingly popular. There are also some purely engineering reasons, that is, when regression prediction is made, logarithmic processing of the target can not only reduce the data range, but also compress the variable scale to make the data more stable. This conversion method is only a special case, usually driven by the desire to adapt the data set to the requirements of the algorithm.

However, data requirements are not only imposed by parameterization methods. If features are not normalized, for example, when the distribution of one feature is located near 0 and the range does not exceed $(-1,1)$, while the distribution range of another feature can reach the order of magnitudes of hundreds of thousands, it will cause features whose distribution is located near 0 to become completely useless.

Here is a simple example: suppose the task is to predict the cost of apartments based on two variables: the number of rooms and the distance to the city center. The number of apartments rarely exceeds 5, and the distance to the city center can easily reach several kilometers. At this moment, it is not possible to use models such as linear regression or KNN, and these two variables need to be normalized.

- **Standardization.** The simplest conversion is standardization (or zero—mean normalization). Standardization requires calculating the mean and standard deviation of a feature, and its formula is expressed as formula (4.1) where μ is the mean and σ is the standard deviation.

$$x' = \frac{x - \mu}{\sigma} \tag{4.1}$$

- **Interval scaling.** There are many ideas for interval scaling. The common one is to use two maximum values to scale, so that all points can be scaled within a predetermined range, i.e. [0, 1]. The formula for interval scaling is expressed as formula (4.2):

$$X_{\text{norm}} = \frac{X - X_{\text{min}}}{X_{\text{max}} - X_{\text{min}}} \tag{4.2}$$

4.2.2 Data Transformation of Continuous Variables

4.2.2.1 log Transformation

Performing log transformation can make the skew data close to the normal distribution, because most machine learning models cannot handle non-normally distributed data well, such as right skewed data. You can apply log $(x + 1)$ conversion to correct the skew, where the purpose of adding 1 is to prevent the data from equaling 0 while ensuring that x is all positive. Taking logarithms does not change the nature and correlation of the data but compresses the scale of variables. Doing so not only makes the data more stable, but also weakens the collinearity, heteroscedasticity, etc. of the model.

> **Extended Learning**
> cbox-cox transformation—a method to automatically find the best normal distribution transformation function. This method is not commonly used in competitions, and readers who are interested in it can learn about it.

4.2.2.2 Discretization of Continuous Variables

The discretized features have strong robustness to abnormal data, making it easier to explore the correlation of the data. For example, the result of discretizing the age feature is: if the age is greater than 30, it is 1; otherwise, it is 0. If this feature is not discretized, then an abnormal data "age 300 years old" will cause great interference

Fig. 4.3 GBDT + LR
Models

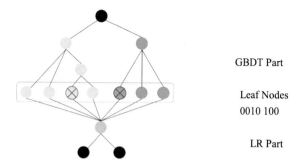

GBDT Part

Leaf Nodes
0010 100

LR Part

to the model. After discretization, we can also cross-combine the features. Commonly used discretization is divided into unsupervised type and supervised type.

- **Unsupervised discretization.** The bucketing operation can discretize continuous variables and smooth the data at the same time, that is, reduce the influence of noise, generally being divided into two bucketing methods: equifrequency and isometry.
- **Equifrequency**. The boundary value of the interval is selected so that each interval contains approximately the same number of variable instances. For example, if it is divided into 10 intervals, each interval should contain about 10% of the instances. This bucketing method can transform the data into a uniform distribution.
- **Isometry**. Divide the instances from the minimum to the maximum into N equal parts, and the spacing of each part is equal. Only the boundary is considered here, and the number of instances per equal part may vary. Isometry can maintain the original distribution of the data, and the more intervals, the better the original appearance of the data will be.
- **Supervised discretization.** This type of method has a good ability to distinguish between targets. It is commonly applied to return leaf nodes for discretization by using the tree model. In the GBDT + LR classic model shown in Fig. 4.3, GDBT is used first to convert continuous values into discrete values. The specific method is to use all continuous values and label output in the training set to train LightGBM, and train two decision trees in total; the first tree has 4 leaf nodes, and the second tree has 3 leaf nodes. If a sample falls on the third leaf node of the first tree or falls on the first leaf node of the second tree, then its code is 0010 100, with a total of 7 discrete features, of which there will be two positions with a value of 1, corresponding to the position of the sample placement in each tree. Eventually we will get num_trees * num_leaves dimensional features.

4.2.3   Category Feature Transformation

In real data, features are not always numerical values, but may also be categories. For discrete category features, there are generally two situations: natural number encoding (features are meaningful) and one-hot encoding (features are meaningless).

- **Natural number encoding.** A list of meaningful category features (that is, with sequential relationships) can be coded with natural numbers, and their relation of sequence can be preserved by referring to order of natural numbers. In addition, when features are not represented by numbers but by letters or symbols, etc., they cannot be directly "fed" into the model as training labels, such as age, education, etc. At this time, it is necessary to convert the feature values into numbers first. If there are K values in a list of category features, then after natural number coding, you can get numbers with values of $\{0, 1, 2, 3, \ldots, K - 1\}$, that is, each category is assigned with a number. The advantages of doing so are low memory consumption and fast training time. The disadvantage is that some feature information may be lost. The following two common ways of encoding natural numbers are given.
- Call a Function in sklearn:

```
from sklearn import preprocessing
for f in columns:
  le = preprocessing.LabelEncoder()
  le.fit(data[f])
```

- Custom Implementation (fast):

```
for f in columns:
    data[f] = data[f].fillna(-999)
    data[f] = data[f].map(dict(zip(data[f].unique(), range(0,
data[f].nunique()))))
```

- **One-hot encoding.** When the category feature is meaningless (that is, there is no order relationship), you need to use one-hot encoding. For example, red > blue > green does not represent anything. After one-hot encoding, the value of each feature corresponds to a one-dimensional feature, and the final result is a 0~1 matrix with the number of samples × the number of categories. You can directly call the API in sklearn.

4.2.4 Irregular Feature Transformation

In addition to numerical features and category features, there is another type of irregular feature that may contain a lot of information about the sample, such as the ID card number. According to the provisions on citizenship ID numbers in the "National Standard GB 11643-1999" of the People's Republic of China, the citizenship identification number is a feature combination code, consisting of a 17-digit ontology code and a one-digit verification code. The order from left to right is: six-digit address code, eight-digit date of birth code, three-digit sequence code and one-digit verification code. Among them, the odd number of the sequence code is distributed to men and the even number is distributed to women. The verification code is a verification code calculated according to ISO 7064:1983. MOD 11-2 verification code based on the previous 17 digits. Therefore, we can obtain the user's birthplace, age, gender, and other information from the ID number. Of course, the ID number involves the user's privacy, and it is impossible for the organizer to provide this information in the competition, so it is only an example.

4.3 Feature Extraction

Machine learning models are difficult to identify complex patterns; it is especially difficult to learn information that interacts with different combinations of features, so we can create some features based on intuitive analysis of the data set and business understanding to help the model learn effectively. We will introduce the feature extraction method of structured data below. (Structured data consists of clearly defined data types, while unstructured data consists of data that is not easy to search, such as audio, video, and pictures.)

4.3.1 Statistics Features Related to Categories

Category features can also be called discrete features. In addition to the specific meaning of each category attribute, continuous statistical features can also be constructed to mine more valuable information, such as constructing features such as target coding, count, nunique, and ratio. In addition, it is also possible to construct more fine grained features through cross combinations between category features.

4.3.1.1 Target Coding

Target coding can be understood as encoding category features with the statistics of target variables (labels), that is, supervised feature construction according to target

Training Set Five-fold
Cross Construction Training Set

Complete Training Set
Construction Verification Set

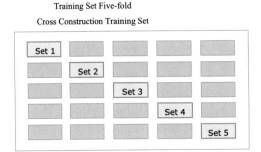

Fig. 4.4 Five-fold Cross Statistics Construction Features

variables. If it is a classification problem, you can count the number of positive samples, the number of negative samples, or the proportion of positive and negative samples; if it is a regression problem, then you can count the target mean, median, and extreme values. Target coding can be a good substitute for category features or as a new feature.

When using target variables, it is very important not to reveal any information about the verification set. All features based on the target code should be calculated on the training set, and the testing set should be constructed from the complete training set. More strictly, when we count the features of the training set, we need to use the K-fold cross statistics method to construct the target code features, so as to prevent information leakage to the greatest extent. As shown in Fig. 4.4, we divide the sample into five parts. For each part of the data, we will use the other four parts to calculate the frequency, proportion, or mean value of the target variable corresponding to the value of each category. Simply put, the unknown data (one part) takes features from the known data (four parts).

Target coding methods are usually effective for low-cardinality category features, but for high-cardinality category features, there may be a risk of over-fitting. Because there will be some categories with very low frequency, the statistical results are not representative. Generally, we will add smoothness to reduce the risk of over-fitting. When handled properly, whether it is a linear model or a nonlinear model, target coding is the best coding method and feature construction method. In order to help you better understand, the five-fold cross statistics is implemented by the following code:

```
folds = KFold(n_splits=5, shuffle=True, random_state=2020)
for col in columns:
  colname = col+'_kfold'
  for fold_, (trn_idx, val_idx) in enumerate(folds.split(train,
  train)):
    tmp = train.iloc[trn_idx]
    order_label = tmp.groupby([col])['label'].mean()
    train[colname] = train[col].map(order_label)
order_label = train.groupby([col])['label'].mean()
test[colname] = test[col].map(order_label)
```

4.3.1.2   count, nunique, ratio

These three categories are often used to construct category features in competitions. The count (counting feature) is used to count the frequency of occurrence of category features. The structure of nunique and ratio is relatively complicated, and it often involves a joint structure of multiple category features. For example, in the advertisement click-through rate forecast problem, for the user ID and advertisement ID, the use of nunique can reflect the range of the user's interest in the advertisement, that is, how many kinds of advertisement IDs the user ID has seen; the use of ratio can indicate the user's preference level for a certain type of advertisement; that is, the ratio of the frequency of user ID clicking on a certain type of advertisement ID to the frequency of user clicking on all AD IDs is calculated. Of course, this also applies to other problems, such as malicious attacks, anti-fraud, and credit scores, which need to construct behavior information or describe distribution information.

4.3.1.3   Cross Combination Between Category Features

Cross combinations can describe more fine grained content. Crossover combination of category features is a very important task in competitions, so that good nonlinear feature fitting can be performed. As shown in Fig. 4.5, user age and user gender can be combined into new features such as "age _ gender". Generally, we can combine two categories or three category features, also known as second-order combinations or third-order combinations. Simply put, it is to perform Cartesian product operations on two category features to generate new category features. In the real data, there may be many category features. If you have 10 category features and consider all the second-order cross combinations, you can produce 45 combinations.

Not all combinations need to be considered. We will analyze from two aspects. The first is the business logic. For example, the combination of the user's operating system version and the user's city is meaningless. Then there is the cardinality of

Fig. 4.5 Cross Combination between Category Features

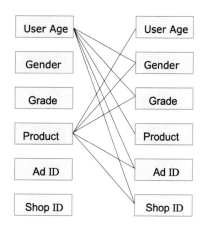

category features. If the cardinality is too large, many categories may appear only once. In a round of training, each category will only be trained once. Obviously, the confidence level of the weight corresponding to the feature is very low.

4.3.2   Numerical Correlation of Statistical Features

The numerical features mentioned here, we think, are continuous, such as housing price, sales volume, number of clicks, number of comments, and temperature, etc. Unlike category features, the value of a numerical feature is meaningful, and usually can be directly "fed" to the model for training without processing. In addition to various transformations of numerical features in advance, there are some other common ways to construct numerical features.

- **Cross combination between numerical features.** Unlike the cross combination between category features, typically, the cross combination of arithmetic operations such as addition, subtraction, multiplication, and division will be performed on numerical features. This requires us to combine business understanding and data analytics to construct, rather than make a violent construction without thinking twice. For example, if you give the size of the housing (in square meters) and the selling price, you can construct the average price per square meter. Or given the user's monthly consumption amount in the past three months, you can construct the total consumption amount and average consumption amount in these three months to reflect the user's overall consumption ability.
- **Cross combination between category features and numerical features.** In addition to the cross combination between different category features and that between different numerical features, cross combination between category features and numerical features can also be constructed. Such features usually calculate some statistics of numerical features in a certain category in category features, such as mean, median, and extreme values.
- **Row-by-row statistics of related features.** This approach is somewhat similar to feature crossover, where information about multi-column features is combined. However, row statistics will contain more columns when constructed, and directly count multiple columns by row, such as the number of 0, null values, positive and negative values, or mean, median, extreme values, or sum, etc. Multi-column features may be the amount of consumption and electricity consumption per month, and in industrial data those can be the temperature and concentration of each stage of chemical experiments. For these data containing multi-column related features, we all need to analyze the changes of multi-column values and extract valuable features from them.

4.3 Feature Extraction   69

4.3.3 Time Features

In real data, the time characteristic usually given is the timestamp attribute, so you need to separate it into multiple dimensions first, such as year, month, day, hour, minute, and second. If your data source comes from different geographic data sources, you also need to use time zones to standardize the data. In addition to the basic time features separated, the time difference feature may also be constructed, i.e., the calculated numerical difference between the time of each sample and a future time, so that the gap is the time difference of the UTC, thereby converting the time characteristic into a continuous value, such as the time difference between the date of the user's first behavior and the date of registration of the user, and the time difference between the user's current behavior and the user's last behavior.

4.3.4 Multiple-Valued Features

In an actual competition, you may encounter a situation where each row in a column of features contains multiple attributes, which is a multiple-valued feature. For example, the interest category in the 2018 Tencent Advertising Algorithm Competition contains five interest feature groups, and each interest feature group contains several interest IDs. For multiple-valued features, sparsity or vectorization can usually be performed. This operation generally occurs in natural language processing. For example, after text segmentation, TF-IDF, LDA, NMF, etc. are used for processing. Here, multiple-valued features can be regarded as the results after the word segmentation of the text, with the same processing afterwards.

As shown in Fig. 4.6, the most basic way to deal with multiple-valued features is to expand completely, that is, to expand the *n* attributes contained in this list of features into an n dimensional sparse matrix. Using the CountVectorizer function in

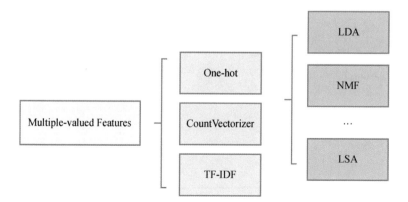

Fig. 4.6 Multiple-valued Feature Processing Method

sklearn, multiple-valued features can be easily expanded, by only taking into account the frequency of each attribute in this feature.

There is also another case. For example, in the 2020 Tencent advertising algorithm contest, it is necessary to predict the user's attribute tab according to the user's historical click behavior. At this time, the user's click sequence is particularly important. When we construct the user's corresponding historical click sequence, in addition to using the above-mentioned TF-IDF and other methods, we can also extract the embedded representation of goods or advertisements in the click sequence, such as using Word2Vec, DeepWalk, and other methods to obtain the embedding vector representation. Because we want to extract a single feature of the user, we can aggregate statistics on the embedded vectors in the sequence. This method is essentially based on the assumption that the goods or advertisements clicked by the user are equally important, which is a relatively crude way of processing. We can introduce time attenuation factors, or use sequence models such as RNN, LSTN, GRU, and apply NLP methods to solve them.

So far, the construction method of basic type features has been given. Of course, there are still many types that have not been mentioned, such as spatial features, time series features, and text features, as well as clustering and dimensionality reduction methods, which we will introduce in detail in the following chapters while dealing with specific problems.

4.4   Feature Selection

As shown in Fig. 4.7, when we add new features, we need to verify whether it can indeed improve the accuracy of model prediction to determine whether useless features are not added, because this will only increase the complexity of the algorithm operation. At this time, it is necessary to automatically select the optimal

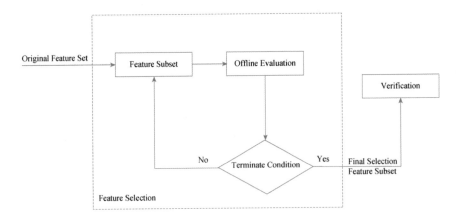

Fig. 4.7 Feature Selection Process

subset of the feature set through the feature selection algorithm to help the model provide better performance. The feature selection algorithm is used to identify and delete unnecessary, irrelevant, and redundant features from the data, which may reduce the accuracy and performance of the model. The methods of feature selection mainly include prior feature correlation analysis and posterior feature importance analysis.

4.4.1 Feature Correlation Analysis

Feature correlation analysis uses statistics to score the correlation between features. Features are sorted according to scores, either retained or deleted from the data set. Correlation analysis methods are usually for single variables and independently consider features or dependent variables. Common feature correlation analysis methods include Pearson correlation coefficient, chi-square test, mutual information method, and information gain. These methods are very fast and convenient to use but ignore the relationship between features and the relationship between features and models.

- **Pearson correlation coefficient.** This method can not only measure the linear correlation between variables and solve the problem of collinear variables, but also measure the correlation between features and labels. Collinear variables refer to the high correlation between variables, which reduces the learning usability, interpretability, and generalization performance of the testing set of the model. Obviously, these three features are all we want to add, so removing collinear variables is a valuable step. We will establish a basic threshold for deleting collinear variables (depending on the number of features you want to retain), and then delete one from any pair of variables higher than this threshold.

The following code is used to solve the problem that features do not have correlation with labels, extracting similar features of top 300 according to the calculation of Pearson correlation coefficient:

```
def feature_select_pearson(train, features):
    featureSelect = features[:]
    # Perform Calculation of Pearson Correlation Coefficient
    corr = []
for feat in featureSelect:
    corr.append(abs(train[[feat, 'target']].fillna(0).corr().values
    [0][1]))

se = pd.Series(corr, index=featureSelect).sort_values
(ascending=False)
feature_select = se[:300].index.tolist()
# Return Training Set after Feature Selection
return train[feature_select]
```

- **Chi-square test**. It is used to test the correlation between the characteristic variable and the dependent variable. For classification problems, it is generally assumed that the feature independent of the label is an independent feature, and the chi-square test happens to be able to perform the independence test, so it is suitable for feature selection. If the test result is that a certain feature is independent of the label, the feature can be removed. Chi-square formula is as formula (4.3):

$$\chi^2 = \sum \frac{(A - E)^2}{E} \tag{4.3}$$

- **Mutual information method**. Mutual information is a measure of the relationship between two variables in a joint distribution. It can also be used to evaluate the correlation between two variables. The reason why the mutual information method can be used for feature selection can be explained from two aspects: based on KL divergence and based on information gain. The greater the mutual information, the higher the correlation between the two variables will be. Mutual information formula is shown as formula (4.4):

$$\mathrm{MI}(x_i, y) = \sum_{x_i \in \{0,\, 1\}} \sum_{y \in \{0,\, 1\}} p(x_i, y) \log\left(\frac{p(x_i, y)}{p(x_i)p(y)}\right) \tag{4.4}$$

The $p\,(x_i, y)$, $p\,(x_i)$, and $p\,(y)$ here are all obtained from the training set. It is not very convenient to use mutual information directly for feature selection for the following two main reasons.

- It is not a measure method, and there is no way to normalize it, so the results on different data sets cannot be compared.
- It is not very convenient to calculate continuous variables (X and Y are sets, x_i and y are discrete values). Usually, continuous variables need to be discretized first, and the results of mutual information are very sensitive to the way of discretization.

4.4.2   Feature Importance Analysis

A feature selection method often used in real-world competitions is to evaluate the importance score of features based on the tree model. The higher the importance score of the feature, the more times the feature is used in the model to build a decision tree. Here we take XGBoost as an example to introduce three calculation methods (weight, gain, and cover) for the tree model to evaluate the importance of features. (LightGBM can also return feature importance.)

- **The weight calculation method.** This method is relatively simple. It calculates the number of times a feature is selected as a split feature in all trees and uses this as a basis for evaluating the importance of the feature. The code example is as follows:

```
params = {
    'max_depth': 10,
    'subsample': 1,
    'verbose_eval': True,
    'seed': 12,
    'objective':'binary:logistic'
}

xgtrain = xgb.DMatrix(x, label=y)
bst = xgb.train(params, xgtrain, num_boost_round=10)
importance = bst.get_score(fmap = '',importance_type='weight')
```

- **The gain calculation method.** gain represents the average gain. When evaluating the importance of a feature, gain is used to represent the sum of the information gains of the feature as a split node in all trees and then divided by the frequency of occurrence of the feature. The code example is as follows:

```
importance = bst.get_score(fmap = '',importance_type='gain')
```

- **The cover calculation method.** cover is more complicated. Its specific meaning is the coverage rate of the feature to each tree, that is, the sum of the second derivatives of the sample where the feature is assigned to the node, and the standard of feature measurement is the average coverage value. The code example is as follows:

```
importance = bst.get_score(fmap = '',importance_type='cover')
```

Using Skills

Although feature importance can help us quickly analyze the importance of features in the model training process, it cannot be used as an absolute reference basis. In general, as long as features do not lead to over-fitting, we can select features with high importance for analysis and expansion. For features with low importance, we can consider removing them from the feature set, then observe the offline effect and make further judgment.

4.4.3   Encapsulation Methods

The encapsulation method is a time-consuming feature selection method. The selection of a set of features can be regarded as a search problem, in which different combinations are prepared, evaluated, and compared to find the optimal feature subset. The search process, such as the best priority search, can be systematic; it can also be random, such as random hill climbing algorithm, or heuristic methods, such as adding and deleting features by searching forward and backward (similar to pre-pruning and post-pruning algorithms). The following describes two commonly used encapsulation methods.

- **Heuristic method.** Heuristic methods are divided into two types: forward search and backward search. Forward search, to put it bluntly, is to incrementally select one of the remaining unselected features and add it to the feature set. When the number of features in the feature set reaches the initial threshold, it means greedily selecting the feature subset with the smallest error rate. Since there is incremental addition, there will be incremental subtraction. The latter is called backward search; that is, starting from the complete set of features, one feature is deleted and evaluated at a time until the number of features in the feature set reaches the initial threshold, and the best feature subset is selected.

   We can also expand on this basis. Because the heuristic method will lead to local optimization, a simulated annealing method is added to improve it. This method does not discard the newly added feature because it cannot improve the effect but adds weights to it and puts it into the selected feature set.

   This heuristic method has been carried out in the competition and is a time-consuming and resource-consuming operation. In most cases, it can be used when the online and offline gains are the same and the data set magnitude is not large.

- **Recursive elimination feature method.** Recursive elimination feature method uses a base model for multiple rounds of training. Each round of training will first eliminate the features of several weight coefficients, and then conduct the next round of training based on the new feature collection. You can use the RFE class of the feature_selection library for feature selection. The code example is as follows:

```
from sklearn.feature_selection import RFE

from sklearn.linear_model import LogisticRegression

# Recursive Elimination Feature Method, Return Data after Feature Selection
# Parameter estimator is the Base Model
# Parameter n_features_to_select is the Number of Features Selected
RFE(estimator=LogisticRegression(),n_features_to_select=2).
fit_transform(data, target)
```

Using Skills

When using encapsulation methods for feature selection, training with full data is not the wisest choice. Big data should be sampled first, and then encapsulation methods should be used for small data.

The above three feature selection methods need to be selected or combined in accordance with real problems. It is recommended to give priority to feature importance, followed by feature relevance. In addition, there are some uncommon feature selection methods, such as the very classic null importance feature selection method on Kaggle.

Models are sometimes really stupid. Many features that are not associated with the target tag at all can also be associated with the target tag. Features that are falsely associated with the testing set will cause over-fitting, which will have a negative impact. After that, feature importance analysis will become less reliable. So, how can we distinguish whether a feature is useful in feature importance analysis?

The idea of null importance is actually very simple, that is, to feed the constructed features and correct labels into the tree model to obtain a feature importance score, then to feed the features and disrupted labels into the tree model to obtain a feature importance score, and then to compare the two scores. If the former does not exceed the latter, then this feature is a useless feature.

4.5 Practical Cases

With the content foreshadowing of Chaps. 2 and 3, we can then carry out the actual practice operation of the feature engineering part. Here is mainly a brief review of what we have learned in this chapter, but there are also many feature engineering skills that are not usually used, but please rest assured that in the link for later actual practice in competitions, we will conduct more detailed application practice.

4.5.1 Data Preprocessing

The main work at this stage might be data cleaning, and timely processing of missing values and outliers. Execute the following code to perform basic data reading, delete feature columns with missing values greater than 50%, and fill in the missing features of the object type:

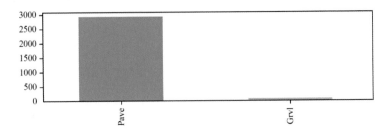

Fig. 4.8 Display of Street Distribution

```
test = pd.read_csv("../input/test.csv")
train = pd.read_csv("../input/train.csv")
ntrain = train.shape[0]
ntest = test.shape[0]

data = pd.concat([train, test], axis=0, sort=False)
# delete feature columns with missing values greater than 50%
missing_cols = [c for c in data if data[c].isna().mean()*100 > 50]
data = data.drop(missing_cols, axis=1)

# fill in the missing features of the object type
object_df = data.select_dtypes(include=['object'])
numerical_df = data.select_dtypes(exclude=['object'])

object_df = object_df.fillna('unknow')
```

Next, fill in the numerical features with the median:

```
missing_cols = [c for c in numerical_df if numerical_df[c].isna().sum
     () > 0]
for c in missing_cols:
   numerical_df[c] = numerical_df[c].fillna(numerical_df[c].median
   ())
```

For the extremely uneven distribution of attributes in the features, such as the existence of an attribute that accounts for more than 95%, it is also necessary to consider whether to delete it. The Street shown in Fig. 4.8 is one such feature, with others such as Heating, RoofMatl, Condition2, and Utilities also belonging to this type.

The following is the feature deletion code:

```
object_df = object_df.drop
([ 'Heating','RoofMatl','Condition2','Street','Utilities'],axis=1)
```

4.5.2 *Feature Extraction*

The feature extraction stage will construct specific features from multiple angles, and each feature constructed has practical significance.

4.5.2.1 Basic Feature Construction

When the housing was built is also a factor that affects the price of it. When constructing this feature, it is found that there is data with a sales date (YrSold) smaller than the construction date (YearBuilt), which needs to be adjusted for this anomaly. Specifically, change the sales date of the anomaly data to the maximum year (2009) of the sales date in the data set. Here is the specific code for constructing the feature:

```
numerical_df.loc[numerical_df['YrSold'] < numerical_df
['YearBuilt'], 'YrSold'] = 2009
numerical_df['Age_House'] = (numerical_df['YrSold'] - numerical_df
['YearBuilt'])
```

Next, construct business-related features, such as summing BsmtFullBath and BsmtHalfBath in the original feature, summing FullBath and HalfBath, summing the first floor area (1stFlrSF), the second floor area (2ndFlrSF), and the basement area to represent the structural information of the housing. The code is as follows:

```
numerical_df['TotalBsmtBath'] = numerical_df['BsmtFullBath'] +
   numerical_df['BsmtHalfBath']*0.5
numerical_df['TotalBath'] = numerical_df['FullBath'] + numerical_df
['HalfBath']*0.5
numerical_df['TotalSA'] = numerical_df['TotalBsmtSF'] + numerical_df
['1stFlrSF'] +
   numerical_df['2ndFlrSF']
```

4.5.2.2 Feature Encoding

Features of object class cannot directly participate in model training, they need to be encoded first, and there are many coding methods for category features. So, how to choose the coding method becomes the key.

First, it is necessary to distinguish category features: for ordinal features with big-small relationship, *0-N* mapping conversion can be performed, that is, natural number coding; for features without such relationship, one-hot coding, or frequency (count) coding can be performed. The code for feature extraction is as follows:

```
bin_map = {'TA':2,'Gd':3, 'Fa':1,'Ex':4,'Po':1,'None':0,
   'Y':1,'N':0,'Reg':3,'IR1':2,'IR2':1,
```

```
    'IR3':0,"None" : 0,"No" : 2, "Mn" : 2,
    "Av": 3,"Gd" : 4,"Unf" : 1, "LwQ": 2,
    "Rec" : 3,"BLQ" : 4, "ALQ" : 5, "GLQ" : 6}
object_df['ExterQual'] = object_df['ExterQual'].map(bin_map)
object_df['ExterCond'] = object_df['ExterCond'].map(bin_map)
object_df['BsmtCond'] = object_df['BsmtCond'].map(bin_map)
object_df['BsmtQual'] = object_df['BsmtQual'].map(bin_map)
object_df['HeatingQC'] = object_df['HeatingQC'].map(bin_map)
object_df['KitchenQual'] = object_df['KitchenQual'].map(bin_map)
object_df['FireplaceQu'] = object_df['FireplaceQu'].map(bin_map)
object_df['GarageQual'] = object_df['GarageQual'].map(bin_map)
object_df['GarageCond'] = object_df['GarageCond'].map(bin_map)
object_df['CentralAir'] = object_df['CentralAir'].map(bin_map)
object_df['LotShape'] = object_df['LotShape'].map(bin_map)
object_df['BsmtExposure'] = object_df['BsmtExposure'].map(bin_map)
object_df['BsmtFinType1'] = object_df['BsmtFinType1'].map(bin_map)
object_df['BsmtFinType2'] = object_df['BsmtFinType2'].map(bin_map)
PavedDrive = {"N" : 0, "P" : 1, "Y" : 2}object_df['PavedDrive'] =
object_df['PavedDrive'].map(PavedDrive) # select the remaining object features
rest_object_columns = object_df.select_dtypes(include = ['object'])
# perform one-hot coding
object_df = pd.get_dummies(object_df, columns = rest_object_columns.
columns)
data = pd.concat([object_df, numerical_df], axis=1, sort=False)
```

We have not carried out all kinds of violent extraction operations, but mainly guided everyone to extract features from the business, in order to enable everyone to master the skills and usage methods of feature extraction, and there will be more feature extraction methods in the following cases.

4.5.3   Feature Selection

This section will use the correlation evaluation method for feature selection. This method is a kind of correlation analysis, which can filter out features whose similarity is greater than a certain threshold and reduce feature redundancy. The following creates an auxiliary function for correlation evaluation:

```
def correlation(data, threshold):
    col_corr = set()
    corr_matrix = data.corr()
    for i in range(len(corr_matrix.columns)):
        for j in range(i):
            if abs(corr_matrix.iloc[i, j]) > threshold: # comparison of similarity
            score and threshold
                colname = corr_matrix.columns[i] # get column name
                col_corr.add(colname)
    return col_corr
```

```
all_cols = [c for c in data.columns if c not in ['SalePrice']]
corr_features = correlation(data[all_cols], 0.9)
```

It is generally difficult to quantify the determination of the threshold, which can be considered from the perspective of the total feature quantity and the overall similarity score. For example, if our machine does not allow too many features to be used, we can use this as a basis to determine the feature retention quantity, which is also a relatively flexible value. Feature sets under different thresholds can be retained for training to assist in improving the fusion results of the model.

4.6 Thinking Exercises

1. How can we choose from the average, median, and mode when filling the missing values of numerical features?
2. In feature selection, there is another type of feature selection method based on penalty terms, which includes L1 regularity and L2 regularity. So, what is the difference between these two regularities? How can you choose from them?
3. How many parts is the whole feature project mainly divided into? What are the main contents of each part?
4. How are numerical features, category features, and irregular features respectively defined? Please give examples.
5. In the process of data cleaning, how could we select the method of outlier processing?
6. Why should memory be optimized and what optimization methods are available?
7. Please give examples to illustrate the meaning of count, nunique, and ratio.
8. Under what circumstances do we need to carry out feature transformation?
9. When selecting features, should the feature correlation analysis method or the feature importance analysis method be selected?

Chapter 5
Model Training

This chapter will introduce the common models used in algorithm competitions. Good models can help us approach the upper limit of scores. The optional models are mainly divided into three types: linear models, tree models, and neural networks. There is a saying that there is no best model, but only the most suitable model, so in this chapter, we will illustrate the application scenarios suitable for their corresponding models and give the using skills and application codes.

5.1 Linear Models

This section will demonstrate two linear reduction methods: Lasso regression and Ridge regression. The difference between these two linear regression models only lies in how to penalize and how to solve the overfitting problem. Then the mathematical forms, advantages and disadvantages, and application scenarios of the two models will be explained accordingly.

5.1.1 Lasso Regression

The full name of Lasso is least absolute shrinkage and selection operator, which is to optimize ordinary linear regression using L1 regularization. By penalizing or limiting the sum of the absolute values of the estimated values, some coefficients can be made zero, so as to achieve the effect of feature coefficients and feature selection. This is convenient when we need some automatic selection for features and variables, or when we deal with highly relevant predictors, because the regression coefficients of a standard regression are usually too large. The mathematical form of Lasso regression is as in the formula (5.1):

© The Author(s), under exclusive license to Springer Nature Singapore Pte Ltd. 2023
W. He et al., *Machine Learning Contests: A Guidebook*,
https://doi.org/10.1007/978-981-99-3723-3_5

$$\min\left(\|Y - X\theta\|_2^2 + \lambda\|\theta\|_1\right) \tag{5.1}$$

In this formula, λ is the coefficient of the regularization item (penalty term). By changing the value of λ, the penalty term can be more or less controlled, that is, the L1 norm of θ. The bigger the value of λ, the greater the influence of the penalty term will be, and vice versa.

5.1.1.1   Code Implementation

You can directly call the sklearn library to implement Lasso regression, where the L1 regular parameter selected is 0.1.

```
from sklearn.linear_model import Lasso
lasso_model = Lasso(alpha = 0.1, normalize = True)
```

5.1.2   Ridge Regression

Ridge regression is an optimization of ordinary linear regression using L2 regularization, and a penalty term is set for the weight coefficient of features.

Its mathematical form is as formula (5.2):

$$\min\left(\|Y - X\theta\|_2^2 + \lambda\|\theta\|_2^2\right) \tag{5.2}$$

It is basically the same as the loss function of Lasso regression, except that the penalty term is modified to the L2 norm of the θ.

5.1.2.1   Code Implementation

Ridge regression can be implemented directly by calling the sklearn library:

```
from sklearn.linear_model import Ridge
Ridge_model = Ridge(alpha=0.05, normalize=True)
```

5.1.2.2   Problem Discussion

Now that we have a preliminary understanding of Lasso regression and Ridge regression, let's consider an example. Suppose we now have a very large data set containing 10,000 features. Only some of these features are related. Then think about which regression model should be used for training, Lasso regression or Ridge regression?

If we use Lasso regression for training, then the difficulty encountered is mainly that when there are relevant features, Lasso regression retains only one of the features while setting other relevant features to zero. This may lead to the loss of some information, thus reducing the accuracy of model prediction.

If you choose Ridge regression, although it can reduce the complexity of the model, it will not reduce the number of features, because Ridge regression will never make the coefficient zero, but only minimize the coefficient. Nevertheless, this is not conducive to feature reduction. In the face of 10,000 features, the model will still be very complex, so it may lead to poor model performance.

All things considered, what is the solution to this problem? You can choose Elastic Net Regression for extended learning.

5.2 Tree Models

This section will illustrate the common tree models in the competition. These models are simple to use and can bring high returns. The tree model can be divided into random forest (RF) and gradient boosting decision tree (GBDT). The biggest difference between the two is that the former is parallel while the latter is serial. In the gradient boosting decision tree section, we will introduce three most popular tree models in competitions now: XGBoost, LightGBM, and CatBoost. Being able to use these three models flexibly is a necessary skill in the competition. Next, the mathematical form, advantages and disadvantages, details of usage, and application scenarios of various tree models will be explained in detail.

5.2.1 Random Forest

In short, the random forest algorithm is to integrate multiple decision trees together through the idea of integrated learning. The decision tree here can be a classifier, and there is no correlation between each decision tree. The random forest algorithm votes on the results of multiple decision trees to get the final result, which is also the simplest bagging idea. Random forest is a model based on nonlinear trees, which can usually provide accurate results.

5.2.1.1 Construction Process of Random Forest

The construction process of random forest is shown in Fig. 5.1.

The specific process is as follows.

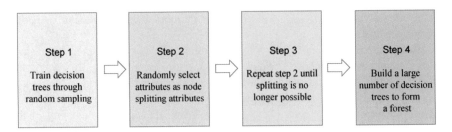

Fig. 5.1 Construction Process of Random Forest

1. Suppose the number of samples in the training set is N. N samples are randomly selected with return; that is, the samples can be repeated, and then a decision tree is trained with the N samples obtained.
2. If there are M input variables (features), specify a number m (column sampling) much smaller than M, so that m input variables will be randomly selected from M input variables for each split. Then select the best splitting variable from the m variables chosen (split based on information gain, information gain ratio, Gini index, etc.).
3. Repeat step (2) to split each node without pruning to maximize the growth of nodes (the condition for stopping splitting is that all samples of nodes belong to the same class).
4. Construct a large number of decision trees through steps (1) to (3), and then summarize these decision trees to predict new data (vote and select on classification problems and calculate the mean value of regression problems).

5.2.1.2   Advantages and Disadvantages of Random Forest

Random forest has obvious advantages: it can not only solve classification and regression problems, but also deal with category features and numerical features at the same time; it is not easy to overfit, and the risk of overfitting is reduced by averaging the decision tree; it is very stable—even if a new data point appears in the data set, the whole algorithm will not be affected too much. The new data point will only affect one decision tree, and it is difficult to affect all decision trees.

Many shortcomings are relatively speaking. Although the random forest algorithm is more complex and costly than the decision tree algorithm, it has natural parallel property and can be trained quickly in a distributed environment. Gradient boosting trees need to continuously train residuals, so the results are more accurate, but random forest is less easy to overfit and more stable, which is also due to its bagging characteristics.

5.2.1.3 Code Implementation

```
from sklearn.ensemble import RandomForestClassifier
rf = RandomForestClassifier(max_features='auto', oob_score=True,
random_state=1, n_jobs=-1)
```

5.2.2 *Gradient Boosting Decision Tree*

Gradient boosting decision tree (GBDT) is based on the boosting improvement. In the boosting algorithm, a series of base learners need to be generated serially, learning one tree at a time, and the learning goal is the residual of the previous tree. Like AdaBoost, the gradient boosting decision tree is also based on the gradient descent function. The gradient boosting decision tree algorithm has proved to be one of the most mature algorithms in the boosting algorithm set. It is characterized by increased estimation variance, more sensitive to noise in the data (both of which can be reduced by using subsamples), and due to non-parallel operations, the computational cost is significant, so it is much slower than random forest.

The gradient boosting decision tree is the foundation of XGBoost, LightGBM, and CatBoost, and its principle will be briefly introduced here. We know that the GBDT is an additive model of boosting, which is a combination of K models. Its form is as formula (5.3):

$$\widehat{y}_i = \sum_{k=1}^{K} f_k(x_i), f_k \in F \tag{5.3}$$

Generally speaking, the loss function describes the relationship between the predicted value y and the real value $\widehat{y}$. The gradient boosting decision tree is based on the residual ($y_i - F_{x_i}$, F_{x_i} is the previous model) to continuously fit the training set. The square loss function is used here. Then for n samples, it can be written as formula (5.4):

$$L = \sum_{i=1}^{n} l(y_i, \widehat{y}_i) \tag{5.4}$$

Still further, the objective function can be written as in the formula (5.5):

$$\text{Obj} = \sum_{i=1}^{n} l(y_i, \widehat{y}_i) + \sum_{k=1}^{K} \Omega(f_k) \tag{5.5}$$

Here Ω represents the complexity of the base model. If the base model is a tree model, the depth of the tree, the number of leaf nodes, and other indicators can reflect the complexity of the tree.

For boosting, it uses a forward optimization algorithm, that is, to gradually establish a base model from the front to the back to approximate the objective function; the specific process is as formula (5.6):

$$
\begin{aligned}
\widehat{y}_i^0 &= 0 \\
\widehat{y}_i^1 &= f_1(x_i) = \widehat{y}_i^0 + f_1(x_i) \\
\widehat{y}_i^2 &= f_1(x_i) + f_2(x_i) = \widehat{y}_i^1 + f_2(x_i) \\
&\vdots \\
\widehat{y}_i^t &= \sum_{k=1}^{t} f_k(x_i) = \widehat{y}_i^{t-1} + f_t(x_i)
\end{aligned}
\tag{5.6}
$$

Then how can a new model be learned in each step of the approximation process? The key to the answer is still in the objective function of the gradient boosting decision tree, that is, the addition of new models is always for the purpose of optimizing goal function. Rewrite the objective function as formula (5.7):

$$
\begin{aligned}
\text{Obj}^t &= \sum_{i=1}^{n} l(y_i, \widehat{y}_i^t) + \sum_{i=1}^{t} \Omega(f_i) \\
\text{Obj}^t &= \sum_{i=1}^{n} l\left(y_i, \widehat{y}_i^{t-1} + f_t(x_i)\right) + \Omega(f_t) + \text{constant}
\end{aligned}
\tag{5.7}
$$

Expand Taylor formula (5.6) to second order as formula (5.8):

$$
\text{Obj}^t = \sum_{i=1}^{n} \left[l\left(y_i, \widehat{y}_i^{t-1}\right) + g_i f_t(x_i) + \frac{1}{2} h_i f_t^2(x_i) \right] + \Omega(f_t) + \text{constant}
\tag{5.8}
$$

Remove the constant term to obtain formula (5.9):

$$
\text{Obj}^t \approx \sum_{i=1}^{n} \left[g_i f_t(x_i) + \frac{1}{2} h_i f_t^2(x_i) \right] + \Omega(f_t)
\tag{5.9}
$$

The reason for removing the constant term is that the constants in the function do not work in the process of function minimization. As a result, the optimization objective function of the gradient boosting decision tree becomes very unified. It only depends on the first and second derivatives of each data point on the error function, and then obtains an overall model according to the addition model.

5.2.3   *XGBoost*

XGBoost is an integrated machine learning algorithm based on decision trees, and it uses gradient boost as the framework. At the SIGKDD 2016 conference, Chen Tianqi and Carlos Guestrin published the paper "XGBoost: A Scalable Tree Boosting System", which has caused a sensation throughout the field of machine learning and has gradually played a dominant role in Kaggle and data science field. XGBoost also introduces the boosting algorithm.

In addition to its successful performance in accuracy and computational efficiency, XGBoost is also a scalable solution. Due to important adjustments to the initial tree boost GBM algorithm, XGBoost represents a new generation of GBM algorithms.

5.2.3.1   Main Characteristics

- Using the sparse sensing algorithm, XGBoost can take advantage of sparse matrices, saving memory (no dense matrices are required) and reducing computation time (zero values are handled in a special way).
- Approximate tree learning (weighted quantiles sketches) is a kind of learning method that can obtain approximate results, but it saves a lot of time than complete branch cutting exploration does.
- Perform parallel computing on one machine (using multithreading in the search for the best segmentation phase) and similar distributed computing on multiple machines.
- Use an optimization method called out-of-core computing to solve the problem of taking too long to read data on disks. Divide the data set into multiple blocks and store them on disks. Use a separate thread to read data from disk and load it into memory. In this way, reading data from disks and completing data calculation in memory can run in parallel.
- XGBoost can also effectively deal with missing values, and automatically learn the segmentation direction for the missing values during training. The basic idea is to let the missing values be segmented to the left and right nodes of the decision tree respectively in each segmentation, then select the segmentation direction with large gain by calculating the gain score to split, and finally learn an optimal default segmentation direction for the missing values of each feature.

5.2.3.2   Code Implementation

Input: training set X_train, training set label y_train
  verification set X_valid, verification set label y_valid
  testing set X_test
  Output: trained model model, testing set result y_pred

```
import xgboost as xgb
params = {'eta': 0.01, 'max_depth': 11,'objective': 'reg:linear',
'eval_metric': 'rmse' }
dtrain = xgb.DMatrix(data=X_train, label=y_train)
dtest = xgb.DMatrix(data=X_valid, label=y_valid)
watchlist = [(train_data, 'train'), (valid_data, 'valid_data')]
model=xgb.train(param, train_data,
        num_boost_round=20000,
        evals=watchlist,
        early_stopping_rounds=200,
        verbose_eval=500)
y_pred = model.predict(xgb.DMatrix(X_test), ntree_limit=model.
best_ntree_limit)
```

5.2.4   LightGBM

LightGBM is an open-source project developed by a Microsoft team on Github. The LightGBM algorithm with high-performance has the advantages of being able to be distributed and quickly process large amounts of data. Although LightGBM is based on decision trees and XGBoost, it also follows other different strategies. XGBoost uses a decision tree to split a variable and explore different cut points on that variable (tree growth strategy by level), while LightGBM focuses on splitting by leaf node in order to get a better fit (this is tree growth strategy by leaf). This allows LightGBM to quickly get a good data fit and generate a solution that can replace XGBoost. Algorithmically, XGBoost calculates the split structure of the decision tree as a graph, using breadth-first search (BFS), while LightGBM uses depth-first search (DFS).

5.2.4.1   Main Characteristics

- Higher accuracy and shorter training time than XGBoost.
- Supporting parallel tree enhancement, providing better training speed than XGBoost even on large datasets.
- By using histogram algorithm to extract continuous features into discrete features, amazing fast training speed and low memory usage are realized.
- Achieve higher accuracy by using segmentation by leaves instead of by levels, speed up the process of objective function convergence, and capture the underlying patterns of training data in a very complex tree. Use num_leaves and max_depth hyperparameters to control overfitting.

5.2.4.2   Code Implementation

```
import lightgbm as lgb
params = {'num_leaves': 54,'objective': 'regression','max_depth':
18,
    'learning_rate': 0.01,'boosting': 'gbdt','metric':
'rmse','lambda_l1': 0.1}
model = lgb.LGBMRegressor(**params, n_estimators = 20000, nthread =
4, n_jobs = -1)
model.fit(X_train, y_train,
    eval_set=[(X_train, y_train), (X_valid, y_valid)],
    eval_metric='rmse',
    verbose=1000, early_stopping_rounds=200)
y_pred = model.predict(X_test, num_iteration=model.best_iteration_)
```

5.2.5   CatBoost

CatBoost is a GBM algorithm open-sourced by the Russian search engine Yandex in July 2017. Its most powerful point is that it can use a strategy that mixes one-hot coding and average coding to deal with category features.

The method used by CatBoost to encode category features is not a new method. It is mean-value coding, which has become a feature engineering method and is widely used in various data science competitions, such as Kaggle. Mean-value coding, also known as likelihood coding, impacts coding, or target coding, can converts labels to numbers based on them, and associate them with target variables. If it is a regression problem, the label is converted based on the typical average target value of the level; if it is a classification problem, then only the target classification probability of the label (the target probability depends on the value of each category) is given. Mean-value coding may seem like a simple and smart feature engineering trick, but in fact it also has side effects, mostly overfitting, because it will bring target information into the prediction.

5.2.5.1   Main Characteristics

- Class features are supported, so we do not need to preprocess class features (e.g. by label encoding or one-hot encoding). In fact, CatBoost doc says not to use one-hot coding during preprocessing because "it will affect the training speed and result quality".
- A new gradient boosting mechanism (Ordered Boosting) is proposed, which can not only reduce the risk of overfitting, but also greatly improve the accuracy.
- Support GPU training out of the box (just set task_type = "GPU").
- Combination category features are used in the training, and the connection between features is used to greatly enrich the feature dimensions.

5.2.5.2   In-depth Comprehension of Feature Combinations

Another powerful feature of CatBoost is that when the tree splits to select nodes, the combination between all category features can be taken into account; that is, the combination of two category features can be combined. The specific approach is: not to consider the combination of category features during splitting for the first time, and then consider the combination between category features during later splitting, using the greedy algorithm to generate the best combination, and then converting the combined category features into numerical features. CatBoost will use the two sets of values obtained by splitting as category features to participate in the subsequent feature combination to achieve a more fine-grained combination.

5.2.5.3   Code Implementation

```
from catboost import CatBoostRegressor
params = {'learning_rate': 0.02,'depth': 13,'bootstrap_type':
'Bernoulli',
     'od_type': 'Iter', 'od_wait': 50, 'random_seed': 11}
model = CatBoostRegressor(iterations=20000, eval_metric='RMSE',
**params)
model.fit(X_train, y_train, eval_set=(X_valid, y_valid),
     cat_features=[], use_best_model=True, verbose=False)
y_pred = model.predict(X_test)
```

Each type of tree model has its own distinctive features. Next, we will deeply understand these tree models from four aspects—the growth strategy of decision trees, gradient deviation, category feature processing, and parameter comparison—to help participants better apply them to competitions.

5.2.5.4   More Functions

CatBoost currently also supports input text features, so there is no need to perform cumbersome operations to obtain standardized input and then feed it to the model as before. Text features are marked the same way as category features—just assign a list of text variable names to the text_features during training. So, how does CatBoost handle text features internally? In fact, the operation is very conventional. CatBoost converts the input text features into numerical features internally. The specific process is word segmentation, dictionary creation, and text features converted into multi-valued numerical features. The following processing method can choose more items, such as fully expanding into Boolean 0/1 features, or performing word frequency statistics.

5.2.6 In-Depth Comparison of Models

XGBoost, LightGBM, and CatBoost are three very core tree models. This section will analyze them because there are countless relationships between the three. Only by clarifying the relationships can we make better use of these three models.

5.2.6.1 Decision Trees Growth Strategy

Figure 5.2 lists the three ways a decision tree grows.

XGBoost uses level-wise level-by-level growth, which can split leaves of the same layer at the same time, thus performing multi-threaded optimization. It is not easy to over-fit, but many leaf nodes have low split gain, which will affect performance.

LightGBM uses leaf-wise splitting method. Each time, the node with the greatest gain is selected from the current leaf for splitting, and the loop iterates, but it will grow a very deep decision tree, resulting in over-fitting. At this time, you can adjust the parameter max_depth to prevent over-fitting.

CatBoost uses an oblivious-tree (symmetric tree), which makes the nodes growth mirrored. Compared with traditional growth strategies, oblivious-tree can easily fit the scheme and quickly generate models. This tree structure plays a regularization role and is not easy to over-fit.

5.2.6.2 Gradient Bias

The boosting tree algorithms in XGBoost and LightGBM are biased gradient estimates, and the data used in the gradient estimates are the same as the data used in the current model, which will lead to data leakage and overfitting.

CatBoost has improved the boosting tree algorithm to convert the original biased gradient estimate to the unbiased gradient estimate. The specific approach is to use all training sets (except article i) to build model M_i, and then use data from article 1 to

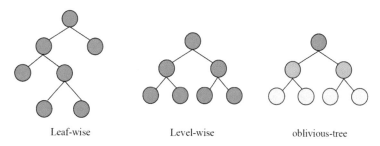

Leaf-wise   Level-wise   oblivious-tree

Fig. 5.2 Growth Pattern of a Decision Tree

article $i - 1$ to build a correction tree M, which is accumulated to the original model M_i.

5.2.6.3   Category Features Processing

XGBoost does not deal with category features, so we need to perform one-hot coding, count coding, and target coding according to the actual situation of the data.

LightGBM directly supports category features without the need for one-hot expansion. Here, the segmentation method of many-vs-many is used to deal with category features, and the time complexity of searching for the best segmentation point can be controlled to the linear level, which is almost the same as the original one-vs-other method. The algorithm first sorts according to the label mean value corresponding to each category (i.e. avg(y) = Sum(y)/Count(y)), and then enumerates the best segmentation points in turn according to the sorting results. Different from the segmentation method of numerical features, it regards a certain category as one category and then regards all other categories as one category.

CatBoost takes a more nuanced approach to category features. While LightGBM may require more coding for category features, CatBoost can choose not to use redundant coding.

The implementation process is to randomly order the input sample set first, and then, for a certain value in the category feature, average it based on the class label that precedes the sample while converting the feature of each sample to a numerical type. All class feature results are performed as shown in the formula (5.10) operation, so that they are converted into numerical results.

$$\frac{\sum_{j=1}^{p-1}\left[x_{\sigma_j,k}=x_{\sigma_p,k}\right] \times Y_{\sigma_j} + a \times P}{\sum_{j=1}^{p-1}\left[x_{\sigma_j,k}=x_{\sigma_p,k}\right] + a} \tag{5.10}$$

Here [] is the indicator function; take 1 when two elements in square brackets are equal and take 0 when not equal. a $(a > 0)$ is the weight of a priori value p, and it is a common practice to add a priori value, which helps to reduce noise obtained from low-frequency categories and reduce overfitting. For regression problems, take the average value of the label as a priori value; for classification problems, take the probability of the occurrence of a positive class as a priori value.

5.2.6.4   Parameters Comparison

As shown in Fig. 5.3, the parameters of the tree model are compared from three aspects, namely, three types of parameters used to control overfitting, to control training speed, and to adjust category characteristics. Here only some important parameters are enumerated, and there are a large number of useful parameters that will not be introduced one by one.

Parameter Function/ Tree Models	XGBoost	LightGBM	CatBoost
Parameters used to control overfitting	1. **learning_rate / eta:** reduce the weight of each step (Shinkage method); generally between 0.01 and 0.2 2. **max_depth:** max depth of tree division 3. **min_child_weight:** default 1; weight sum of mini leaf node samples	1. **learning_rate:** learning rate 2. **max_depth:** default 20; max depth of trees; num_leaves = 2 ^ (max_depth) representing the max number of leaf nodes 3. **min_data_in_leaf:** default 20; Each leaf node corresponds to the mini amount of data	1. **learning_rate:** learning rate 2. **Depth:** the depth of the tree 3. no parameters similar to min_child_weight 4. **l2-leaf-reg:** L2 regularization coefficient, which is the constraint term of leaf node weight
Parameters used to control training speed	1. **colsample_bytree:** ratio of random column samples 2. **subsampling:** the proportion of random samples 3. **n_estimators:** the max number of decision trees; higher values may lead to over-fitting	1. **feature_fraction:** random column sampling ratio per iteration 2. **bagging_fraction:** ratio of random sample per iteration 3. **num_iterations:** default 100, iteration times	1.**rsm:** random subspace, proportion of features selected per split 2. no parameters similar to sample sampling 3. **iterations:** max number of trees that can be built
Parameters used to adjust category features	No such parameters	1. **categorical_feature:** positional indexes corresponding to category features	1. **cat_features:** positional indexes corresponding to category features 2. **one_hot_max_size:** used to limit the length of the one-hot feature vector; the default is False

Fig. 5.3 Comparison of Core Parameters

Fig. 5.4 Neural Networks

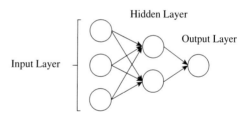

5.3   Neural Networks

If you want to go further in the competition, then neural networks are also models that must be mastered. Generally speaking, as the amount of data that we have continues to increase, the possibility of neural networks defeating traditional machine learning models will also increase.

First, there is an example of the cost of a house to show the functional details of neural networks. Specifically, the price of a house should be estimated according to certain functions of the house. If detailed information such as the area of the house, its location, and the number of bedrooms is provided, and the task is to estimate the price of the house, then neural networks will be the most appropriate method under this situation.

The simple structure of neural networks is described in Fig. 5.4. It has three different types of layers: the input layer, the hidden layer, and the output layer. Each hidden layer can contain any number of nerve cells (nodes). The number of nodes in

Fig. 5.5 Structural Details
of Input Layer Nodes

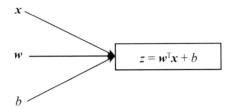

Fig. 5.6 Using Activation
Functions for Input Layer
Results

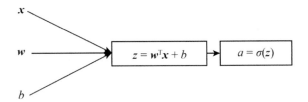

the input layer is equal to the number of features used in the prediction problem (there are 3 features in the above example, namely, housing area, location, and number of bedrooms). The number of nodes in the output layer is equal to the number of values to be predicted (there is 1 value to be predicted in the above example, that is, housing price). Next, let's try to go a little deeper and understand what is happening in each node.

As shown in Fig. 5.5, in each node of the input layer, the input elements x, weight w, and bias b are used as inputs, and z is calculated and output. Constructing these values into a matrix can make the calculation easier and more efficient. What are weights and biases? These two values are first randomly initialized values from Gaussian distribution, which are used to calculate the output of input nodes. We adjust these values to make neural networks fit the input data.

As shown in Fig. 5.6, after calculating the value z, we use the activation function σ for z. The activation function is used to introduce some nonlinearity to the model. If we do not apply any activation function, the output result can only be a linear function, and may not be able to successfully map complex inputs to outputs.

The input of the node in the hidden layer is the output of the node in the previous layer. The final output layer predicts a value and compares the value with the known value (the real value) to calculate the loss. Intuitively, the loss represents the error between the predicted value and the real value.

The whole process first calculates each variable and the weight of each layer and calculates the error (forward propagation), then traverses each layer through back propagation to measure the error contribution of each connection, and finally slightly adjusts the weight and bias of the connector to reduce the error to ensure correct prediction of the output results.

So far, we have had a basic understanding of neural networks. Next, we will introduce multilayer perceptrons, convolutional neural networks, and recurrent neural networks.

5.3.1 Multilayer Perceptions

The multilayer perceptron (MLP) can also be called deep neural networks (Deep Neural Networks, DNN), which is a neural network with multiple hidden layers. Even if a single perceptron has a certain fitting ability, the multilayer perceptron will definitely have a stronger fitting ability and can be used to solve more complex problems.

As shown in Fig. 5.7, the most basic structure of the multilayer perceptron is given, mainly divided into the input layer, the hidden layer, and the output layer. The different layers of the multilayer perceptron are all fully connected (fully connected means that any single nerve cell in the l layer must be connected to any nerve cell in the $l + 1$ layer). Next, three main parameters need to be studied in detail: weight, bias, and activation function.

5.3.1.1 Weight and Bias

The weight is used to indicate the connection strength between nerve cells, and the size of the weight indicates the possibility. The bias is set to correctly classify samples, which is an important parameter in the model, that is, to ensure that the output value calculated through the input cannot be activated casually.

5.3.1.2 Activation Function

The activation function can play the role of non-linear mapping and can limit the output amplitude of nerve cells to a certain range, generally between $(-1,1)$ or $(0,1)$. Commonly used activation functions are sigmoid, tanh, ReLU, etc. Among them, the sigmoid function can map the number between $(-\infty,+\infty)$ to the range of $(0,1)$, and the rest of the functions are not introduced too much here.

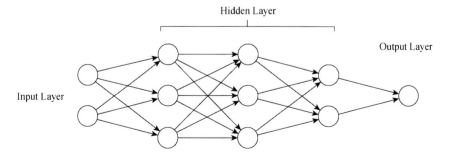

Fig. 5.7 Structure of the Multilayer Perceptron

5.3.1.3   Code Implementation

```
def create_mlp(shape):
  X_input = Input((shape, ))
  X =Dense(256, activation='relu')(X_input)
  X = Dense(128, activation='relu')(X)
  X = Dense(64, activation='relu')(X)
  X = Dense(1, activation='sigmoid')(X)
  model = Model(inputs=X_input, outputs=X)
  model.compile(optimizer='adam', loss='binary_crossentropy',
  metrics=['accuracy'])
  return model
mlp_model = create_mlp(x_train.shape[0])
mlp_model.fit(x=x_train, y=y_train, epochs=30, batch_size=512)
```

5.3.2   Convolutional Neural Networks

Convolutional neural networks (CNN) are similar to multi-layer perceptrons, and the difference between the two lies in the different network structures. The proposal of convolutional neural networks is inspired by biological processing processes, and its structure is similar to that of animal visual cortex. Convolutional neural networks are widely used in the field of computer vision, such as facial recognition, automatic driving, image segmentation, etc., and have achieved excellent results in various competition cases. Convolutional neural networks have two major characteristics.

- It can effectively reduce the dimension of a large amount of data into a small amount of data and simplify complex problems. In most scenarios, dimensionality reduction will not affect the result. For example, if there is a cat in an image, after reducing the pixels of the image from 1000 to 200, even the naked eyes will not recognize the cat as a dog by mistake, nor does the machine.
- It can effectively retain image features and conform to the principles of image processing. When flipping, rotating, or changing the position of an image, convolutional neural networks can effectively identify which images are similar.

As shown in Fig. 5.8, the most basic structure of convolutional neural networks is given. It is mainly divided into three layers: convolution layer, pooling layer (sampling layer), and fully connected layer. The three layers perform their respective duties. The convolution layer is responsible for extracting features, the pooling layer is responsible for feature selection, and the fully connected layer is responsible for classification. The fully connected layer is the neural network we mentioned earlier, so only the convolution layer and the pooling layer will be introduced in detail.

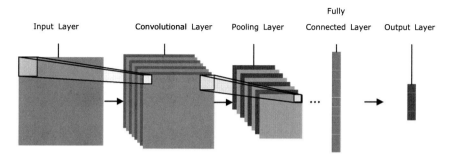

Fig. 5.8   Structure of Convolutional Neural Networks

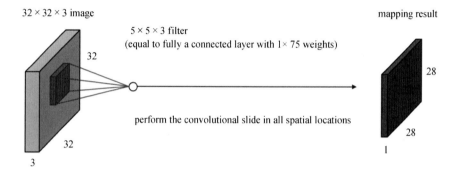

Fig. 5.9   Convolution Layer Operation

5.3.2.1   Convolutional Layer

The convolution layer is used to extract local features in the image, which can effectively reduce the data dimension. As shown in Fig. 5.9, suppose that the RGB image size of our input is 32×32, the actual size of the input will be $32 \times 32 \times 3$ because there are channels for 3 colors. Then select a 5×5 convolution kernel for convolution calculation, and contain three channels, so each time you extract a $5 \times 5 \times 3$ size square, set the extraction step (stride) to 1, and do not fill to the outside (padding $= 0$), and finally you can get a feature map of $28 \times 28 \times 1$.

5.3.2.2   Pooling Layer

The pooling layer is used for feature selection. Compared with the convolution layer, the data dimension can be reduced more effectively, which can not only greatly reduce the amount of computation, but also effectively avoid overfitting. For example, when the data passes through the convolution layer to obtain a $28 \times 28 \times 1$ feature map, we set the stride size to 2 and the convolution kernel size to 2×2, and then obtain a new feature map of $14 \times 14 \times 1$ through the maximum pooling layer/

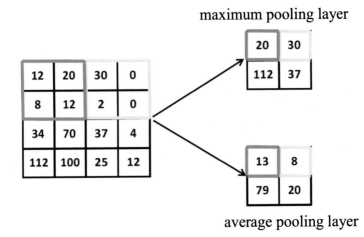

Fig. 5.10 Maximum/Average Pooling Layer (convolution kernel size is 2 × 2; stride size is 2)

average pooling layer, as shown in Fig. 5.10. With the introduction to the basic structure above, we can use keras to construct our own convolutional neural networks.

5.3.2.3   Code Implementation

```
def create_cnn():
    X_input = Input((28,28,1))

    X = Conv2D(24,kernel_size=5,padding='same',activation='relu')
    (X_input)
    X = MaxPooling2D()(X)
    X = Conv2D(48,kernel_size=5,padding='same',activation='relu')
    (X_input)
    X = MaxPooling2D()(X)
    X = Flatten()(X)

    X = Dense(128, activation='relu')(X)
    X = Dense(64, activation='relu')(X)
    X = Dense(1, activation='sigmoid')(X)

    model = Model(inputs=X_input, outputs=X)
    model.compile(optimizer='adam', loss='binary_crossentropy',
    metrics=['accuracy'])

    return model

cnn_model = create_cnn()
cnn_model.fit(x=x_train, y=y_train, epochs=30, batch_size=64)
```

5.3.3   Recurrent Neural Networks

The recurrent neural network (RNN) is an extension of neural networks and excels better at modeling and processing sequence data. For traditional feedforward neural networks, the input is generally a fixed-length vector and cannot process variable-length sequence information. Even if the sequence is processed into a fixed-length vector through some methods, the model is difficult to capture the long-distance dependencies in the sequence. Recurrent neural networks process serialized data by serializing nerve cells. Since each nerve cell can use its internal variables to store the sequence information previously entered, the entire sequence is condensed into an abstract representation, which can be classified, or new sequences can be generated.

When processing sequence data and completing classification decisions or regression estimates with sequence data, recurrent neural networks are very effective, and they are usually used to solve tasks related to sequence data, mainly including natural language processing, speech recognition, machine translation, time series forecast, etc. Of course, recurrent neural networks can also be used for non-sequence data.

The basic structure of recurrent neural networks is shown in Fig. 5.11. It consists of a nerve cell receiving input, generating an output, and returning the output to itself, as shown in Fig. 5.11 (1). At each time step t (also called a frame), the circulating nerve cell receives input x_t and its own previous time stride h_{t-1}. We can expand (1) into a network according to the timeline, as shown in (2) in Fig. 5.11

Expressed by the formula as follows:

$$y_t = g(V \cdot h_t)$$
$$h_t = f(U \cdot X_t + W \cdot h_{t-1})$$

(5.11)

In the formula, V is the weight matrix from the hidden layer to the output layer, U is the weight matrix from the input layer to the hidden layer, and W is also the weight matrix, which represents that the last value of the hidden layer is the weight of this input. In addition, X is the input layer, h is the hidden layer, and y is the output layer.

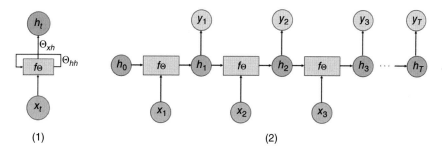

Fig. 5.11   Recurrent Neurons (1); Expand in Time Sequence (2)

Sequence related models extended by recurrent neural networks include LSTM, GRU, etc., which will also be introduced and applied in later chapters.

5.3.3.1   Code Implementation

Different from other models, the input data of recurrent neural networks contain sequence information. For example, if we want to classify a movie review, we need to preprocess it. If the text length is inconsistent, we need to use pad_sequences to truncate it to ensure that the length of each sample sequence is consistent. The following gives a simple implementation code, in which X_train is pre-processed training set.

```
def create_rnn():

emb = Embedding(10000, 32) # 10000 means the total number of words, 32 means
the output dimension
X = SimpleRNN(32)(emb)
X = Dense(256, activation='relu')(X)
X = Dense(128, activation='relu')(X)
X = Dense(1, activation='sigmoid')(X)

model = Model(inputs=X_input, outputs=X)
model.compile(optimizer='adam', loss='binary_crossentropy',
metrics=['accuracy'])

return model

rnn_model = create_rnn()
rnn_model.fit(x=x_train, y=y_train, epochs=30, batch_size=64)
```

5.4   Practical Cases

This section only needs to select multiple models to run the results. The models given above are not complete. Verification methods will also be added here to make the results more reliable. The results of multiple models can be compared and analyzed to help optimize the integration part of the model.

```
# following on to the code, construction training set, and testing set of the practical case in Chapter 5
x_train = data[:ntrain][all_cols]
x_test = data[ntrain:][all_cols]
# log the selling price
y_train = np.log1p(data[data.SalePrice.notnull()]['SalePrice'].
values)
```

5.4.1   *XGBoost*

The more conventional five-fold cross validation is used here,

```
import xgboost as xgb
from sklearn.model_selection import KFold
kf = KFold(n_splits=5, shuffle=True, random_state=2020)
for i, (train_index, valid_index) in enumerate(kf.split(x_train,
y_train)):
trn_x, trn_y, val_x, val_y = x_train.iloc[train_index], y_train
[train_index],
    x_train.iloc[valid_index], y_train[valid_index]

    params = {'eta': 0.01, 'max_depth': 11,'objective': 'reg:linear',
    'eval_metric': 'mae' }

    dtrain = xgb.DMatrix(data=trn_x, label=trn_y)
    dtest = xgb.DMatrix(data=val_x, label=val_y)

    watchlist = [(dtrain, 'train'), (dtest, 'valid_data')]

    model=xgb.train(params, dtrain,
            num_boost_round=20000,
            evals=watchlist,
            early_stopping_rounds=200,
            verbose_eval=500)
```

5.4.2   *Multilayer Perceptions*

Before constructing a multi-layer perceptron, it is necessary to ensure that there are no missing values in the data and to perform regularization processing. Here, the data set is randomly segmented for offline verification.

```
from sklearn.preprocessing import StandardScaler
x_train = x_train.fillna(0)
x_train = StandardScaler().fit_transform(x_train)

from sklearn.model_selection import train_test_split
trn_x, val_x, trn_y, val_y = train_test_split(x_train, y_train,
random_state = 2020)

def create_mlp(shape):
    X_input = Input((shape,))

    X = Dropout(0.2)(BatchNormalization()(Dense
    (256, activation='relu')(X_input)))
    X = Dropout(0.2)(BatchNormalization()(Dense
    (128, activation='relu')(X)))
```

```
X = Dropout(0.2)(BatchNormalization()(Dense
(64, activation='relu')(X)))
X = Dense(1)(X)

model = Model(inputs=X_input, outputs=X)
model.compile(optimizer='adam', loss='mse', metrics=['mae'])

return model
mlp_model = create_mlp(trn_x.shape[1])
mlp_model.fit(x=trn_x, y=trn_y, validation_data = (val_x, val_y),
    epochs=30, batch_size=16)
```

The models given so far are relatively easy to implement, which helps to quickly feed the results back. Comparing with the offline results of XGBoost (average absolute error: 0.08×, after the result is logarithmic) and the multilayer perceptron (average absolute error: 0.21×), it is found that the effect of the latter is much worse, and it is difficult for the multilayer perceptron to achieve a better result with more than 2000 training data.

5.5   Thinking Exercises

1. In the Lasso regression and Ridge regression section, we know that L1 and L2 can reduce the risk of overfitting. So, what is the appropriate value of this parameter?
2. When the tree model splits, it can actually look at the cross-combination stage of features. Is it necessary to construct a cross feature to feed the tree model?
3. This chapter introduces the core parameters of the tree model, and there are still many that have not been introduced. Please try to analyze the relationship between the parameters and in which step of the algorithm the specific parameters appear, so as to deepen the understanding of the parameters.
4. There are still many commonly used activation functions. When training deep learning related models, different activation functions still have a great impact on the results. Try to sort out the advantages and disadvantages of activation functions such as sigmoid, tanh, ReLU, leaky ReLU, SELU, and GELU and their corresponding applicable scenarios.

Chapter 6
Model Integration

This chapter will introduce to you the key steps to improve performance in the algorithm competition. This is also the typical method in the final stage, that is, model integration (or integrated learning). Model integration is carried out by combining the advantages of different sub-models. Of course, this is in an ideal state.

This chapter is mainly divided into three parts: building diversity, training process integration, and training result integration. Model integration is often the key to winning the competition. In contrast, model integration with differences can often bring great improvement to the results. Although model integration cannot always play a great role every time you use it, but in terms of usual competition experience, we have come to a conclusion that model integration will bring more or less help in most cases. In the competitions, especially when the final results are not much different, the method of model integration will often become one of the keys to success. In different types of competitions, it is hard for us to guarantee which method will definitely be better than the others; evaluation indicators often have to be based on online results. It just implies that the more model integration methods you know, the higher the probability of winning in the end. Therefore, in this chapter, we will also introduce the application scenarios of different model integration methods from these three parts, and at the same time give the using skills and application codes.

6.1 Building Diversity

This section will introduce three ways to construct diversity in model integration, namely feature diversity, sample diversity, and model diversity. Diversity refers to the difference between sub-models, which can be constructed by reducing the homogeneity of sub-model integration. Good diversity helps to improve the effect of model integration.

© The Author(s), under exclusive license to Springer Nature Singapore Pte Ltd. 2023
W. He et al., *Machine Learning Contests: A Guidebook*,
https://doi.org/10.1007/978-981-99-3723-3_6

6.1.1   Characteristic Diversity

Constructing multiple feature sets with differences and establishing models separately can make the features exist in different hyperspaces, so that the multiple models established will have different generalization errors, and the final model integration can play a complementary role. In the competition, the feature sets between teammates are often different, and in the case of little difference in their scores, direct model integration will usually ensure helpful effects or results.

In addition, max_features in random forest, colsample_bytree in XGBoost, and feature_fraction in LightGBM are all used to sample features in the training set, which is to build feature diversity in essence.

6.1.2   Sample Diversity

Sample diversity is also a common model integration method in competitions. The diversity here mainly comes from different sample sets. The specific method is to cut the data set into multiple parts, and then build models separately. We know that many tree models will be sampled during training; the main purpose of doing so is to prevent overfitting, so as to improve the accuracy of prediction.

Sometimes the data set is not divided into multiple parts randomly but is divided according to the specific competition data. It is necessary to consider how to divide the data to construct the maximum data difference and use the segmented data to train the models respectively.

For example, in the Tianchi "Global Urban Computing AI Challenge", the competition training set contains a total of 25-day subway records of card swiping data from January 1 to January 25, 2019, and it is required to predict the average passenger flow volume per ten minutes at each subway station on January 26 (January 26, 2019 is Saturday). Obviously, there is a big difference in the distribution of traffic volume between weekdays and weekends. At this time, there will be a problem. If you only keep the weekend data for training, you will waste a lot of data; if you keep all the data for a week, it will have a certain impact on the workday data. At this time, you can try to build two sets of different samples to train the model separately; that is, the overall data is kept as one group and the weekend data as the other group. Of course, the score after the model integrated will be greatly improved.

6.1.3   Model Diversity

Different models have different ability to express data. For example, FM can learn the cross information between features and has strong memory; the tree model can handle continuous features and discrete features (such as LightGBM and CatBoost) well, and it is also robust for outliers. The integration of these two types of models with different data assumptions and representation capabilities will definitely achieve certain results.

For competitions, traditional tree models (XGBoost, LightGBM, CatBoost) and neural networks all need to be tried one by one, and then the models already tried are integrated together as the model with differences.

More Diversity Methods
In addition to what is mentioned in this section, there are many other ways to build diversity, such as training target diversity, parameter diversity, and loss function selection diversity, which can produce very good results.

6.2 Training Process Integration

There are two ways to model integration. The first is training process integration, such as the random forest and XGBoost we know. Based on these two models, multiple decision trees are constructed in training for integration. The multiple decision trees here can be regarded as multiple weak learners. Among them, random forest is aggregated by Bagging, and XGBoost is fused by Boosting.

6.2.1 Bagging

The idea of bagging is very simple, that is, to take data with return (Bootstrapping) from the training set; these data form a sample set, which also ensures that the size of the training set remains unchanged, and then the sample set is used to train the weak classifier. Repeat the above process many times and take the average value or use the voting mechanism to obtain the final result of model integration. The schematic diagram of the above process is shown in Fig. 6.1.

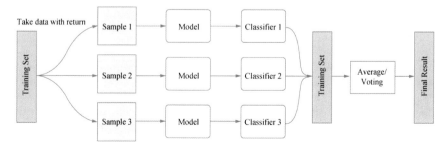

Fig. 6.1 Bagging Process

When we train models on different sample sets, Bagging reduces the variance of the classifier by reducing the difference between errors. In other words, Bagging can reduce the risk of overfitting. The efficiency of the Bagging algorithm comes from the different training data. There are great differences between the models, and the errors of the training data can cancel each other out in the process of weighted fusion. Of course, here you can choose the same classifier for training, or you can choose different classifiers. In addition, Bagging-based algorithms include Bagging meta-estimator and random forest.

6.2.2 Boosting

It is no exaggeration to say that the idea of boosting is in fact not difficult to understand. First train a weak classifier and record the samples of the wrong class of this weak classifier, and at the same time give this weak classifier a certain weight; then establish a new weak classifier. The new weak classifier trains based on the previously recorded error samples. Similarly, we also give this classifier a weight. Repeat the above process until the performance of the weak classifier reaches a certain index. For example, when the newly established weak classifier does not significantly improve the accuracy rate, the iteration will stop. Finally, multiply these weak classifiers by the corresponding weights and add them all together to get the final strong classifier. In fact, there are many algorithms based on Boosting, including Adaboost, LightGBM, XGBoost, and CatBoost.

6.3 Training Result Integration

The second way of model integration is the training result fusion, which is mainly divided into weighting method, stacking, and blending. These methods can effectively improve the overall prediction ability of the model and are also methods that participants must master in the competition.

6.3.1 Weighting Method

The weighting method is very effective for a series of tasks (such as classification and regression) and evaluation indicators (such as AUC, MSE, or Logloss). For example, we have 10 algorithm models that all predict the results. The results are directly averaged, or given different weights to each algorithm, and then the fusion result is obtained. The weighting method can usually reduce overfitting because the results of each model may have some noise. The weighting method can smooth the noise and improve the generalization of the model.

6.3.1.1 Classification Problems

For classification problems, it is important to note that the output ranges of different classifiers are consistent because the prediction results output can be 0/1, or probabilities between 0 and 1. In addition, voting is also a special weighting method, assuming that three models each output three sets of results:

1010110011
1110110011
1110110011

As long as the weights of the three results are consistent, the final fusion result is 1110110011 no matter the voting method (the minority is subordinate to the majority) or the weighting method (fixing 0.5 as the threshold) is used.

6.3.1.2 Regression Problems

For regression problems, if the weighting method is used, it will be very simple. Here we mainly introduce the algorithm average and geometric average. So, why are there two choices? It is mainly because of the evaluation index. In the 2019 Tencent Advertising Algorithm Competition, the effect of choosing geometric average is far better than that of choosing arithmetic average. This is because the scoring rule is the symmetric mean absolute percentage error (SMAPE). At this time, if you choose arithmetic average, the result of model integration will be too large. This does not conform to the intuition of the mean absolute percentage error. The smaller the value, the greater the impact on the score will be, and the arithmetic average will lead to greater errors. Therefore, choosing geometric average can make the result biased towards small values.

$$\text{SMAPE} = \frac{1}{n} \sum_{t=1}^{n} \frac{|F_t - A_t|}{(F_t + A_t)/2} \tag{6.1}$$

- **Arithmetic average.** The integration method based on arithmetic average is the most commonly used in the algorithm, because it is not only simple, but also has a high probability of obtaining good results every time the algorithm is used. The formula is as follows (6.2):

$$\text{pred} = \frac{\text{pred}_1 + \text{pred}_2 + \cdots + \text{pred}_n}{n} \tag{6.2}$$

- **Geometric average.** According to many contestants, weighting based on geometric average is not used much in the algorithm, but in actual situations, sometimes the model integration effect based on geometric average is slightly better than that based on arithmetic average.

$$\text{pred} = \sqrt[n]{\text{pred}_1 + \text{pred}_2 + \cdots + \text{pred}_n} \tag{6.3}$$

6.3.1.3   Sorting Problems

The main task in a general recommendation question is to sort the recommendation results. Common evaluation indicators include mAP (mean Average Precision), NDCG (Normalized Discounted Cumulative Gain), MRR (Mean Reciprocal Rank), and AUC. MRR and AUC will be explained here.

MRR

Given the recommendation result q, if the position of q in the recommendation sequence is r, then MRR (q) is $1/r$. It can be seen that if the product recommended to the user hits in the recommendation sequence, the higher the position of the hit, the higher the score. Obviously, the significance of ranking results is not the same for different positions, so we not only need to carry out weighted fusion, but also need to make the results biased towards small values. Under such conditions, the results must be converted, and then integrated by using the weighting method. In general, the conversion method used is log transformation. The basic idea is as follows.

First, enter three prediction result files. Each prediction result file contains M records. Each record corresponds to N prediction results, and finally the integrated results of the three prediction result files will be output. The internal details can be divided into the following two steps.

Step 1: Count the positions of all recommended commodities (a total of N commodities) recorded in the three prediction result files; For instance, the recommended position of commodity A in the first file is 1, the recommended position of it in the second file is 3, and it does not appear in the third file. In this context, we calculate the score of commodity A as $\log 1 + \log 3 + \log(N + 1)$, where we use $N + 1$ to represent the condition of not appearing. In other words, in the N recommended products, the product A cannot be found, so it can only be $N + 1$.

Step 2: Sort the commodities in each record by the scores calculated, from small to large, and take the top N as the final recommended result of this record.

AUC

As a ranking index, AUC generally uses the integration idea of ranking means, using relative order to replace the original probability value. Many competitions with AUC as the index have achieved very good results. The following two steps are what in a using process.

Step 1: Sort the probability of classification in each classifier, and then use the ranking value (rank) obtained after ranking each sample as the new result.

Step 2: Calculate the arithmetic average value of the ranking value of each classifier as the final result.

6.3.2   Stacking Integration

Although it is simple to use the weighting method for integration, it requires manual labor to determine the weights, so a more intelligent way shall be considered to learn the weights of each classifier through a new model. Here we assume that there are two layers of classifiers. If a specific base classifier in the first layer mistakenly learns a certain area of the feature space, this wrong learning behavior may be detected by the second layer classifier, which, like the learning behavior of other classifiers, can correct inappropriate training. The above process is the basic idea of stacking integration.

Two points should be paid attention to here: first, the new models constructed are generally simple models, such as linear models like logistic regression; second, using multiple models for stacking integration will have better results.

Stacking integration uses the prediction results of the base model as the input of the second-level model. However, we cannot simply use the complete training set data to train the base model, which will create the risk that the base classifier will already "see" the testing set during prediction, resulting in an over-fitting problem when providing the prediction results. Thus, we should use out-of-fold to predict the results, that is, making use of K-fold cross validation to predict the results. Here we divide the stacking integration into two parts: the training phase and the testing phase and will show the specific operation of each part in the form of a flow chart. The training phase is shown in Fig. 6.2.

In Fig. 6.2, we use a five-fold cross validation method for each model, then we can obtain the prediction probability results of the complete verification set, and finally splice the N-column probability results and the training set labels obtained into a second-layer training sample, so that the second-layer model can be trained. After that, we use the model trained during the five-fold cross validation (such as model 1, which can be trained to obtain 5 different model 1) as the training of the testing set.

As shown in Fig. 6.3, the testing phase will use the model trained in the training phase. First, the testing set will be predicted using the 5 models obtained from model 1, and then the 5 probability results will be obtained by weighted average to obtain a probability result—probability 1. Then the above operations are also performed on model 2 to model N in turn, and finally N probability results are obtained. Regard these N results as the testing samples of the second layer, and then use the model obtained from the second layer training to predict the testing samples of the second layer to get the final result.

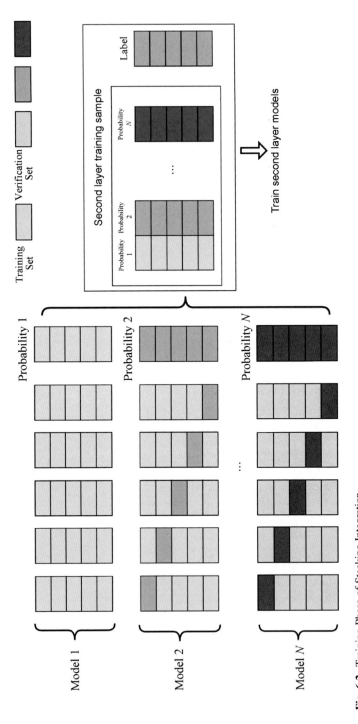

Fig. 6.2 Training Phase of Stacking Integration

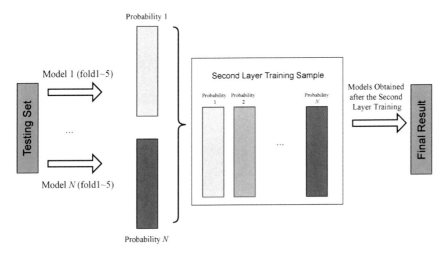

Fig. 6.3 Testing Phase of Stacking Integration

Extended Learning
Please think about the linear stacking of feature weighting; you can refer to the corresponding paper "Feature-Weighted Linear Stacking two layer stacking", which is actually an extension of the traditional stacking integration method in depth. The probability value is obtained through the traditional stacking integration method, and then this value is spliced with the basic feature set to reconstitute a new feature set for a new round of training.

6.3.3   Blending Integration

Unlike stacking integration, which uses K-fold cross validation to obtain prediction results, blending integration is to build a holdout set and use disjoint data sets for different layers of training, which can greatly reduce the risk of overfitting. Suppose we construct two layers of blending, divide the training set into two parts (train_one and train_two) at a ratio of 5:5, and the testing set is test.

The first layer uses train_one to train multiple models and merges the prediction results of this model for train_two and test into the original feature set, as the feature set of the second layer. The second layer uses train_two feature set and labels to train the new model, and then predicts the test to obtain the final integration result.

6.4 Practical Cases

This section will take you through the operation of Stacking integration, which requires the construction of multiple model prediction results, generally more than 3. Here we choose ExtraTreesRegressor, RandomForestRegressor, Ridge, and Lasso as the base classifiers, and Ridge as the final classifier. First import some new packages:

```
from sklearn.ensemble import ExtraTreesRegressor
from sklearn.ensemble import RandomForestRegressor
from sklearn.metrics import mean_squared_error
from sklearn.linear_model import Ridge, Lasso
from math import sqrt
# still use five-fold cross validation
kf = KFold(n_splits=5, shuffle=True, random_state=2020)
```

Then build a functional class of the model in sklearn, initialize the parameters and then train and predict. This code is very reusable. We recommend that you continue to improve the construction, constructing your own set of repositories.

```
class SklearnWrapper(object):
  def __init__(self, clf, seed=0, params=None):
    params['random_state'] = seed
    self.clf = clf(**params)

  def train(self, x_train, y_train):
    self.clf.fit(x_train, y_train)

  def predict(self, x):
    return self.clf.predict(x)
```

Then, encapsulate the cross validation function; the reusability of this code is also very frequent:

```
def get_oof(clf):
  oof_train = np.zeros((x_train.shape[0],))
  oof_test = np.zeros((x_test.shape[0],))
  oof_test_skf = np.empty((5, x_test.shape[0]))

  for i, (train_index, valid_index) in enumerate(kf.split(x_train,
  y_train)):
    trn_x, trn_y, val_x, val_y = x_train.iloc[train_index],
    y_train[train_index],
      x_train.iloc[valid_index], y_train[valid_index]
    clf.train(trn_x, trn_y)
    oof_train[valid_index] = clf.predict(val_x)
    oof_test_skf[i, :] = clf.predict(x_test)
```

```
oof_test[:] = oof_test_skf.mean(axis=0)
return oof_train.reshape(-1, 1), oof_test.reshape(-1, 1)
```

Next is the part of the code trained and predicted by the base classifier, which can predict the verification set results and testing set results of the four models, and assist the last step of the stacking integration operation:

```
et_params = {
  'n_estimators': 100,
  'max_features': 0.5,
  'max_depth': 12,
  'min_samples_leaf': 2,
}
rf_params = {
  'n_estimators': 100,
  'max_features': 0.2,
  'max_depth': 12,
  'min_samples_leaf': 2,
}
rd_params={'alpha': 10}
ls_params={'alpha': 0.005}
et = SklearnWrapper(clf=ExtraTreesRegressor, seed=2020,
params=et_params)
rf = SklearnWrapper(clf=RandomForestRegressor, seed=2020,
params=rf_params)
rd = SklearnWrapper(clf=Ridge, seed=2020, params=rd_params)
ls = SklearnWrapper(clf=Lasso, seed=2020, params=ls_params)

et_oof_train, et_oof_test = get_oof(et)
rf_oof_train, rf_oof_test = get_oof(rf)
rd_oof_train, rd_oof_test = get_oof(rd)
ls_oof_train, ls_oof_test = get_oof(ls)
```

The final part is the Stacking part. It uses the ridge model. Of course, you can also try more complex models such as the tree model:

```
def stack_model(oof_1, oof_2, oof_3, oof_4, predictions_1, predictions_2,
predictions_3, predictions_4, y):
  train_stack = np.hstack([oof_1, oof_2, oof_3, oof_4])
  test_stack = np.hstack([predictions_1, predictions_2, predictions_3,
predictions_4])

  oof = np.zeros((train_stack.shape[0],))
  predictions = np.zeros((test_stack.shape[0],))
  scores = []

  for fold_, (trn_idx, val_idx) in enumerate(kf.split(train_stack, y)):
    trn_data, trn_y = train_stack[trn_idx], y[trn_idx]
    val_data, val_y = train_stack[val_idx], y[val_idx]
```

```
clf = Ridge(random_state=2020)
clf.fit(trn_data, trn_y)

oof[val_idx] = clf.predict(val_data)
predictions += clf.predict(test_stack) / 5

score_single = sqrt(mean_squared_error(val_y, oof[val_idx]))
scores.append(score_single)
print(f'{fold_+1}/{5}', score_single)
print('mean: ',np.mean(scores))

return oof, predictions

oof_stack , predictions_stack = stack_model(et_oof_train, rf_oof_train,
    rd_oof_train, ls_oof_train, et_oof_test, rf_oof_test, rd_oof_test,
    ls_oof_test, y_train)
```

After comparing the final effect, it can be seen that the stacking integration is 0.13157, and the optimal base classifier is 0.13677, which has an improvement of about five thousandths, indicating that the model integration still has a certain effect. In addition, we also carried out a common weighted average fusion scheme, with a score of only 0.13413, which showed that the effect was relatively poor.

6.5   Thinking Exercises

1. There are also many methods to construct diversity, such as training target diversity, parameter diversity, and loss function selection, which can all produce very good results. Please sort out and summarize more methods.
2. Intuitively, stacking integration can bring good results. However, why will the effect of stacking integration sometimes deteriorate? Is it a problem of base model selection, or is it because the number of layers is not enough? Please analyze what factors will affect the final fusion results.
3. Try to build the framework of stacking integration and make it reusable, which is convenient for participants to call flexibly in the competition.

Part II
Birds of a Feather Flock Together

Chapter 7
User Profiles

As the old saying goes, thousands of people have thousands of quite different faces. It is true that everyone in the world is a unique individual. Just as no two leaves of any plants are identical in the world, there are no identical people in the world either. After the birth, people begin to have their own unique labels, such as their last name, who their parents are, where their home is, when they come to the world, and what kind of life journey they are about to embark on. The more you experience, the more obvious your uniqueness becomes. Even if among siblings, there will always be times when they live independently, and differences will emerge. Loneliness, a description of the state of mind, is an eternal theme in one's life. As long as you find a soul mate, you will no longer be alone. However, not being alone is not equal to not feeling lonely. Due to the huge differences and individual uniqueness between individuals, the analysis and study of individuals might be very complicated, even impossible, and unnecessary. Therefore, psychology and sociology are mostly about group characteristics, and occasionally the study of individual abnormal behaviors is just to trace the source based on the existing results.

Even today when it has witnessed explosion of Internet information and the era of so-called big data artificial intelligence, the information that can be obtained from a person's digital records is still only a part of their life, and others simply cannot fully know what they are thinking and what they will do in the future. Of course, there is no need to be too pessimistic. If you can make good use of the recorded information, you can also understand a person to a certain extent. Therefore, based on only a certain level of data, some individual profiles can be produced, which can be used to describe the general differences between groups and individuals from a specific perspective. Having said so much, what does this have to do with the user profile discussed in this chapter? First of all, it is necessary to clarify who the users in the user profile are: data collectors (that is, product providers) often develop a product for people to use, and these users are the users of the data collectors. In order to promote the product while continuing to maintain and improve the user experience, data collectors will need to mine the data generated by user operations, so as to discover the behavioral preferences of groups and even individuals, forming a

W. He et al., *Machine Learning Contests: A Guidebook*,
https://doi.org/10.1007/978-981-99-3723-3_7

so-called profile at the data level. More broadly speaking, an image can be formed of any group, such as a region, an era, a community of groups, etc.

The popularity of smart phones has attracted people to play with them during most of their leisure time. Various apps have generated a large amount of recorded data about user behavior, which has laid the foundation for the formation of user profiles. Therefore, machine learning algorithms based on user profiles have many application scenarios, and the competitions of such scenarios also become important events. This chapter will explain from five parts: what the user profile is, tagging system, user profile data features, user profile application, and thinking exercises.

7.1   What Are User Profiles

There is no denying the fact that when thinking of a person, whether on the surface or subconsciously, one will actually have a general impression for the person, such as the figure and face, social attributes, personality cultivation, interests and hobbies, etc. Although the general impression in one's mind can also be regarded as a portrait with relatively subjective consciousness, it is obviously not the user profile for business analysis and data mining to be discussed in this section.

The user profile mentioned in machine learning is usually based on the given data to describe the user portraits and behaviors, then extract the user's personalized indicators, next analyze the possible group commonalities, and apply them to various business scenarios. In the Internet era, there are a lot of user-oriented products and data collection is relatively easy, which promotes the application of machine learning in user-oriented aspects, and user profile is the most important part. In all kinds of machine learning algorithm competitions, the mining of user data always occupies a place, so user profiles often haunt the competitions and play an indispensable role. Next, we will introduce the composition of user profiles and how to use user profiles in the competitions.

7.2   Tagging System

The core of user profiles is actually to "tag" users, which means to tag the behavioral characteristics of users, making it possible for enterprises to take advantage of the tags in a user profile to analyze the user's social attributes, living habits, consumer behavior, and other information, and then apply such information into commercial use. Building a tagging system has become the key for enterprises to empower more business, and the tagging system is also the content to be introduced in detail in this section, specifically from three aspects, namely, the tag classification method, multi-channel access to tags, and the tagging system framework.

7.2.1 Tag Classification

As shown in Fig. 7.1, the tag classification method is shown by analyzing the features of a user.

7.2.2 Multi-Channel Access to Tags

According to the tag acquisition method, tags can be divided into three types: fact tag, rule tag, and model tag. The tag acquiring method can also be regarded as a feature obtaining method. With the help of these three methods, features that can represent user characteristics, commodity features, or data characteristics can be extracted from the original data source.

7.2.2.1 Fact Tags

Fact tags are the easiest to obtain, directly from the original data source, such as gender, age, membership level, and other fields. Of course, it is also possible to extract fact labels after simple statistic work on the original data source, such as the number of user behaviors, and total consumption.

7.2.2.2 Rule Tags

Rule tags are widely used. They are tags generated by multiple rules set by operators and data personnel through joint negotiation. They are characterized by direct, effective, and flexible, low computational complexity, and high interpretability. They are mainly used for more intuitive and clearer user-related tags, such as

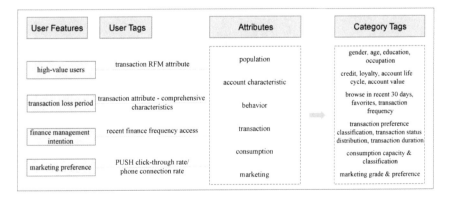

Fig. 7.1 Tag Classification Method

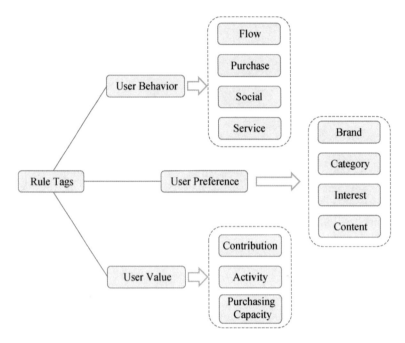

Fig. 7.2 Rule Tags Classification

geographical affiliation, family type, age, etc. The technical knowledge used is mainly mathematical statistics knowledge, such as basic statistics, numerical stratification, probability distribution, mean analysis, variance analysis, etc.

As shown in Fig. 7.2, the final rule tag is mainly generated by logical operation and function operation on a single or multiple indicators, which is divided into three parts: user behavior, user preference, and user value. If you are interested, you can conduct deeper operations on users.

7.2.2.3   Model Tags

Model tags are insight tags generated by secondary processing after model processing such as machine learning and deep learning, for example, predicting user status, predicting user credit score, dividing interest groups and classifying comment text, etc. The results obtained through these processes are model tags. It is characterized by high degree of synthesis and complexity. Most tags need to construct corresponding mining index system in a targeted way first, and then rely on classical mathematical algorithms or models to carry out comprehensive calculation among multiple indexes to obtain model tags, which often requires a combination of multiple algorithms to build models.

As shown in Fig. 7.3, based on the model tags, the RFM model can be used to measure user value and user profitability, build user behavior information modeling

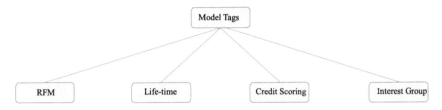

Fig. 7.3 Model Tags Classification

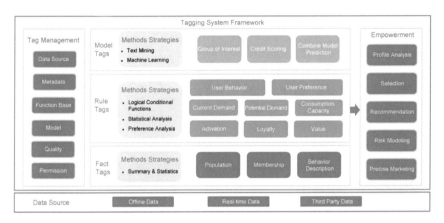

Fig. 7.4 Tagging System Framework

to predict user lifetime change, forecast user credit scores through models, and use graph embedding or user hierarchical model to divide interest groups. In addition, there are many ways to get tags through the model.

7.2.3   Tagging System Framework

With a preliminary comprehension of tagging classification and acquisition, we can connect them together to form a basic tagging system framework, including extracting bottom layer data to empower business applications.

As shown in Fig. 7.4, the whole tagging system framework is divided into four parts: data source, tag management, tag hierarchy classification, and tag service empowerment.

7.3 User Profile Data Features

Whether it is building a user profile or conducting an algorithm competition, data is the core of creating benefits. Broadly speaking, the data source of user profiles is user data, commodity data, and channel data. For example, user transaction data and behavior data can be obtained from e-commerce websites and WeChat platform (WeChat is a Chinese multi-purpose instant messaging, social media, and payment app), and user attributes data can be obtained from the user system on the platform. These data exist in various forms, and by understanding of the data form, statistics, coding, and dimension reduction can be carried out to extract effective features. Then these features can be used to construct the tags we need. In this section, we will introduce common data forms and some feature extraction methods for competitions related to user profiles.

7.3.1 Common Data Forms

In a variety of competitions, data forms and formats are diverse. This section takes the user profile as an example to roughly divide the relevant fields of data into four common data forms: numerical variables, category variables, multi-value variables, and text variables. Each variable has a corresponding processing method. It should be emphasized that these variables are aimed at the user level; that is, all sample data is distinguished by using the user as the unique primary key, and each user has only one record. The reason for using this example is that the data required for machine learning models based on user profiles are usually presented in the form of a user pool, and the user's tags are subjected to corresponding feature learning. The data given in the actual competition may be very complex, and even describes the user's behavior in the form of dotting records. At this time, participants are often required to construct and extract user attributes, which involves better application skills.

7.3.1.1 Numerical Variables

The most common numerical variable is continuous variable, which refers to variables with numerical meaning, such as age, height, weight, etc. shown in Fig. 7.5, and others such as consumption expenditure, traffic accumulation, etc.

7.3.1.2 Category Variables

Category variables refer to variables with category identification, such as gender, nationality, city, etc. These variables record the inherent attributes of users, as shown in Fig. 7.6.

User ID	Age	Height	Weight
1001	25	175	80
1002	18	170	68
1003	26	172	72

Fig. 7.5   Continuous Variables

User ID	Gender	Nationality	City
1001	Male	Tianjin	Beijing
1002	Male	Sichuan	Hong Kong
1003	Male	Hubei	Shanghai

Fig. 7.6   Category Variables

User ID	Hobbies	Dressing Styles	Favorite Movies
1001	Food, Reading, Movies	High Necks	*Parasite, The Attorney*
1002	Fitness, Photography	Sweatshirts	*Waiting Alone, Heidi*
1003	Cycling, Basketball	Jackets	*Spiderman*

Fig. 7.7   Multi-valued Variables

7.3.1.3   Multi-valued Variables

Multi-valued variables refer to variables for which users have multiple values in a certain dimension, such as hobbies, dressing styles, movies they have watched, etc. Due to their special structure, such variables cannot be directly applied to the model and need to be processed with special data structures such as sparse matrix, as shown in Fig. 7.7.

7.3.1.4   Text Variables

Text variables (shown in Fig. 7.8) are variables that use text records, such as a user's comment on an item or a purchase. Handling such variables requires some tools of natural language processing, such as the Chinese word segmentation tool jieba.

User ID	Commodity	Star Rating	Comment
1001	Large Teppanyaki	4	Satisfying in taste and portion size
1002	TNT Hip Hop Center	5	A great dancing experience
1003	xx	3	xx

Fig. 7.8 Text Variables

Next, we will introduce some common feature extraction methods to enable readers to better solve problems when facing competitions related to user profiles, specifically, text mining algorithms, magical embedded representations, and similarity calculation methods.

7.3.2   Text Mining Algorithm

For basic raw data such as frequent user tag sets, shopping evaluation, etc., in addition to common statistical features, it is also possible to extract features based on text mining algorithms, and at the same time preprocess and clean the raw data sources, so as to achieve the effect of matching and identifying user data. This section will introduce common text mining algorithms LSA, PLSA, and LDA, all of which are unsupervised learning methods.

7.3.2.1   LSA

LSA (Latent Semantic Analysis) is a non-probabilistic topic model related to word vectors, which is mainly used for topic analysis of docs. Its core idea is to discover the topic-based semantic relationship between docs and words through matrix decomposition. Specifically, the doc set is represented as a word - doc matrix, and the word - doc matrix will undergo SVD (singular value decomposition) to obtain the topic vector space and the representation of doc in the topic vector space.

The specific use of LSA is also very simple. We will take the data in the 2020 Tencent Advertising Algorithm Contest as an example. First, construct the ID series (creative_id) of the advertising material clicked by the user, and then perform TF-IDF calculations. Finally, the results are obtained through SVD. The implementation code is as follows:

```
from sklearn.feature_extraction.text import TfidfVectorizer
from sklearn.decomposition import TruncatedSVD
from sklearn.pipeline import Pipeline
# extract the point-and-click series by users
docs=data_df.groupby(['user_id'])['creative_id'].agg(lambda x:"
   ".join(x)).reset_index()['creative_id']
```

```
# tfidf + svd
tfidf = TfidfVectorizer()
svd = TruncatedSVD(n_components=100)
svd_transformer = Pipeline([('tfidf', tfidf), ('svd', svd)])
lsa_matrix = svd_transformer.fit_transform(documents)
```

7.3.2.2   PLSA

The PLSA (Probability Latent Semantic Analysis) model is actually proposed to overcome some of the potential shortcomings of the LSA model. The PLSA model gives LSA a probabilistic interpretation through a generative model. The model assumes that every doc contains a series of possible potential topics, and every word in the doc is not dreamt up but generated through a certain probability under the guidance of potential topics.

7.3.2.3   LDA

LDA (Potential Dirichlet Allocation) is a probabilistic topic model, which has nothing to do with word vectors. The topic of each doc in the doc set can be given in the form of a probability distribution. By analyzing a batch of doc sets and extracting their topic distribution, topic clustering or text classification can be carried out according to the topic distribution. At the same time, it is a typical bag-of-words model, that is, a doc is composed of a group of mutually independent words, and there is no sequence relationship between words.

7.3.3   Magic Embedded Representation

It is no exaggeration to say that anything that can form a network structure can have an embedding representation, and the embedded representation can convert high-dimensional sparse feature vectors into low-dimensional dense feature vectors to represent. The concept of embedding was originally widely used in the field of NLP, and has now been extended to other applications, such as e-commerce platforms. E-commerce platforms regard a user's behavior sequence as a sentence composed of a series of words, such as a user clicking sequence and a purchase sequence and obtain embedded vectors about goods after training. This article mainly introduces the classic Word2Vec and the DeepWalk method in network representation learning.

Input Layer   Hidden Layer   Output Layer

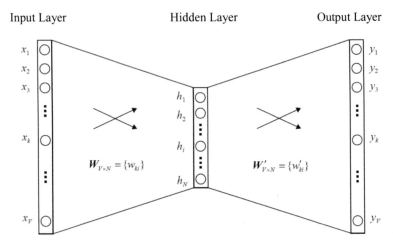

Fig. 7.9 Simple Version of Word2Vec Model

7.3.3.1   Word Embedding Word2Vec

Word2Vec is often used in competitions and can bring unexpected effects. It is very important to master its principle. Word2Vec trains word vectors according to the relationship between contexts. There are two training modes, namely Skip-Gram (Skip-Gram model) and CBOW (continuous bag-of-words model). The main difference between the two lies in the difference between the input layer and the output layer. Simply put, Skip-Gram uses a word as input to predict its context; CBOW uses the context of a word as input to predict the word itself.

As shown in Fig. 7.9, Word2Vec is essentially a fully connected neural network with only one hidden layer, which is used to predict words with a large degree of association with a given word. The size of the model glossary is V, the dimension of each hidden layer is N, and the connection between adjacent nerve cells is fully connected.

The input layer in Fig. 7.9 is a one-hot vector that converts a word into a given word, and then converts the word into a $\{x_1, x_2, x_3, \ldots x_v\}$ sequence. Only one value in this sequence is 1, and the others are all 0; simply map the sequence in the hidden layer through the weight matrix $W_{V \times N}$ between the input layer and the hidden layer; there is a weight matrix $W'_{N \times V}$ between the hidden layer and the output layer, and the score of each word in the glossary is obtained by calculating the weight; finally the probability result of each word is output using the softmax activation function. Next, let's look at the specific model structure of Skip-Gram and CBOW.

As shown in Fig. 7.10, Skip-Gram predicts a given sequence or context based on the current word. Assume that the target word entered is x_k, the defined context window size is c, and the corresponding context is $\{y_1, y_2, y_3, \ldots y_c\}$. These y are independent of each other.

Fig. 7.10 Skip-Gram

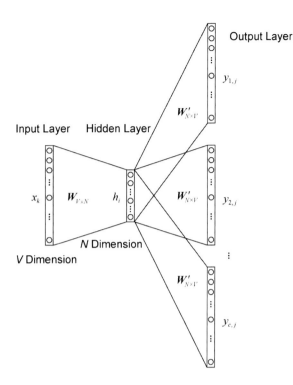

Figure 7.11 is the model structure of the CBOW, which defines the context window size as c, and the context word as $\{x_1, x_2, x_3, \ldots x_c\}$. The current word is predicted through the word in the context, and the corresponding target word is y.

Word2Vec is very convenient to use—just call the gensim package directly, and pay attention to several specific parameters, such as window size, model type selection, and vector length of the generated word.

The choice of Skip-Gram and CBOW can be roughly based on the following three points: CBOW is much faster than Skip-Gram in training because CBOW predicts the word itself based on context, only requiring adding other words in the window as input to implement prediction. No matter how big the window is, it only needs one calculation; compared to Skip-Gram, CBOW can better represent common words; Skip-Gram can also represent rare words or phrases in a small number of training sets.

7.3.3.2   Graph Embedding DeepWalk

In many scenarios, data objects have not only sequence relationships, but also graph or network structure relationships, while Word2Vec can only extract embedding representations under sequence relationships but cannot obtain graph-related information crossing the original sequence relationships. In order to span from "one-

Fig. 7.11 CBOW

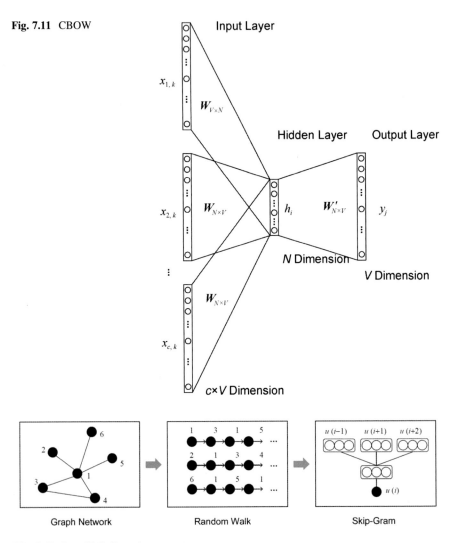

Fig. 7.12 DeepWalk Procedure

dimensional" relationships to "two-dimensional" relationships, graph embedding has become a new research direction, the most influential of which is DeepWalk.

As shown in Fig. 7.12, the DeepWalk algorithm mainly includes three parts. First, the graph network is generated. For example, the commodity graph network is constructed via the user clicking sequence or purchasing commodity sequence. The commodities in the sequence are connected by lines and used as the edge of the graph. The commodities are used as points, so that a certain commodity point can be associated with a large number of domain commodities; then a random walk is performed based on the graph network to obtain the co-occurrence relationship between nodes in the network, generating a commodity sequence in accordance with the walk length; finally, the generated sequence is input as a sentence to Skip-Gram for word vector training, and the graph embedding representation of the commodity is obtained.

For ease of understanding, the code description of DeepWalk is given below:

```
def deepwalk_walk(walk_length, start_node):
    walk = [start_node]
    while len(walk) <= walk_length:
        cur = walk[-1]
        try:
            cur_nbrs = item_dict[cur]
            walk.append(random.choice(cur_nbrs))
        except:
            break
    return walk

def simulate_walks(nodes, num_walks, walk_length):
    walks = []
    for i in range(num_walks):
        random.shuffle(nodes)
        for v in nodes:
            walks.append(deepwalk_walk(walk_length=walk_length,
            start_node=v))
        return walks

if __name__ == "__main__":
# Step 1: generate the graph network (omitted)
# Construct item_dict and save nodes relationship between nodes, namely,
          dictionary structure storage, with key as the node, value as the
          field
# Step 2: generate commodities sequence through DeepWalk
nodes = [k for k in item_dict] # Node Sets
num_walks = 5 # Rounds of random walk
walk_length = 20 # Length of random walk
sentences = simulate_walks(nodes, num_walks, walk_length) # Sequence set

# Step 3: train commodities word vectors through Word2Vec
model = Word2Vec(sentences, size=64, window=5, min_count=3,
seed=2020)
```

Extended Learning
The derivation of Item2Vec for Word2Vec and more graph embedding method, such as LINE, Node2Vec, and SDNE, are all worth studying. From the traditional Word2Vec to the embedded representation in the recommendation system, and then to the gradual transition to graph embedding today, these embedding methods are all widely used.

7.3.4   Similarity Calculation Method

Feature extraction based on similarity calculation includes Euclidean distance, cosine similarity, Jaccard similarity, etc. It is helpful to extract the similarity of users, products, and text. After the embedding representation of users and products, the word segmentation representation of text, and various sparse representations have been obtained, the similarity calculation of these vector representations can be performed. Similarity-based calculation has been widely used in applications such as user hierarchical clustering, customized recommendation, or advertising.

7.3.4.1   Euclidean Distance

Euclidean distance is the easiest way to understand distance calculation. It is the distance formula between two points in two-dimensional, three-dimensional or multi-dimensional space. In n-dimensional space, for vectors $A = [a_1, a_2, \ldots, a_n]$, $B = [b_1, b_2, \ldots, b_n]$, the formula used is formula (7.1):

$$d(A, B) = \sqrt{\sum_{i=1}^{n} (a_i, b_i)^2} \tag{7.1}$$

The code for implementing Euclidean distance is as follows:

```
def EuclideanDistance(dataA, dataB):
    # np.linalg.norm for norm calculations, the default is two-norm, which is equivalent to
finding the square root of the sum of squares
    return 1.0 / ( 1.0 + np.linalg.norm(dataA - dataB))
```

7.3.4.2   Cosine Similarity

First of all, the included angle cosine of the sample data is not an included angle cosine in the true geometric sense. In fact, the former is just borrowing the name of the latter and becomes an algebraic sense of the "included angle cosine", which is used to measure the difference between sample vectors. The smaller the included angle, the closer the cosine value is to 1, and vice versa, approximating -1. See formula (7.2) for the included angle cosine between vector A and vector B above:

$$\cos \theta = \frac{\sum_{i=1}^{n} (a_i \times b_i)}{\sqrt{\sum_{i=1}^{n} a_i^2} \times \sqrt{\sum_{i=1}^{n} b_i^2}} \tag{7.2}$$

The codes for cosine similarity are implemented as follows:

```
def Cosine(dataA, dataB):
  sumData = np.dot(dataA, dataB)
  denom = np.linalg.norm(dataA) * np.linalg.norm(dataB)
  # regularize to the range of [0,1]
  return ( 1 - sumData / denom ) / 2
```

7.3.4.3   Jaccard Similarity

Jaccard similarity is generally used to measure the difference between two sets. The idea is that the more elements two sets share, the more similar they will be. In order to control their value range, we can add a denominator, that is, all the elements owned by the two sets. The Jaccard similarity formula for set C and set D is shown in formula (7.3):

$$J(C,D) = \frac{|C \cap D|}{|C \cup D|} = \frac{|C \cap D|}{|C| + |D| - |C \cap D|} \qquad (7.3)$$

The code for Jaccard similarity is implemented as follows:

```
def Jaccard(dataA, dataB):
    A_len, B_len = len(dataA), len(dataB)
    C = [i for i in dataA if i in dataB]
    C_len = len(C)
    return C_len / ( A_len + B_len - C_len)
```

Extended Learning
More similarity calculation methods include Pearson Correlation Coefficient, Adjusted Cosine Similarity, Hamming Distance, Manhattan Distance, Levenshtein Distance, etc.

7.4   Application of User Profiles

The reason why user profiles are worth studying and learning is that they have a wide range of application scenarios, and the user behavior in the Internet era can generate a large amount of data for analysis and modeling, which also provides good conditions for user profiles. Although many companies have different emphases in considering user profiles, after all abstracted for analysis, these profiles can be divided into several categories as shown in Fig. 7.13.

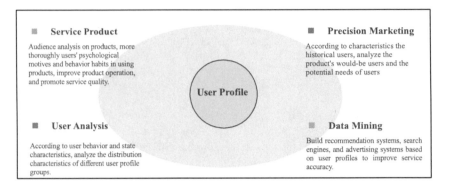

Fig. 7.13 Application scenario of user profile

7.4.1 User Analysis

DAU and MAU are two words that can be heard from time to time. They are usually used to describe the number of people who are active online every day or every month. They are important evaluation indicators for Internet products. Although each product has more or less targeted user groups at the beginning of the design, the product will always face a variety of challenges after going online, including users' referral, activation, retention, figuring out what the characteristics of new users, and whether the attributes of core users have changed, which are all difficult problems that need to be analyzed and studied. Therefore, it is necessary to constantly do user profile analysis, refine crowd characteristics, and then continuously optimize product performance and UI interaction.

1. The competition "User's Purchase Forecast of Stores under the Category" of JD. com on JDATA platform in 2019 provided data information from users, merchants, commodities, and other aspects, including information of merchants and the content of the goods themselves, comment information, and rich interaction between users and merchants. Players were required to build prediction models of relevant categories in the user's purchase of merchant's goods through data mining technology and machine learning algorithms, output matching results of users, shops, and commodity categories, and provide high-quality target groups for users to invite new users. In essence, it is the matching work of goods searching for users, providing suitable goods for would-be users, enlarging the conversion rate of user types, and facilitating the promotion of platform GMV.

2. Tencent Advertising "2020 Tencent Advertising Algorithm Contest" was also focused on the user's behavior information, that was, using the user's interaction behavior in the advertising system as input to predict the user's demographics attributes (such as age, gender, occupation, etc.). For example, for practitioners who lacked user information, inferring user attributes based on their own system data could help them achieve intelligent targeting or audience protection on a wider population. Although the information filled in by the user might be

falsified, it was difficult to falsify the user's behavior habits. Fully mining the user's interaction behavior was helpful to verify the user's attribute characteristics.

7.4.2 Precision Marketing

E-commerce is always among the top ten apps for smartphone downloads. The convenience of online shopping and the wide variety of products greatly enhance users' shopping experience. Using users' historical consumption behavior for user profiling can show users' consumption preferences, so that e-commerce platforms can quickly and accurately respond to users' needs when users need to purchase items, and merchants can easily find their own seed users. In addition to e-commerce search, recommendation systems and advertising also belong to the category of precision marketing. On the major competition platforms of machine learning, competitions on user-oriented precision marketing emerge one after another. The following is a brief introduction to the algorithm competition examples of user profile in precision marketing.

1. The first championship of the competition career of the two co-authors of this book, Wang He and Liu Peng, was won in the "2018 HKUST iFLYTEK AI Marketing Algorithm Competition" held on the DC competition platform. The organizer is iFLYTEK AI Marketing Cloud, which is affiliated to iFLYTEK CO., LTD. Based on years of deep cultivation of artificial intelligence technology and big data accumulation, it endows marketing intelligence and innovative brain, and helps advertisers with a sound product matrix and a full range of services. AI realizes the overall improvement of marketing efficiency and creates a new ecology of digital marketing. With the rapid development of iFLYTEK AI Marketing Cloud, it has accumulated a large amount of advertising data and user data. How to effectively use these data to predict the user's advertising click probability is a key issue in the application of big data in precision marketing, and it is also the core technology that all intelligent marketing platforms must have. This competition provides a large amount of advertising data from iFLYTEK AI Marketing Cloud. The contestants need to build a prediction model through artificial intelligence technology to estimate the user's advertising click probability, that is, to predict the advertising click probability under the condition of given advertising, media, users, contextual content, and other information related to advertising click. Although this competition also belongs to the field of advertising, the two co-authors have made good achievements by digging deep into the user profile through the user tag set given, accurately linking users with advertisements.

2. Tencent Advertising "2018 Tencent Advertising Algorithm Contest" is based on Look-alike (similar crowd expansion), based on seed users, that is, advertisers' existing consumers, through a certain algorithm evaluation model, to find out potential consumers similar to existing consumers, in order to effectively help advertisers to tap new customers and expand business. This topic will provide

participants with hundreds of seed groups, user characteristics corresponding to a large number of candidate groups, and advertising characteristics corresponding to seed groups. Participants need to predict whether the promising user of the seed package belongs to the score of the seed package. The higher the score, the more likely the promising user is to be a potential user to be extended in a certain package. In the process of mining similar groups, Look-alike mainly relies on the user's basic attributes and the behavioral information they have, which requires a huge data stock as the source of analysis.

7.4.3 Areas of Risk Control

In addition to the application mentioned in the previous two sections, user profile also has a special type of application scenario, which is the risk control problem in the financial field. This type of scenario mainly focuses on the user's economic situation and evaluates the user's ability to repay loans in combination with dimensions such as credit registry. In the era of mobile payment, electronic payment has replaced cash flow, and at the same time enables people's various consumption behaviors to be automatically recorded, which generates a large amount of transaction statement and provides support for evaluating the consumer's consumption ability and credit. Related competition cases are as follows.

1. "Group Image of Consumers - Intelligent Scoring of Credits" of the DF competition platform. Under the big background of the rapid development of the social credit standard system, China Mobile, as a communication operator, has massive, extensive, high-quality, and time-sensitive data. Breaking the traditional credit scoring method, how to intelligently score customers based on rich big data is a difficult problem for China Mobile and Newland Technology Group to tackle key problems at present. China Mobile Fujian Company has provided sample data (desensitized) for a certain month in 2018, including multi-dimensional data such as customers' various communication expenditures, arrears, travel conditions, consumption places, social contacts, personal interests, etc. Participants are required to accurately evaluate users' consumption credit scores through analysis and modeling, using machine learning and deep learning algorithms.

2. "PPDAI 4th Magic Mirror Data Application Contest". This competition is based on the Internet financial credit business as the background, considering that the amount of a single underlying asset in the Internet financial credit business is small and complex and diverse, which brings huge fund management pressure to lenders or institutions, the contestants need to use the data provided to predict the daily repayment amount of the asset portfolio in the future for a period of time. The competition topic covers common problems in the financial field such as credit default prediction and cash flow prediction and is also a complex timing problem and multi-objective prediction problem. The organizer provides information such as basic information of users borrowing money, user profile tag list, borrowing and repayment behavior log of users, attribute table of borrowing, etc.

The above two competitions are about typical risk control problems on credit scoring and cash flow prediction respectively, and the characteristics of problems in the field of risk control are very obvious. First, the business has high explanatory requirements on the model and certain requirements on timeliness, which requires participants to learn to weigh the complexity and precision of the model and optimize the algorithm kernel appropriately in actual modeling. Second, there are various business models, each of which has a very high connection with the business objectives, and it is often necessary to build appropriate models according to the business. Third, the proportion of negative samples is very small, which is one of the main fields of balanced learning algorithm.

7.5   Thinking Exercises

1. Do you think the user profile is used to reflect the commonness or individuality of users? Why?
2. Think about how the algorithm and operation team will draw a profile of you with the app you use every day.
3. There are also many text mining algorithms. Try to sort out the calling methods of these algorithms and get familiar with the parameter setting according to the principle.
4. Embedding methods are widely used. In addition to Word2Vec and DeepWalk, what are other embedding algorithms? What is the specific principle?
5. There are many methods to calculate similarity, but it is not easy to retrieve what is the most similar or the top N similar from a large amount of data. Is there any good retrieval algorithm then?

Chapter 8
Case Study: Elo Merchant Category Recommendation

This chapter will take the Kaggle platform's 2019 Elo Merchant Category Recommendation competition (as shown in Fig. 8.1) as an example to explain the real-world practice related to user profiles and illustrate the complete process and precautions of real cases end to end. This chapter is mainly divided into the following parts: question understanding, data exploration, feature engineering, model selection, model integration, efficient scoring, and summarizing competition questions, the common organizational structures of all chapters involving case studies in this book, as well as important components of a competition process. I believe that under the guidance of this book, readers can quickly become familiar with the competition process and apply it in practice.

8.1 Understanding the Competition Question

There's a saying that goes, "Sharpening your axe will not delay your job of chopping wood." Before the competition, we should fully understand the relevant information of the competition questions and know the needs behind them, so as to achieve the purpose of correctly examining the questions.

8.1.1 Competition Background

Imagine that when you are hungry in an unfamiliar place and want to find something delicious, will you get a restaurant recommendation that is exclusively recommended based on your personal preferences, and will the recommendation also be accompanied by the discount information provided by your credit card provider for nearby restaurants?

© The Author(s), under exclusive license to Springer Nature Singapore Pte Ltd. 2023   137
W. He et al., *Machine Learning Contests: A Guidebook*,
https://doi.org/10.1007/978-981-99-3723-3_8

Fig. 8.1 Elo Merchant Category Recommendation Competition

Currently, Elo, one of Brazil's largest payment brands, has established a partnership with merchants to provide customers with information on promotions or discounts. But are these promotions beneficial to both customers and merchants? Do customers like their event experience? Can merchants see duplicate transactions? Personalization is the key to answer these questions.

Elo has built machine learning models to understand customers' preferences, ranging from food to shopping, in the most important aspects in a customer's lifetime. But so far, those learning models have not been specifically tailored for individuals or personal data, which is why this competition is held.

In this competition, participants are required to develop algorithms to identify and provide individuals with the most relevant opportunities by finding signs of customer loyalty. Your opinions will improve the lives of customers, help Elo reduce unnecessary activities, and create accurate and correct experiences for customers.

8.1.2 Competition Data

In order to ensure privacy and information security, all data in this competition are simulated and fictitious data or desensitized data, not real customer data. Specifically, the following data files are included.

- **train.csv**: training set.
- **test.csv**: testing set.
- **sample_submission.csv**: an example of a correct and standardized submission document, containing all the card_id that the contestant needs to predict
- **historical_transactions.csv**: the transaction history of credit cards (card_id) at a given merchant; for each credit card, it contains up to three months of its transaction history.
- **merchants.csv**: additional information for all merchants (merchant id) in the dataset
- **new_merchant_transactions.csv**: the shopping data of each credit card at the new merchant, including up to two months of data
- **Data_D ictionary.xlsx**: the description file of the data dictionary that provides the field meanings of the above sheets, including the corresponding instructions

for train, historical_transactions, new_merchant_period, and merchant; I believe the contestants are as confused as the author about what this new_merchant_period is, and will continue to read.

8.1.3 Competition Task

The task is to train the model by using the customer's historical transaction records and the information data of the customer and the merchant, and to finally predict the loyalty scores of all credit cards in the testing set.

8.1.4 Evaluation Indicators

In this competition, the root mean square error (RMSE) is used as the evaluation indicator to calculate the results submitted by the participants. The specific calculation method is as formula (8.1):

$$\text{RMSE} = \sqrt{\frac{1}{n} \sum_{i=1}^{n} (y_i - \hat{y}_i)^2} \tag{8.1}$$

Here $\hat{y}_i$ is the loyalty score predicted by the participants for each credit card, and y_i is the true loyalty score of the corresponding credit card.

8.1.5 Competition FAQ

Q There are so many data files provided in the competition. Which are indispensable for completing the modeling?

A At least train.csv and test.csv are required. These two files contain card_id of all credit cards that will be used for training and testing. In addition, historical_transactions.csv and new_merchant_transactions.csv contain the transaction records of every single credit card.

Q How can participants use the rest of the data?

A train.csv and test.csv contain the card_id of all credit cards and information about the credit card itself (such as when the card is activated in the first month, etc.). In addition, train.csv also includes the target value of some customers, that is, the exact

loyalty   score   of   these   customers.   historical_transactions.csv   and
new_merchant_transactions.csv are designed to be combined with train.csv, test.
csv and merchants.csv, because as described above, these two files contain the
transaction records of each credit card, so by combing transaction records with the
merchant, additional information such as the merchant level can be provided.

8.2   Data Exploration

I believe that many contestants, like the writer, still feel a little confused even after
reading all the contents of Sect. 8.1. As the old saying goes, "Talk is cheap; show me
the data". No words are as realistic as directly understanding the data. I believe many
problems can be solved by observing and analyzing the data. In the field of data
mining, there is a proper term called exploratory data analysis (EDA, which is called
"data exploration" in this book). This can not only help participants understand the
real meaning of the topic of the competition and achieve a grasp of the general
situation of the data, but also play a guiding role in the following feature engineering
and modeling, further enhancing participants' comprehension of the business and
application of technology. Therefore, the first thing we need to do is to explore the
data set. After reading this, some readers may have begun to try to actuate "code
power". Before writing the code, I would like to suggest, if possible, take advantage
of Excel, a powerful spreadsheet tool to open various files provided by the competition and get an intuitive feeling. This topic enables us to directly view train.csv,
test.csv, sample_submission.csv, and Data_D ictionary.xlsx. Generally speaking,
files over 50 MB are not convenient to be opened directly with Excel, because it
will cause computer lagging. Of course, it should be noted that the data format of
Excel itself will also affect the presentation of files, such as scientific counting, text,
and date.

8.2.1   Field Category Meaning

Before exploring the data, participants should first clarify the introduction to each
data file and the meaning of the fields in the file in order to understand the
competition questions and build analysis logic. By referring to the field information
table Data_D ictionary.xlsx provided by the organizer of the competition, you can
see the fields and their meanings in the five data files as follows.

8.2.1.1   Fields and Meanings in train.csv and test.csv

- card_id: a unique credit card ID, for example C_ID_92a2005557;
- first_active_month: the month of the first purchase with a credit card, in the
  format of YYYY-MM, such as 2017-04;

- feature_1/2/3: anonymous credit card discrete features 1/2/3, for example, 3;
- target: loyalty numerical score calculated 2 months after historical and evaluation period; loyalty score target column, for example, 0.392913.

By looking at the meaning of the above fields, it can be seen that all three features are anonymous credit card discrete fields, and there is a month of first purchase, and target is the loyalty score calculated quantitatively two months after the historical and evaluation period. It should be noted that the evaluation period here should refer to the information in the new_merchant_transactions.csv, and it also corresponds to the new_merchant_period field in Data_Dictionary.xlsx. At the same time, please check the correctness of the data. Then you will find that the card_id of the training set and the testing set are unique values, and the card_id of the training set and the testing set are not repeated.

8.2.1.2   Fields and Meanings in historical_transactions.csv and new_merchant_transaction.csv

- card_id: a unique credit card identifier, i.e., credit card ID, for example C_ID_415bb3a509;
- month_lag: the month from the reference date, for example, $[-12, -1]$, $[0,2]$;
- purchase_date: shopping date (time), such as 2018-03-11 14:57:36;
- category_3: anonymous category feature 3, such as A/B/C/D/E;
- installments: the quantity of goods purchased, for example, 1;
- category_1: anonymous category feature 1, such as Y/N;
- merchant_category_id: product type ID (anonymized), for example, 307;
- subsector_id: product category group ID (anonymized), for example, 19;
- merchant_id: product ID (anonymized), such as M_ID_b0c793002c;
- purchase_amount: standardized purchase amount, for example, -0.557574;
- city_id: city ID (anonymized), for example, 300;
- state_id: state ID (anonymized), for example, 9;
- category_2: anonymous category feature 2, for example, 1.

8.2.1.3   Fields and Meanings in merchants.csv

- **merchant_id**: a unique product identifier, i.e., the product ID, such as M_ID_b0c793002c;
- **merchant_group_id**: commodity group (anonymized), for example, 8353;
- **merchant_category_id**: product type ID (anonymized), for example, 307;
- **subsector_id**: product category group ID (anonymized), for example, 19;
- **numerical_1/2**: anonymous numerical features 1/2, for example, -0.057471;
- **category_1**: anonymous category feature 1, such as N/Y;
- **most_recent_sales_range**: the sales level in the most recent active month, such as A, B, C, D, E (The level decreases in turn);

- **most_recent_purchases_range**: the level of number of transactions in the most recent active month, such as A, B, C, D, E (The level decreases in turn);
- **avg_sales_lag3/6/12**: average monthly income for the past 3, 6, and 12 months divided by the income in the previous active month, for example, −82.13;
- **avg_purchases_lag3/6/12**: average monthly trading volume in the past 3, 6, and 12 months divided by the transaction volume in the previous active month, for example, 9.6667;
- **active_months_lag3/6/12**: the number of active months in the past 3, 6, and 12 months, for example, 3;
- **category_4**: anonymous category feature 4, such as Y/N.

8.2.2 Field Value Status

After combing the fields and meanings of each table file, participants can specifically view the specific value status of each field in each table. In general, in addition to the meaning of the field, the value type of the field should be determined in combination with the meaning of the field. There are two types: characters (object) and numerical values (int, float). Attention should be paid to whether the meaning of the field is discrete or not and whether the value of the field is a numerical value. There is no inevitable connection. Because the value of a discrete field may be a numerical value, such as a city_id field, although its values are all numerical types, there is no relationship of comparing which is big and small between them; the value of a numerical field may also be a character, such as a most_recent_sales_range field. Although its values are of character types, you can obviously feel which is greater than another, that is the relationship between them.

No matter what type the field is, participants need to be mainly concerned about two aspects: one is the missing value situation, and the other is the approximate value range and distribution of the field. The focus of discrete features is the number distribution of eigenvalues, while numerical features need to pay attention to their value range and abnormal values, off-group points, etc. Here, the target column of this competition is listed as an example, and the target column is of continuous values. The method describe of pandas.series can be used to analyze its value range and interval.

What is interesting is if the value_counts method of analyzing discrete feature distribution is adopted at the same time, the contestants will be pleasantly surprised to find that the target column has an extreme outlier −33.219281, accounting for approximately 1%. In the following modeling tasks, the contestants will gradually realize the importance and particularity of this discovery.

8.2.3   Difference in Data Distribution

There are three special data set titles in the field of machine learning, namely training set, verification set, and testing set. The model learns the correlation between features and tags from the training set, and uses the verification set to evaluate, so as to avoid over-fitting and under-fitting. After learning to an appropriate extent, the model can be applied to the testing set for prediction. In order to make the prediction effect of the model excellent, one of the prerequisites is that the data distribution of the training set, verification set, and testing set should be similar, especially the joint distribution of features and tags being consistent, so that the correlation learned by the model can be generalized. The verification set has a variety of selection ways according to different modeling tasks. Usually, modeling tasks that do not involve time sequence can randomly divide a data set into training sets and verification sets; the testing set can only be divided by the distribution or joint distribution of some features because there is no way to know the tag, which is the target column. This question will take train.csv and test.csv as examples to explore and analyze the differences in data distribution.

As shown in Fig. 8.2, the univariate distribution display of the fields first_active_month, feature_1, feature_2, and feature_3—in train.csv and test.csv shows that the absolute quantity distribution shapes on all univariates are extremely similar in the training set and the testing set. Further examination of the relative proportion distribution is needed to obtain a more accurate conclusion. The

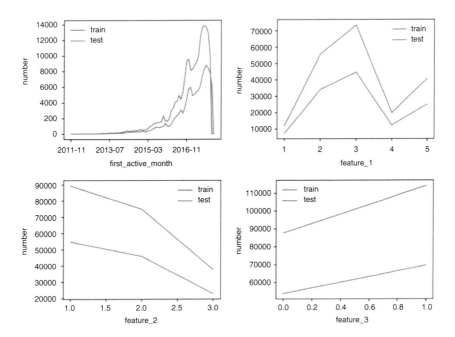

Fig. 8.2 Feature Distribution Difference between Training Set and Testing Set

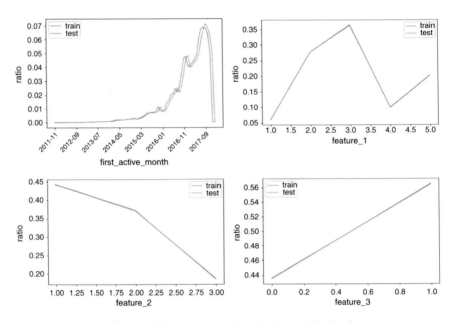

Fig. 8.3 Feature Distribution Difference between Training Set and Testing Set

univariate proportion distribution of these four fields will continue to be displayed as shown in Fig. 8.3.

As can be seen from Fig. 8.3, the relative proportion distribution shape on all univariates of the training set and the testing set is basically the same, so it is assumed that the training set is generated in the same way as the testing set, and the joint distribution can continue to be verified as a factual basis for strengthening this conjecture.

It should be noted that there is an imprecision in the above analysis by the means of drawing—that is, the univariate value range of the training set and the testing set may not be exactly the same—so drawing two lines on the same graph may make mistakes, such as offset. Curious readers can verify by themselves whether the abscissas of the two are exactly the same. If not, what will happen when you run the same drawing code? In the following joint distribution verification, we will solve this problem.

When looking at the joint distribution of multi-variables, a scatter plot can usually be used, but the four fields here are all discrete features, and the scatter plot is not suitable for continuing the idea of drawing univariate diagram above, so contestants can splice the two variables together to transform the joint distribution of multi-variables into univariate distribution, and the results are shown in Fig. 8.4.

After correcting the omission, participants can find that the joint distribution of the two variables of the training set and the testing set is also generally the same, which implies that the training set and the testing set are generated in the same way; that is, the training set and the testing set are the result of randomly dividing the same

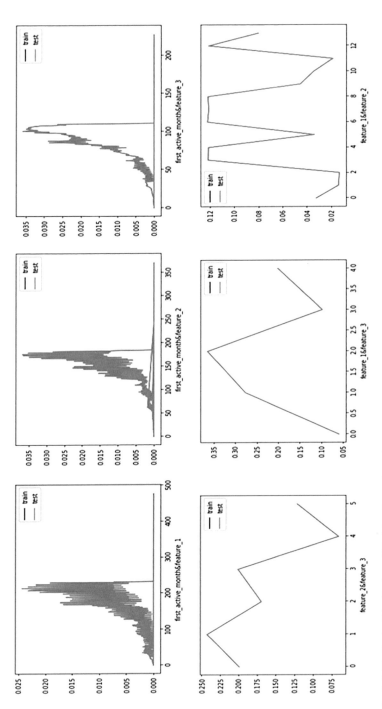

Fig. 8.4 Verify the Distribution of Training Set and Testing set

batch of data. Participants who are interested in this can keep verifying the three-variable and four-variable distributions. Assuming that this conjecture about the training set and the testing set is true, it will greatly increase participants' confidence in subsequent feature engineering, enabling them to have an overall grasp of the modeling method.

8.2.4   Table Correlation

From the above exploration, it can be seen that train.csv and test.csv help participants define the training set, testing set, and modeling objectives; historical_transactions.csv and new_transactions.csv have the same fields, but they are different in time, providing participants with rich customer transaction information; merchants.csv describes the business situation of the merchants. Participants need to combine the merchant's basic information table and the customer transaction records table to conduct data mining on the user's consumption behavior, in order to find as rich relevant information as possible in the target column, thus achieving excellent prediction effect.

8.2.5   Data Preprocessing

In order to facilitate the subsequent feature extraction and keep the data clean and tidy, participants can complete the corresponding data cleaning while exploring the data. For different data, the processing skills are also varied, but the purpose is always to clear the obstacles for the subsequent feature extraction. Only the detailed steps are given here. Please refer to the eda.ipynb in the attached resources of this book for the specific code.

8.2.5.1   train.csv and test.csv

These two tables only have one missing value in the first_active_month field of the test.csv. Generally, having only one missing value has little influence, and this field is of character type, so it needs to be encoded. Considering that it has a sequence relationship in essence, it can be encoded by dictionary ordering.

8.2.5.2   merchants.csv

The processing steps are as follows:

1. Divide discrete field category_cols and continuous field numeric_cols according to business meaning;
2. Perform dictionary ordering and encoding of discrete fields in character type;
3. In order to make it more convenient to statisticize, the missing values are processed, and the discrete fields are uniformly filled with −1;
4. Positive infinite values are found in exploring discrete fields, which is unacceptable for feature extraction and models, so infinite values need to be processed, and the maximum value is used to replace them here;
5. There are many ways to deal with the missing values of discrete fields. Here, the average value is used for filling first, and then optimization is needed later;
6. Remove columns that duplicate those in the transaction record table and the duplicate records of the merchant_id.

8.2.5.3 new_merchant_transactions.csv and historical_transactions.csv

The processing steps are as follows:

1. In order to unify the processing, the two tables are first spliced together, and then be further distinguished by the condition—month_lag $> = 0$;
2. Divide discrete fields, continuous fields, and time fields;
3. Carry out dictionary ordering and encoding of character discrete fields and filling of missing values by imitating the processing method of merchants.csv;
4. Process the time period. For simplicity, extract the information of month, day of the week (weekdays and weekend), and time period (morning, afternoon, evening, and before dawn);
5. Perform dictionary ordering and coding of newly generated discrete fields of months with purchase;
6. After processing the tables of merchant information and transaction records, these tables are merged for the convenience of unified calculation of features, and then the corresponding field types are re-divided.

8.3 Feature Engineering

After basic data exploration, I believe the contestants already have a good comprehension to the data and the tasks of the competition. The focus of this competition is to mine the relationship between various transaction behaviors of users and target columns, thus achieving a good model learning effect and enabling the model to accurately predict the loyalty scores of users in testing sets. Therefore, this is a topic that pays attention to the profile of credit card users' local consumption preferences. By finding similar users in training sets to analogize the loyalty scores of users of testing sets, high-value groups can be distinguished to provide decision support for merchants and credit card banks, and the shopping experience of consumers can be improved at the same time. Therefore, feature engineering can focus on the portrait of users' transaction behavior, that is, the quantification of users' shopping behavior in various dimensions, such as the expense and purchase quantity in the latest month, etc.

In the field of profiles for evaluating user values, there is a classic RFM theory, namely Recent, Frequency, and Money. Combined with the previous data exploration, participants should be able to clarify the feasibility of this theory. Here we will simulate frequency with the purchase quantity and use the expense as money. This question not only has a wide range of modeling objectives, but also its data structure has typical characteristics; that is, it mainly uses tables recording user behavior (historical_transactions.csv, merchants.csv, and new_merchant_transactions.csv) for information mining. Next, we will introduce two methods of feature extraction, one is to use python's native dictionary structure to extract general features, and the other is to use pandas, the powerful data processing tool's statistical function to extract business features.

8.3.1   General Features

The key-value structure of the dictionary provides a good mapping relationship that is easy to use. The feature extraction here can take the user as the key-value of the first layer, the feature field as the key-value of the second layer, and then convert the dictionary into pandas. DataFrame format after the statistics are completed; in short, it is to know the purchase quantity and expense of the user for every value-taking in each category field.

First, create a dictionary to store the generated statistical features and assign values to each card_id:

```
features = {}
card_all = train['card_id'].append(test['card_id']).values.tolist()
for card in card_all:
    features[card] = {}
```

Second, record the index of each field so that the target value can be directly obtained when processing by rows:

```
columns = transaction.columns.tolist()
idx = columns.index('card_id')
category_cols_index = [columns.index(col) for col in category_cols
numeric_cols_index = [columns.index(col) for col in numeric_cols]
```

Then, extract and update the features of the corresponding fields by rows:

```
# record the running time
s = time.time()
num = 0
```

```
    for i in range(transaction.shape[0]):
    va = transaction.loc[i].values
    card = va[idx]
    for cate_ind in category_cols_index:
        for num_ind in numeric_cols_index:
            col_name = '&'.join([columns[cate_ind], va[cate_ind],
columns[num_ind]])
            features[card][col_name] = features[card].get(col_name, 0)
+ va[num_ind]
    num += 1
    if num%1000000==0:
        print(time.time()-s, "s")
    del transaction
    gc.collect()
```

Finally, convert the dictionary into a characteristic DataFrame table structure and reset the column name of the table.

```
df = pd.DataFrame(features).T.reset_index()
del features
cols = df.columns.tolist()
df.columns = ['card_id'] + cols[1:]
```

After the table is generated, the training set and testing set can be spliced for subsequent model training. In order to distinguish from subsequent features, the feature set is named dict here, and the complete code is dict.ipynb in the code resources.

8.3.2   Business Features

The advantage of universal feature extraction based on dictionary structure is that it can be read and processed by row, regardless of speed or memory, and it can also quantify the user behavior under each subclass. However, its disadvantages are also obvious; that is, it requires a fixed data structure and will produce higher-dimensional results. Another solution is to use the groupby method of the pandas tool for statistics. This method is much simpler but needs higher memory performance because all data needs to be loaded. It should be noted that in order to meet the statistical needs of pandas, missing values and discrete fields are no longer converted.

At the same time, two features are added. These two features are related to the time interval between the user's two purchases. They are depicted in terms of day and month respectively. The code is as follows:

```
transaction['purchase_day_diff'] = transaction.groupby("card_id")
['purchase_day'].diff()
transaction['purchase_month_diff'] = transaction.groupby
("card_id")['purchase_month'].diff()
```

First, set the corresponding statistics you want to obtain according to the type of fields and give the corresponding field list to prepare for subsequent calculations. This method has clear logic and more comprehensive feature structure:

```
aggs = {}
for col in numeric_cols:
    aggs[col] = ['nunique', 'mean', 'min', 'max','var','skew', 'sum']
for col in categorical_cols:
    aggs[col] = ['nunique']
    aggs['card_id'] = ['size', 'count']
cols = ['card_id']
for key in aggs.keys():
    cols.extend([key+'_'+stat for stat in aggs[key]])
```

Then, perform calculations and statistics separately for new_merchant_transactions.csv, historical_transactions.csv, and the whole time period, to obtain statistical features from multiple angles:

```
df = transaction[transaction['month_lag']<0].groupby('card_id').
agg(aggs).reset_index()
df.columns = cols[:1] + [co+'_hist' for co in cols[1:]]

df2 = transaction[transaction['month_lag']>=0].groupby('card_id').
agg(aggs).reset_index()
df2.columns = cols[:1] + [co+'_new' for co in cols[1:]]
df = pd.merge(df, df2, how='left',on='card_id')

df2 = transaction.groupby('card_id').agg(aggs).reset_index()
df2.columns = cols
df = pd.merge(df, df2, how='left', on='card_id')
```

It can be seen that the number of features counted by the groupby method will be much less, and the statistics of various user behaviors will be concentrated. In order to distinguish it from subsequent features, the feature set here is named groupby.

8.3.3   Text Features

In addition to the conventional features mentioned above, this competition question can also extract a class of features, which is based on the TF-IDF vector features in the field of CountVector and NLP. Different from the previous dict and groupby, here only part of the discrete fields is used for word frequency statistics. CountVector is similar to the features in the dict part, while TF-IDF is a supplement to the multivariate joint distribution.

First, the corresponding fields are processed into a standard input format, and then the relevant methods in sklearn are called for calculation. It should be noted that this part of the feature uses the sparse matrix structure of scipy, so it is different from dict and groupby while processing.

8.3.4   Feature Selection

There are two common feature selection methods, one is filtering feature selection, and the other is feature importance selection. The former uses some statistical correlation coefficients for filtering, while the latter selects through the feature importance in the model evaluation process. Broadly speaking, the function of feature selection is mainly to improve the speed and accuracy of model training. In Sect. 8.4, model training will be conducted for different feature selection methods and the final offline and online results will be compared.

8.4   Model Training

After preparing the basic features, participants can begin to try the whole process of model training and prediction. In order to introduce some processing skills to participants as much as possible, this section will introduce the whole process of three models (random forest, LightGBM, and XGBoost) and combine different feature selection methods and parameter tuning methods at the same time.

8.4.1   Random Forest

The first is the random forest model in the sklearn library. The full process of this model is divided into four modules: reading data, feature selection, parameter tuning, and training prediction. The elements of the model are composed of dict and groupby in Sect. 8.3.4. The feature selection uses the Filter method based on Pearson correlation coefficient calculation to obtain the first 300 features, and the parameter tuning uses the sklearn library's GridSearch.

First of all, read the specified feature set and testing set that have been constructed in advance and splice the data set. The specific code is as follows:

```
def read_data(debug=True):
  NROWS = 10000 if debug else None
  train_dict = pd.read_csv("preprocess/train_dict.csv", nrows=NROWS)
  test_dict = pd.read_csv("preprocess/test_dict.csv", nrows=NROWS)
  train_groupby = pd.read_csv("preprocess/train_groupby.csv",
nrows=NROWS)
  test_groupby = pd.read_csv("preprocess/test_groupby.csv",
nrows=NROWS)
```

```
# remove duplicate columns
for co in train_dict.columns:
    if co in train_groupby.columns and co!='card_id':
        del train_groupby[co]
for co in test_dict.columns:
    if co in test_groupby.columns and co!='card_id':
        del test_groupby[co]

train = pd.merge(train_dict, train_groupby, how='left',
on='card_id').fillna(0)
test = pd.merge(test_dict, test_groupby, how='left', on='card_id').
fillna(0)
return train, test
```

Then use the Filter method based on Pearson correlation coefficient calculation to take the first 300 features for selection. 300 here is a number taken at will. Participants can try several more numbers to choose the best one. The specific code is as follows:

```
def feature_select_pearson(train, test):
features = [f for f in train.columns if f not in ["card_id","target"]]
featureSelect = features[:]
# remove those with missing values exceeding 99%
for fea in features:
    if train[fea].isnull().sum() / train.shape[0] >= 0.99:
        featureSelect.remove(fea)
    # perform Pearson correlation coefficient calculation
    corr = []
    for fea in featureSelect:
        corr.append(abs(train[[fea, 'target']].fillna(0).corr().values
[0][1]))

    se = pd.Series(corr, index=featureSelect).sort_values
(ascending=False)
    feature_select = ['card_id'] + se[:300].index.tolist()
    return train[feature_select + ['target']], test[feature_select]
```

Then there is the parameter tuning based on grid search. Grid search is actually a permutation set over different parameters and different values, and it may be necessary to manually iterate the parameter space several times according to the tuning results. Of course, each iteration is adding the unsearched parameter area based on the previous best parameter; the specific code is as follows:

```
def param_grid_search(train):
features = [f for f in train.columns if f not in ["card_id","target"]]
parameter_space = {
    "n_estimators": [80],
    "min_samples_leaf": [30],
    "min_samples_split": [2],
    "max_depth": [9],
    "max_features": ["auto", 80]
}
```

```
# configure as parameter tuning of mse
clf = RandomForestRegressor(
     criterion="mse",
     min_weight_fraction_leaf=0.,
     max_leaf_nodes=None,
     min_impurity_decrease=0.,
     min_impurity_split=None,
     bootstrap=True,
     oob_score=False,
     n_jobs=4,
     random_state=2020,
     verbose=0,
     warm_start=False)
grid = GridSearchCV(clf, parameter_space, cv=2,
scoring="neg_mean_squared_error")
grid.fit(train[features].values, train['target'].values)

  print("best_params_:")
  print(grid.best_params_)
  means = grid.cv_results_["mean_test_score"]
  stds = grid.cv_results_["std_test_score"]
  for mean, std, params in zip(means, stds, grid.cv_results_
["params"]):
        print("%0.3f (+/-%0.03f) for %r"% (mean, std * 2, params))
  return grid.best_estimator_
```

Finally, model training and prediction are carried out according to the best result of parameter tuning. Here, five-fold cross-validation is selected, and attention is paid to saving the cross prediction results of the training set and the prediction results of the testing set, which can be used in Sect. 8.5.

```
def train_predict(train, test, best_clf):
   features = [f for f in train.columns if f not in
["card_id","target"]]
prediction_test = 0
cv_score = []
prediction_train = pd.Series()
kf = KFold(n_splits=5, random_state=2020, shuffle=True)
for train_part_index, eval_index in kf.split(train[features], train
['target']):
   best_clf.fit(train[features].loc[train_part_index].values,
      train['target'].loc[train_part_index].values)
   prediction_test += best_clf.predict(test[features].values)
   eval_pre = best_clf.predict(train[features].loc[eval_index].
values)
   score = np.sqrt(mean_squared_error(train['target'].loc
[eval_index].values,
    eval_pre))
cv_score.append(score)
   print(score)
   prediction_train = prediction_train.append(pd.Series(
      best_clf.predict(train[features].loc[eval_index]),
```

```
index=eval_index))
print(cv_score, sum(cv_score) / 5)
pd.Series(prediction_train.sort_index().values).
    to_csv("preprocess/train_randomforest.csv", index=False)
pd.Series(prediction_test / 5).to_csv("preprocess/
test_randomforest.csv",
index=False)
test['target'] = prediction_test / 5
test[['card_id', 'target']].to_csv("result/
submission_randomforest.csv",
index=False)
return
```

The last step here is to use five-fold cross validation. It not only can avoid the over-fitting of the model to the training set, but also can make the prediction results of the model to the testing set more robust. There is also an incidental benefit that it can be used to generate Stacking integration features, that is, the cross-prediction results of the training set and the model prediction results of the testing set. The two are retained to prepare for subsequent model integration. A total of three files need to be saved: train_randomforest.csv, test_randomforest.csv, and submission_randomforest.csv.

After the prediction result comes out, submit it for test and get the specific score. The cross validation score is 3.68710936, whose public score (public ranking list, commonly known as A list) submitted is 3.75283 (2867/4127), while the private score (hidden list, commonly known as B List) is 3.65493 (2814/4127).

8.4.2 LightGBM

Perform LightGBM modeling by using the same feature set as that of random forest model. The four modules of the whole process of the LightGBM model and the random forest model are same, with the data reading stage being exactly like each other. The difference is that the wrapper method is used in the feature selection stage, and the hyperopt framework is chosen in the parameter tuning stage.

8.4.2.1 Feature Selection

Here, the feature importance is mainly used to select the first 300 features for modeling training, and this number can also be changed according to the modeling effect. The specific code is as follows:

```
def feature_select_wrapper(train, test):
label = 'target'
features = [f for f in train.columns if f not in ["card_id","target"]]
# configure the training parameter of models
```

```
params_initial = {
    'num_leaves': 31,
    'learning_rate': 0.1,
    'boosting': 'gbdt',
    'min_child_samples': 20,
    'bagging_seed': 2020,
    'bagging_fraction': 0.7,
    'bagging_freq': 1,
    'feature_fraction': 0.7,
    'max_depth': -1,
    'metric': 'rmse',
    'reg_alpha': 0,
    'reg_lambda': 1,
    'objective': 'regression'
}
ESR = 30
NBR = 10000
VBE = 50
kf = KFold(n_splits=5, random_state=2020, shuffle=True)
fse = pd.Series(0, index=features)
for train_part_index, eval_index in kf.split(train[features], train
[label]):
    train_part = lgb.Dataset(train[features].loc[train_part_index],
        train[label].loc[train_part_index])
    eval = lgb.Dataset(train[features].loc[eval_index],
        train[label].loc[eval_index])
    bst = lgb.train(params_initial, train_part, num_boost_round=NBR,
        valid_sets=[train_part, eval],
        valid_names=['train', 'valid'],
        early_stopping_rounds=ESR, verbose_eval=VBE)
    fse += pd.Series(bst.feature_importance(), features)

  feature_select = ['card_id'] + fse.sort_values(ascending=False).
index.tolist()[:300]
  print('done')
  return train[feature_select + ['target']], test[feature_select]
```

8.4.2.2   Parameter Tuning

Hyperopt is a sklearn Python library that performs serial and parallel optimization on the search space. The search space can be real values, discrete values, and conditional dimensions, providing an interface for transferring parameter spaces and evaluation functions. Currently supported optimization algorithms are random search, simulated annealing, and TPE (Tree of Parzen Estimators). Compared with grid search, hyperopt can often obtain better parameter results in a relatively short time. The specific code is as follows:

```
def params_append(params):
  params['objective'] = 'regression'
  params['metric'] = 'rmse'
```

```
    params['bagging_seed'] = 2020
    return params

def param_hyperopt(train):
label = 'target'
features = [f for f in train.columns if f not in ["card_id","target"]]
train_data = lgb.Dataset(train[features], train[label], silent=True)
def hyperopt_objective(params):
    params = params_append(params)
    print(params)
    res = lgb.cv(params, train_data, 1000, nfold=2, stratified=False,
shuffle=True,
        metrics='rmse', early_stopping_rounds=20, verbose_eval=False,
        show_stdv=False, seed=2020)
    return min(res['rmse-mean'])
# set the spatial region for parameters
params_space = {
    'learning_rate': hp.uniform('learning_rate', 1e-2, 5e-1),
    'bagging_fraction': hp.uniform('bagging_fraction', 0.5, 1),
    'feature_fraction': hp.uniform('feature_fraction', 0.5, 1),
    'num_leaves': hp.choice('num_leaves', list(range(10, 300, 10))),
    'reg_alpha': hp.randint('reg_alpha', 0, 10),
    'reg_lambda': hp.uniform('reg_lambda', 0, 10),
    'bagging_freq': hp.randint('bagging_freq', 1, 10),
    'min_child_samples': hp.choice('min_child_samples', list(range
(1, 30, 5)))
}
params_best = fmin(
    hyperopt_objective,
    space=params_space,
    algo=tpe.suggest,
    max_evals=30,
    rstate=RandomState(2020))
return params_best
```

For the output of results, grid search is very different from hyperopt. The former outputs the best classifier with parameters, while the latter outputs the best parameter dictionary.

8.4.2.3   Training Prediction

Finally, the training and prediction of the LightGBM model are carried out in the same way, and the prediction result of the final training set and the prediction result of the testing set are also obtained by using the five-fold cross validation method as follows:

```
def train_predict(train, test, params):
    label = 'target'
    features = [f for f in train.columns if f not in ["card_id","target"]]
    params = params_append(params)
```

```
    kf = KFold(n_splits=5, random_state=2020, shuffle=True)
    prediction_test = 0
    cv_score = []
    prediction_train = pd.Series()
    ESR = 30
    NBR = 10000
    VBE = 50

for train_part_index, eval_index in kf.split(train[features], train
[label]):
    train_part = lgb.Dataset(train[features].loc[train_part_index],
      train[label].loc[train_part_index])
    eval = lgb.Dataset(train[features].loc[eval_index],
        train[label].loc[eval_index])
    bst = lgb.train(params, train_part, num_boost_round=NBR,
        valid_sets=[train_part, eval],
        valid_names=['train', 'valid'],
        early_stopping_rounds=ESR, verbose_eval=VBE)
prediction_test += bst.predict(test[features])
prediction_train = prediction_train.append(pd.Series(
    bst.predict(train[features].loc[eval_index]),
index=eval_index))
eval_pre = bst.predict(train[features].loc[eval_index])
score = np.sqrt(mean_squared_error(train[label].loc[eval_index].
values,
    eval_pre))
cv_score.append(score)
print(cv_score, sum(cv_score) / 5)
pd.Series(prediction_train.sort_index().values).to_csv(
    "preprocess/train_lightgbm.csv", index=False)
pd.Series(prediction_test / 5).to_csv("preprocess/test_lightgbm.
csv", index=False)
test['target'] = prediction_test / 5
test[['card_id', 'target']].to_csv("result/submission_lightgbm.
csv", index=False)
return
```

The train_lightgbm.csv, test_lightgbm.csv, and submission_lightgbm.csv files are also saved, and the CV score and online submission score are recorded. The cross-validation score is 3.6773062. The public score is 3.73817 (2786/4127) and the private score is 3.64490 (2719/4127). The scores have been improved compared with those in Sect. 8.4.1, and more optimizations will be made to achieve more breakthrough in the score.

8.4.3   XGBoost

The two models above only use two sets of features—dict and groupby. The XGBoost model in this section will try to add nlp features to the training, and meanwhile skip the feature selection stage to consider using the feature set for modeling; the parameter tuning framework is replaced by beyesian.

The first step is still to read the data; the difference is that the previous feature set and nlp features need to be merged into a sparse matrix; the specific code is as follows:

```
def read_data(debug=True):
    print("read_data...")
    NROWS = 10000 if debug else None
    # read two sets of features - dict and groupby - obtained in the feature
engineering stage
    train_dict = pd.read_csv("preprocess/train_dict.csv", nrows=NROWS)
    test_dict = pd.read_csv("preprocess/test_dict.csv", nrows=NROWS)
    train_groupby = pd.read_csv("preprocess/train_groupby.csv",
nrows=NROWS)
    test_groupby = pd.read_csv("preprocess/test_groupby.csv",
nrows=NROWS)
    # remove duplicate columns
    for co in train_dict.columns:
        if co in train_groupby.columns and co!='card_id':
            del train_groupby[co]
    for co in test_dict.columns:
        if co in test_groupby.columns and co!='card_id':
            del test_groupby[co]
    train = pd.merge(train_dict, train_groupby, how='left',
on='card_id').fillna(0)
    test = pd.merge(test_dict, test_groupby, how='left', on='card_id').
fillna(0)

    features = [f for f in train.columns if f not in ["card_id","target"]]
    # read nlp related features obtained in the feature engineering stage
    train_x = sparse.load_npz("preprocess/train_nlp.npz")
    test_x = sparse.load_npz("preprocess/test_nlp.npz")

    train_x = sparse.hstack((train_x, train[features])).tocsr()
    test_x = sparse.hstack((test_x, test[features])).tocsr()
    print("done")
    return train_x, test_x
```

Then there is the parameter tuning stage. Unlike hyperopt, beyesian parameter adjustment is optimized by maximizing the evaluation score, and the root mean square error of the evaluation indicator should be as small as possible. Therefore, the negative root mean square error is used as the optimization goal. The specific code is as follows:

```
def params_append(params):
    params['objective'] = 'reg:squarederror'
    params['eval_metric'] = 'rmse'
    params["min_child_weight"] = int(params["min_child_weight"])
    params['max_depth'] = int(params['max_depth'])
    return params
```

```
def param_beyesian(train):
  train_y = pd.read_csv("data/train.csv")['target'].values
  train_data = xgb.DMatrix(train, train_y, silent=True)

 def xgb_cv(colsample_bytree, subsample, min_child_weight, max_depth,
  reg_alpha, eta, reg_lambda):
  params = {'objective': 'reg:squarederror',
            'early_stopping_round': 50,
            'eval_metric': 'rmse'}
  params['colsample_bytree'] = max(min(colsample_bytree, 1), 0)
  params['subsample'] = max(min(subsample, 1), 0)
  params["min_child_weight"] = int(min_child_weight)
  params['max_depth'] = int(max_depth)
  params['eta'] = float(eta)
  params['reg_alpha'] = max(reg_alpha, 0)
  params['reg_lambda'] = max(reg_lambda, 0)
  print(params)
  cv_result = xgb.cv(params, train_data, num_boost_round=1000,
                     nfold=2, seed=2, stratified=False, shuffle=True,
                     early_stopping_rounds=30, verbose_eval=False)
 return -min(cv_result['test-rmse-mean'])
xgb_bo = BayesianOptimization(
xgb_cv,
{'colsample_bytree': (0.5, 1),
'subsample': (0.5, 1),
'min_child_weight': (1, 30),
'max_depth': (5, 12),
'reg_alpha': (0, 5),
'eta':(0.02, 0.2),
'reg_lambda': (0, 5)}
)
# init_points means the initial point,  n_iter means iteration times(i.e. sampling number)
xgb_bo.maximize(init_points=21, n_iter=5)
print(xgb_bo.max['target'], xgb_bo.max['params'])
return xgb_bo.max['params']
```

Finally, it is the training prediction part of the XGBoost model. Perform the five-fold cross-training in the same way, and the three files train_xgboost.csv, test_xgboost.csv, and submission_xgboost.csv are saved at the same time. The specific code is as follows:

```
def train_predict(train, test, params):
train_y = pd.read_csv("data/train.csv")['target']
test_data = xgb.DMatrix(test)

  params = params_append(params)
  kf = KFold(n_splits=5, random_state=2020, shuffle=True)
  prediction_test = 0
  cv_score = []
  prediction_train = pd.Series()
  ESR = 30
  NBR = 10000
```

```
  VBE = 50
  for train_part_index, eval_index in kf.split(train, train_y):
        train_part = xgb.DMatrix(train.tocsr()[train_part_index, :],
                train_y.loc[train_part_index])
        eval = xgb.DMatrix(train.tocsr()[eval_index, :], train_y.loc
[eval_index])
        bst = xgb.train(params, train_part, NBR, [(train_part, 'train'),
(eval, 'eval')],
                verbose_eval=VBE, maximize=False,
early_stopping_rounds=ESR, )
        prediction_test += bst.predict(test_data)
        eval_pre = bst.predict(eval)
        prediction_train = prediction_train.append(pd.Series(eval_pre,
index=eval_index))
        score = np.sqrt(mean_squared_error(train_y.loc[eval_index].
values, eval_pre))
        cv_score.append(score)

  print(cv_score, sum(cv_score) / 5)
  pd.Series(prediction_train.sort_index().values).to_csv(
        "preprocess/train_xgboost.csv", index=False)
  pd.Series(prediction_test / 5).to_csv("preprocess/test_xgboost.
csv", index=False)
  test['target'] = prediction_test / 5
  test[['card_id', 'target']].to_csv("result/submission_xgboost.
csv", index=False)
  return
```

8.5   Model Integration

After completing feature engineering and model training, participants may find that their scores are still unsatisfactory. In order to introduce the relevant skills of the machine learning competition as briefly and broadly as possible, this chapter uses the more generally used method (copy and paste) during competitions instead of for-mulating an extremely detailed plan for the competition questions. That is not the purpose of this book. What can be observed is that the score of a single model does seem to be high. This section will try model weighted integration and stacking integration to improve the score. Here is another point the contestants need to know; that is, in machine learning competitions, the power of teams and open source is extremely strong. One individual's thinking, time, and energy are often limited, and there is great difference in modeling methods between different individuals, so they can often bring together great integration benefits. In addition, most competitions, especially the Kaggle competition, allow participants to freely discuss and even open source of codes, which is also a good resource for increasing score. You can integrate your own algorithms and open-source solutions to get better scores.

8.5.1   *Weighted Integration*

Chapter 6 of this book has clearly explained the principle of weighted integration of results. The results obtained by the three single models in Sect. 8.4 can be given weights according to scores and correlations, and then submitted. After submission, specific scores can be obtained: public score being 3.73135 (2741/4127) and private score being 3.63741 (2646/4127).

The specific weight calculation method is data ['randomforest'] * 0.2 + data ['lightgbm'] * 0.3 + data ['xgboost'] * 0.5.

8.5.2   *Stacking Integration*

When training the three models mentioned above, the stacking features of the corresponding models are generated incidentally, that is, the model prediction results of the training set and the testing set. This result can be regarded as the extraction and compression of the feature set information, combined with an open-source scheme of a higher score (note that this open-source code has several bugs, which will not affect the use after making corresponding changes) to perform feature splicing and integrated modeling. Adding the stacking features corresponding to XGBoost, the best performing model in the three models above, to the training process can produce a model with sound scores: specifically, public score—3.68825 (878/4127) and private score—3.60871 (90/4127).

Due to limited pages, this section only lists the weighted integration results with low scores and the stacking integration results with high scores. Readers who want to discover more can try and compare the results of weighted integration and Stacking integration under the same conditions. Since this competition is entitled regression problem and there are outliers, stacking integration is better than weighted integration. For stacking integration code, please refer to the integration techniques in Sect. 8.6.2 below.

8.6   Efficient Scoring

The scores obtained so far are still not very ideal. We need to learn more about some important tools, such as feature selection methods, parameter tuning methods, and some core tree models. In this section, the features and final results will be optimized to efficiently improve the scores.

8.6.1   Feature Optimization

In the final scheme, features are extracted mainly around new_merchant_transactions.csv and historical_transactions.csv. Features include five parts: basic statistical features, global card_id features, card_id in the last two months, second-order features, and supplementary features. These features primarily contain the features included in the feature engineering designed by most contestants. Next, the work done on each feature is introduced in detail.

8.6.1.1   Basic Statistical Features

Features in this part are mainly aggregated (groupby) statistics with card_id as the key, and this part of features are extracted from data sets—new_transactions.csv, historical_transactions.csv (authorized_flag is 1), and historical_transactions.csv (authorized_flag is 0) respectively. The specific aggregation method and statistical dimensions are shown in the following code:

```python
def aggregate_transactions(df_, prefix):

  df = df_.copy()

df['month_diff'] = ((datetime.datetime.today() - df
['purchase_date']).dt.days)//30
df['month_diff'] = df['month_diff'].astype(int)
df['month_diff'] += df['month_lag']

df['price'] = df['purchase_amount'] / df['installments']
df['duration'] = df['purchase_amount'] * df['month_diff']
df['amount_month_ratio'] = df['purchase_amount'] / df['month_diff']

  df.loc[:, 'purchase_date'] = pd.DatetimeIndex(df
['purchase_date']).
        astype(np.int64) * 1e-9
  agg_func = {
            'category_1': ['mean'],
        'category_2': ['mean'],
        'category_3': ['mean'],
        'installments': ['mean', 'max', 'min', 'std'],
        'month_lag': ['nunique', 'mean', 'max', 'min', 'std'],
        'month': ['nunique', 'mean', 'max', 'min', 'std'],
        'hour': ['nunique', 'mean', 'max', 'min', 'std'],
        'weekofyear': ['nunique', 'mean', 'max', 'min', 'std'],
        'dayofweek': ['nunique', 'mean'],
        'weekend': ['mean'],
        'year': ['nunique'],
        'card_id': ['size', 'count'],
        'purchase_date': ['max', 'min'],
        'price': ['mean', 'max', 'min', 'std'],
```

```
        'duration': ['mean','min','max','std','skew'],
        'amount_month_ratio':['mean','min','max','std','skew'],
}
for col in ['category_2','category_3']:
    df[col+'_mean'] = df.groupby([col])['purchase_amount'].
transform('mean')
agg_func[col+'_mean'] = ['mean']

agg_df = df.groupby(['card_id']).agg(agg_func)
agg_df.columns = [prefix + '_'.join(col).strip() for col in agg_df.
columns.values]
agg_df.reset_index(drop=False, inplace=True)

return agg_df
```

8.6.1.2 Global card_id Features

This feature is extracted from data sets of new_transactions.csv, historical_transactions.csv (authorized_flag is 1), and historical_transactions.csv (authorized_flag is 0) respectively. It mainly includes statistics related to the time of user behavior, such as the time difference between the last transaction and the first transaction, the time difference between the credit card activation date and the first transaction; aggregate statistics (mean/sum) of authorized_flag and month_diff statistics, with card_id as the key; use card_id as the key to gather, integrate, and calculate nunique of state_i, city_id, installments, merchant_id, merchant_category_id, etc., and construct the relative value of nunique obtained above and card_id frequency, in order to reflect the behavior purity (range of scatter) of the user card_id; the statistics of variables related with purchase_amount (mean/sum/std/median) are aggregated, with card_id as the key; besides, some pivot-related features are also built.

8.6.1.3 card_id in the Last Two Months

This feature is extracted only for historical_transactions.csv data sets. This part has many similar features to the global card_id feature; the main difference lies in the time range, and here more attention is paid to the recent changes in user behavior.

8.6.1.4 Second-order Features

Extract this part of features only for historical_transactions.csv data set; the premise is to first construct first-order features (nunique, count, sum, etc.); the specific extraction structure is as follows:

```
for col_level1,col_level2 in tqdm_notebook(level12_nunique):
    # first-order extracts nunique features
```

```
level1 = df.groupby(['card_id',col_level1])[col_level2].
    nunique().to_frame(col_level2 + '_nunique')
level1.reset_index(inplace =True)
# construct aggregate statistics features, with card_id as the key (second-order features)
level2 = level1.groupby('card_id')[col_level2 + '_nunique'].agg
(['mean', 'max', 'std'])
level2 = pd.DataFrame(level2)
level2.columns = [col_level1 + '_' + col_level2 + '_nunique_' + col
for col in
    level2.columns.values]
level2.reset_index(inplace = True)
cardid_features = cardid_features.merge(level2, on='card_id',
how='left')
```

8.6.1.5   Supplementary Features

Most of these features have business significance. For example, in order to better find outliers (i.e. labeled -33.219281), the mean coding feature about whether the prediction target is outlier is constructed. There are also some cross statistical features on series features of hist and new.

```
train['outliers'] = 0
train.loc[train['target'] < -30, 'outliers'] = 1
train['outliers'].value_counts()
for f in ['feature_1','feature_2','feature_3']:
    colname = f+'_outliers_mean'
    order_label = train.groupby([f])['outliers'].mean()
    for df in [train, test]:
        df[colname] = df[f].map(order_label)
```

```
# cross statistical features on series features of hist and new, a part of which is presented below
df['card_id_total'] = df['hist_card_id_size']+
df['new_card_id_size']
df['card_id_cnt_total'] = df['hist_card_id_count']+
df['new_card_id_count']
df['card_id_cnt_ratio'] = df['new_card_id_count']/
df['hist_card_id_count']
```

8.6.2   Integration Skills

8.6.2.1   Single Mode Result

After feature optimization, cross validation is used for offline verification, and LightGBM, XGBoost, and CatBoost models are used for training and result prediction to obtain the offline verification score, public score, and private score. Here,

LightGBM, XGBoost, and CatBoost models are packaged into a function with the following code:

```
def train_model(X, X_test, y, params, folds, model_type='lgb',
eval_type='regression'):
    oof = np.zeros(X.shape[0])
    predictions = np.zeros(X_test.shape[0])
    scores = []
    # perform five-fold cross-validation
    for fold_n, (trn_idx, val_idx) in enumerate(folds.split(X, y)):
        print('Fold', fold_n, 'started at', time.ctime())
        # determine the model selected according to model_type
        if model_type == 'lgb':
          trn_data = lgb.Dataset(X[trn_idx], y[trn_idx])
          val_data = lgb.Dataset(X[val_idx], y[val_idx])
          clf = lgb.train(params, trn_data, num_boost_round=20000,
                          valid_sets=[trn_data, val_data],
                       verbose_eval=100, early_stopping_rounds=300)
          oof[val_idx] = clf.predict(X[val_idx], num_iteration=clf.
best_iteration)
          predictions += clf.predict(X_test, num_iteration=clf.
best_iteration) /
                 folds.n_splits
      if model_type == 'xgb':
          trn_data = xgb.DMatrix(X[trn_idx], y[trn_idx])
          val_data = xgb.DMatrix(X[val_idx], y[val_idx])
          watchlist = [(trn_data, 'train'), (val_data, 'valid_data')]
        clf = xgb.train(dtrain=trn_data, num_boost_round=20000,
        evals=watchlist, early_stopping_rounds=200,
        verbose_eval=100, params=params)
        oof[val_idx] = clf.predict(xgb.DMatrix(X[val_idx]),
            ntree_limit=clf.best_ntree_limit)
            predictions += clf.predict(xgb.DMatrix(X_test),
                ntree_limit=clf.best_ntree_limit) / folds.n_splits
            # for CatBoost model, the codes for regression task and
classification task is quite different, and need to be run separately
            if (model_type == 'cat') and (eval_type == 'regression'):
                clf = CatBoostRegressor(iterations=20000,
eval_metric='RMSE', **params)
                clf.fit(X[trn_idx], y[trn_idx],
                    eval_set=(X[val_idx], y[val_idx]),
                    cat_features=[], use_best_model=True, verbose=100)
                oof[val_idx] = clf.predict(X[val_idx])
                predictions += clf.predict(X_test) / folds.n_splits

        if (model_type == 'cat') and (eval_type == 'binary'):
        clf = CatBoostClassifier(iterations=20000,
eval_metric='Logloss', **params)
        clf.fit(X[trn_idx], y[trn_idx],
                eval_set=(X[val_idx], y[val_idx]),
                cat_features=[], use_best_model=True, verbose=100)
        oof[val_idx] = clf.predict_proba(X[val_idx])[:,1]
        predictions += clf.predict_proba(X_test)[:,1] / folds.n_splits
```

```
    print(predictions)
    if eval_type == 'regression': # perform regression scorning
    scores.append(mean_squared_error(oof[val_idx], y[val_idx])
**0.5)
    if eval_type == 'binary': # perform classification scoring
    scores.append(log_loss(y[val_idx], oof[val_idx]))

  print('CV mean score: {0:.4f}, std: {1:.4f}.'.format(np.mean
(scores),
  np.std(scores)))
  return oof, predictions, scores
```

With the above functions, using LightGBM, XGBoost, and CatBoost models will become very convenient. The input parameter model_type decides which model to choose, eval_type decides whether it is a binary classification task or a regression task. Let's look at the code used:

```
lgb_params = {'num_leaves': 63, 'min_data_in_leaf': 32,
'objective':'regression',
    'max_depth': -1,
    'learning_rate': 0.01, "min_child_samples": 20, "boosting":
"gbdt",
    "feature_fraction": 0.9, "bagging_freq": 1, "bagging_fraction":
0.9,
    "bagging_seed": 11, "metric": 'rmse', "lambda_l1": 0.1,
"verbosity": -1}
folds = KFold(n_splits=5, shuffle=True, random_state=4096)
# extract training set and testing set by feature columns
X_train = train[fea_cols].values
X_test = test[fea_cols].values
# use LightGBM model for regression prediction
oof_lgb, predictions_lgb, scores_lgb = train_model(X_train, X_test,
y_train,
    params=lgb_params, folds=folds, model_type='lgb',
eval_type='regression')
```

By doing this, the XGBoost model and the CatBoost model can be trained, and the prediction results can be obtained. The prefix oof represents the verification set prediction results, and the prefix predictions represents the testing set results. These results will be used for model integration.

8.6.2.2   Weighted Integration

Use weighted integration as the basic integration method. The following formula is used to assign the same weight to the three single-model results:

$$\text{Weighted\_average} = + (\text{LightGBM XGBoost} + \text{CatBoost})/3$$

Model	Offline Verification	Public Score and Rank	Private Score and Rank
LightGBM	3.6418	3.68104 (246/4127)	3.61068 (150/4127)
XGBoost	3.6500	3.68879 (910/4127)	3.61062 (144/4127)
CatBoost	3.6481	3.70114 (2175/4127)	3.61492 (820/4127)
Weighted Integration	NULL	3.68667 (460/4127)	3.60832 (82/4127)
Stacking Integration	3.6395	3.67974 (218/4127)	3.60886 (90/4127)

Fig. 8.5 Comparison of Results of Various Schemes

8.6.2.3 Stacking Integration

Stacking integration is only performed on the results of the three models (LightGBM, XGBoost, CatBoost) here. Of course, this is far from enough. In the actual competition, more results with differences can be constructed for integration, not limited to model differences, and there are feature differences. The comparison of results under a single model and two integration methods is shown in Fig. 8.5.

It is obvious that the private score of weighted integration has the best effect, but the public score has not been improved. The relatively stable is stacking integration—both public score and private score of it have been greatly improved. At present, the result of weighted integration can reach 82 in the hidden list, and of course this ranking can be improved.

The specific code for stacking integration of the prediction results of LightGBM model, XGBoost model, and CatBoost model is as follows:

```
def stack_model(oof_1, oof_2, oof_3, predictions_1, predictions_2,
predictions_3, y,
eval_type='regression'):

    train_stack = np.vstack([oof_1, oof_2]).transpose()
    test_stack = np.vstack([predictions_1, predictions_2]).transpose()
    from sklearn.model_selection import RepeatedKFold
    folds = RepeatedKFold(n_splits=5, n_repeats=2, random_state=2020)

oof = np.zeros(train_stack.shape[0])
predictions = np.zeros(test_stack.shape[0])

for fold_, (trn_idx, val_idx) in enumerate(folds.split(train_stack, y)):
  print("fold n° {}".format(fold_+1))
  trn_data, trn_y = train_stack[trn_idx], y[trn_idx]
  val_data, val_y = train_stack[val_idx], y[val_idx]
  print("-" * 10 + "Stacking " + str(fold_) + "-" * 10)
  clf = BayesianRidge()
  clf.fit(trn_data, trn_y)
    oof[val_idx] = clf.predict(val_data)
    predictions += clf.predict(test_stack) / (5 * 2)
if eval_type == 'regression':
    print('mean: ', np.sqrt(mean_squared_error(y, oof)))
```

```
if eval_type == 'binary':
    print('mean: ',log_loss(y, oof))
```

```
return oof, predictions
```

The input parameters oof_1, oof_2, and oof_3 correspond to the prediction results of the verification set of the three models respectively, and the predictions_1, predictions_2, and predictions_3 respectively correspond to the prediction results of the testing set of the three models.

The above code can be used for the integration of regression tasks and the integration of classification tasks at the same time, and the specific input parameters eval_type need to be set. As for the model, BayesianRidge is the final model, because the model structure is very simple and is not easy to over-fit.

8.6.2.4   Trick Integration

First build two models, one is a categorization model that predicts whether a value is an outlier (if it is equal to the extreme outlier −33.219281, it is 1, which means it is an outlier; otherwise, it is 0, which means it is not an outlier), and the other is a regression model after removing outliers. Of course, you can also build these two types of models with single mode first, and then use Stacking integration to get the final result.

Next, the final integration scheme is based on the above categorization model that predicts whether a value is an outlier, the regression model for full data, and the non-outlier regression model; the above three types of models are all obtained by Stacking integration.

Scheme 1: categorization model result × (−33.219281) + (1 − categorization model result) × non-outlier regression model result

Scheme 2: the result of Scheme 1 × 0.5 + regression model for full data × 0.5

The scores of these two schemes have been greatly improved. The public score of scheme 1 is 3.67542 (150/4127) and its private score is 3.60636 (59/4127); the public score of scheme 2 is 3.67415 (136/4127) and its private score is 3.60414 (35/4127).

8.7   A Summary of the Competition Questions

8.7.1   More Options

8.7.1.1   Top 1 Scheme

The top1 scheme demonstrated the trick integration that the champion team used, which could directly improve the score by 0.015 in the local CV. At first, in the discussion section of Kaggle, it was found that more than half of the root mean square error of modeling was caused by extreme outliers −33.219281. Therefore, it could be considered to train a regression model after removing samples containing extreme outliers, and at the same time construct a binary categorization model based on using whether the data was extreme outliers as the objective of prediction. This was also the final scheme adopted by the champion, which mainly integrated the prediction results of the two types of models with the following formula:

```
train['final'] = train['bin_predict'] * (-33.21928) + (1 - train
['bin_predict']) *
train['no_outlier']
```

In this formula, train['bin_predict'] was the probability that the data was an extreme outlier −33.219281; no_outlier was the prediction result of the regression model after removing outliers. It could be specifically explained as: the probability of the binary classification result being an outlier multiplied by the outlier plus the multiplication of the probability that the binary classification result was not an outlier and the prediction result of the model without outliers.

Of course, there is also a more direct method to deal with outliers—after the prediction is completed, the minimum value of the outliers is directly changed before being tackled. However, this method is relatively risky and has no actual value for use, so it is not recommended. Interested readers can have a try to see the effect.

8.7.1.2   Top 5 Scheme

The team ranked 5th introduced the detailed scheme and how each type of model was combined. The team first created thousands of features, and more than 100 remained after feature selection. Figure 8.6 shows the overall scheme framework of the 5th team.

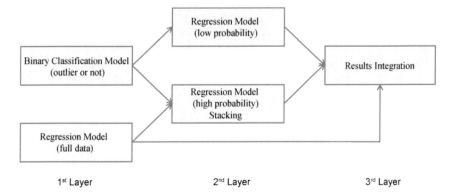

Fig. 8.6 Overall Framework of the Scheme

The following part describes in detail what is involved in the framework.

- Regression model, using full data in training;
- The binary categorization model used to predict whether the data is extremely outliers (bank cards without duplicate transactions in the testing set), whose prediction results will be divided according to a threshold of 0.015, in order to create two regression models (low probability and high probability) of the second layer;
- Low probability regression model—based on the prediction results of the previous layer of binary categorization model, regression modeling is carried out according to the prediction results lower than the 0.015 threshold. Since this is a relatively sparse part of the prediction results, the final integration weight is 0.4.
- High probability regression model—based on the prediction results of the previous layer of binary classification model, perform regression modeling with the prediction results higher than the 0.015 threshold. The main features of this model come from the results predicted by the previous layer of binary classification model and the regression model, plus some other features. This model helps to avoid post-processing.
- In the final submission, the prediction results from the low probability regression model and the high probability regression model are integrated first, and then the result of the regression model in step 1 is integrated.

8.7.2   Knowledge List

8.7.2.1   Feature Engineering

There are three main types of features used in this chapter, namely RFM, groupby, and nlp, using dictionaries, pandas. DataFrame, and sparse matrix respectively to carry out extraction. The RFM model only simulates the use of F and M, and the relevant information in the near future needs to be considered and investigated by readers. To some extent, the second scheme—groupby—contains part of R information, while nlp mainly uses CountVector and TF-IDF. What should be noticed is that the data structures and types required by the three are different, which leads to a slight difference in pre-processing. In the aspect of feature selection, the filter method based on Pearson correlation coefficient and the wrapper method based on the importance of model features are tried. Finally, pandas. DataFrame and scipy. sparse are used during modeling.

8.7.2.2   Parameter Tuning

Overall, this contest question is a relatively standard data mining and machine learning modeling problem. The distribution of the training set and the testing set is highly coupled so that participants only need to focus on characterizing the user's own consumption behavior, and then use the machine learning algorithm to train and predict. The actual practice of this chapter uses three parameter tuning frameworks, namely grid search, hyperopt, and beyesian. They have their own characteristics. In order to quickly find a better parameter combination, participants can only randomly sample part of the data for debugging.

8.7.3 Extended Learning

In view of the limited space, this chapter does not introduce the two algorithms of CatBoost and Word2Vec. CatBoost is also a decision tree model, which is special in that it directly supports the modeling and calculation of discrete fields and text fields, thus saving a lot of preprocessing time. By comparison, Word2Vec is also a classic algorithm in the field of NLP, which also processes text into numerical vectors that the model can understand.

In addition, this section will recommend some similar game questions as an extension of the learning content in order to deepen the understanding of such competition questions.

8.7.3.1 Santander Product Recommendation

The homepage of this competition question is shown in Fig. 8.7.

In order to meet the needs of a series of financial decisions, Santander Bank provides loan services to customers through personalized product recommendations, predicting which products existing customers will use in the next month based on their past behavior and behavior of other similar customers.

The question is based on the purchase records of the Spanish Santander Bank in the first 17 months and the user attributes to predict the most likely items to be purchased by each user in June 2016 (need to give the predicted 7 items and arrange them according to the probability; the evaluation indicator is MAP@7). This is a typical recommendation competition, which builds a model based on the user's historical behavior information. This behavior information is the key information

Fig. 8.7 Santander Product Recommendation

to characterize the user's interests and habits and can also be used to build rich user tags.

Basic ideas: This kind of problem extracts mainly historical features, such as products lag features, products existing period, average purchase time difference of products, time after the last purchase of products, the value of 20 products last month, the number of products purchased last month, etc., and some raw features. The model uses the basic XGBoost and LightGBM (not yet available in 2016), calculates the user's new purchase probability for each product, that is, row_num * n_classes, and finally sorts the probability and extracts the top 7 products for each user. In this way, a baseline scheme can be obtained, and then optimization is mainly carried out for feature amplification, how to model (offline verification scheme), model selection, and post-processing.

8.7.3.2 WSDM—KKBox's Music Recommendation Challenge

The homepage of this competition question is shown in Fig. 8.8.

This is the 11th ACM International Conference on Web Search and Data Mining (WSDM 2018) Challenge, which used data sets provided by KKBOX to build a better music recommendation system. This challenge requires players to predict the possibility of repeated listening to a song when the user's first observable listening event is triggered within a time window. If a repeated listening event is triggered within one month after the user's first observable listening event, the target is marked as 1; otherwise, it is marked as 0. The same rule applies to testing sets.

Basic ideas: The champion team introduced that the contest task was similar to the click-through rate (CTR) prediction. Relistening is like buying, and listening for the first time is like clicking. Compared with "recommendation", "click-through rate prediction" is a more accurate keyword (because CTR prediction and CVR prediction share similar models), so methods such as potential factor-based or neighborhood-based collaborative filtering are not the best way to solve click-through rate prediction.

The model uses common tree models and NN, which are all fine, and then uses the last 20% of the data in the set as the verification set. The characterizing portion is

Fig. 8.8 WSDM—KKBox's Music Recommendation Challenge

the target coding, counting features, the last time an individual listened to a song (categories combined with features such as msno, msno + genre_ids, msno + composer), the next time an individual will hear a song, and the time span between the last time one listened to a song and the present, considering the feature combination construction of different granularities at the same time.

Part III
Learn from History to Create a Bright Future

Chapter 9
Time Series

There is a very old story about time series analysis. In ancient Egypt 7000 years ago, people recorded the ups and downs of the Nile River day by day to form a time series. After long-term observation of this time series, people found that the ups and downs of the Nile River were very regular. Thanks to mastering the laws of ups and downs, agriculture in ancient Egypt developed rapidly. This method of obtaining laws of intuition from observing sequences is the descriptive analysis method. In the development of time series analysis methods, applications in the fields of economy, finance, engineering and so on have always played an important promoting role, and every step of the development of time series analysis is inseparable from the application.

Time series analysis has always received much attention. The most classic contest "Makridakis Competitions" focus on time series forecasting problems. It has been held for five times, the first in 1982 and the fifth in 2020, nearly 40 years apart. This challenge aims to evaluate and compare the accuracy of different prediction methods and solve the problem of times series forecasting.

In competitions, we often encounter tasks related to time series analysis, such as what the index will happen in the next day/week/month, more specifically, predicting the traffic of advertising impressions in the following 2 weeks, and predicting the time of users to buy goods and stock trading time, etc. Of course, it is far more than that; we can use different methods to deal with these prediction tasks after summarizing them, according to the required prediction quality and the length of the prediction cycle.

This chapter will be divided into four parts, which introduce time series analysis, time series patterns, feature extraction methods, and model diversity.

© The Author(s), under exclusive license to Springer Nature Singapore Pte Ltd. 2023   177
W. He et al., *Machine Learning Contests: A Guidebook*,
https://doi.org/10.1007/978-981-99-3723-3_9

9.1   What Is Time Series

This section will give a brief introduction to time series analysis, including the simple definition, common questions, cross-validation, and basic rule methods, to help everyone have a basic understanding of time series analysis and a general idea of solving problems.

9.1.1   Simple Definition

Let's start with the definition of time series analysis, which is a series of data points indexed (or listed or illustrated) in chronological order. Therefore, the data that makes up the time series consists of relatively definitive timestamps. Compared with random sample data, more additional information (such as time trends, change information, etc.) can be extracted from the time series data.

Different from the random observation sample analysis discussed in most other statistical data, the analysis of time series is based on the following assumptions: the data value of the label in the data file represents continuous measurement values at equal intervals, such as traffic within one hour, sales volume within one day, etc. Assuming that there is a correlation in the data, then find the corresponding correlation through modeling, and use it to predict the future data trend.

9.1.2   Common Questions

As one of the most common competition topics, competitions related to times series forecasting can be subdivided into many questions. Through the summary of previous competitions, these questions can be summarized into univariate time series and multivariate time series from the perspective of variables, and then these problems can be concluded into single-step prediction and multi-step prediction according to different prediction objectives.

9.1.2.1   Univariate and Multivariate Time Series

Univariate time series has only a single time-dependent variable, so it is only affected by time factors. This kind of problem focuses on analyzing the changing characteristics of data, which are affected by factors such as correlation, trend, periodicity, and circularity. This type of problem is relatively rare and can generally be regarded as part of multivariate time series i.e., only considering the influence of time on labels. See Fig. 9.1 for an example.

Fig. 9.1 An Example of
Univariate Time Series

Date	Sales
2020-1-1	1937
2020-1-2	2134
2020-1-3	2556
2020-1-4	2209
...	...

Date	Promotion Effort	Visits	Search Volume	...	Sales
2020-1-1	A	32456	73954	...	1937
2020-1-2	A	37984	73954	...	2134
2020-1-3	C	42367	102943	...	2556
2020-1-4	B	40657	94872	...	2209
...	...	...	...	...	...

Fig. 9.2 An Example of Multivariate Time Series

Multivariate time series has multiple time-dependent variables. In addition to being affected by time factors, it is also affected by other variables. For example, the forecast of product sales may be affected by a series of variables such as category, brand, and promotion. This kind of problem is more common and requires more factors to be considered, so the challenge is also greater. See Fig. 9.2 for an example.

9.1.2.2   Single-step Prediction and Multi-step Prediction

The single-step prediction problem is relatively basic. Just adding a time unit to the time basis of the training set can be used as a testing set. In fact, it is a common regression problem, but the input variable is no longer an independent feature variable, but a feature variable that will be affected by historical data over time. As shown in Fig. 9.3, if the testing set only has time of $t + 1$, then it is a single-step prediction; if it has time span of $t + 1$ to $t + n$, it is a multi-step prediction.

The multi-step prediction problem is more complicated. It is to add multiple time units as testing sets on the basis of the time of the training set. There are many solutions to this problem: first, based on single-step prediction, each time the predicted value is added to the training set as the real value to predict the next time unit, this will lead to error accumulation, especially if there is a large error at the beginning, then the prediction effect will become worse and worse; second, directly predict the results of all testing sets, that is, as a multi-output regression problem, so that although the problem of error accumulation is avoided, it will increase the difficulty of model learning because it requires the model to learn a many-to-many system, thus increasing the training difficulty.

t − m to t	t + 1	t + 2	...	t + n

Training Set   Testing Set

Fig. 9.3 Data Set Division

9.1.3   Cross Validation

Before starting to build the model, we should consider how to carry out offline verification. For the stability of the results, we choose cross validation. But how can we perform cross validation for the time series? Since the time series contain a time structure, it is generally necessary to pay attention to the situation that data penetration cannot occur in the fold while retaining this structure. If a randomized cross validation is carried out, then all the time correlation between the tag values will be lost, resulting in data penetration. Fortunately, there is still a suitable method for dealing with time series problems. This approach is called rolling cross validation, as shown in Fig. 9.4.

This cross validation method is quite simple. We first train the model with the data from the initial time to the time t, then perform offline validation with the data from the time t to the time $t + n$, and calculate the score of the evaluation index; next, extend the training sample to the time $t + n$, and verify with the data from the time $t + n$ to the time $t + 2n$; repeat this process continuously until the last available tag value is reached. The number of verification times can be freely controlled, and finally the average value of the verification results is calculated to obtain the final offline verification results.

9.1.4   Basic Rules and Methods

In time series related competitions, rule-based methods are often used to solve problems. Due to noise in the data or some unexpected situations, the model cannot learn all the information. At this time, rule-based methods may be helpful. Here we mainly introduce two common rule-based prediction methods: weighted averaging and exponential smoothing.

9.1.4.1   Weighted Averaging

Weighted averaging is to first obtain the values of the most recent N time units in the data. If the data has strong periodicity (cycles are days, weeks, months, seasons, etc.), you can also consider the extraction under link relative ratio, that is, the values of the corresponding units yesterday, last week, last month, last season; and then make a simple weighting calculation on the extracted subset, usually the closer the data to the current time, the greater the data importance. In terms of how to select the

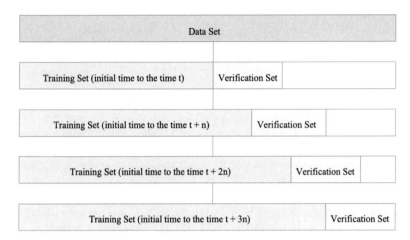

Fig. 9.4   Rolling Cross validation

N value, short-term historical data is generally considered, because the correlation of the data is higher in the short term. The weighted average is calculated by the formula (9.1);

$$y_t = \frac{1}{N}(w_1 \times y_{t-1} + w_2 \times y_{t-2} + w_3 \times y_{t-3} + \cdots + w_N \times y_{t-N}) \qquad (9.1)$$

Here N is the computation period, w is the weight of each time unit, and y is the value of the current time unit.

It is difficult to select the N value and determine the weight, because there are too many possibilities, especially for the weight. Therefore, we can use offline verification to conduct a simple linear search to determine the N value and weight. The following formula (9.2) optimizes the search method for offline verification:

$$\bar{\theta} = \arg\max_{\theta} \sum_{i=1}^{N} score(y_i, pred_i) \qquad (9.2)$$

9.1.4.2   Exponential Smoothing

In the problem of times series forecasting, the closer the time point to the prediction unit is, the more important it is. For example, if there are sales data for nearly 10 days now, the data on the 10th day will have the most impact on the prediction of sales on the 11th day. In addition, the farther the data is from the testing set, the closer its weight is to 0. Decay the weight of each time unit according to the exponential level and perform the final weighting computation. This method is called exponential smoothing. The formula is (9.3):

$$\widehat{y}_t = a \sum_{n=0}^{t} (1-a)^n y_{t-n} \tag{9.3}$$

Wherein $\widehat{y}_t$ is the value obtained by exponential smoothing at the time point of t, and y_{t-n} is the actual value at the time of $t - n$. a is an adjustable hyperparameter value, which is taken between 0 and 1. It can also be called a attenuation memory factor. It can be seen from formula (9.3) that the larger the value of a, the faster the model "forgets" the historical data.

To some extent, exponential smoothing is like a moving average method with infinite memory (the smoothing window is large enough) and exponentially decreasing weights. The calculation results obtained by an exponential smoothing can be expanded beyond the data set and range, so it can also be used for prediction.

Extended Learning

If you think about the two rule-based methods described above, you will find that the forecast result will neither be higher than the historical high nor lower than the historical low. There will obviously be unrealistic occasions. For example, if the automotive industry becomes sluggish and the sales volume of passenger cars decreases year by year, then the sales this year will definitely be lower than that of last year. The reason is that the trend is not taken into account. Therefore, we can use the secondary exponential smoothing linear trend method to predict passenger car sales. There is also the triple exponential smoothing, which can predict time series with both trend and seasonality. It is based on the algorithms of single exponential smoothing and secondary exponential smoothing.

9.2 Time Series Patterns

Solving the time series problem first requires understanding of the key data patterns, and then representing these patterns by extracting features. Four types of time series patterns are mainly introduced here: trend, periodicity, correlation, and randomness. Through the comprehension of these patterns, the direction of feature extraction and model selection can be found.

9.2.1   Trend

Trend is the change of data that continues to rise or fall over a long period of time. Of course, this is not only limited to linear rise or fall, but also can be periodic up or down. Trend appears in many types of time series forecasting problems, such as sales volume forecasting, various types of traffic forecasting, financial related forecasting, etc.

So, how can trends be expressed using features? It should be performing feature construction based on the change of data for the most part, usually starting from the first-order trend and second-order trend to construct. The first-order trend is mainly the data difference and proportion of adjacent time units, which are used to reflect the change degree of data from adjacent time units. The second-order trend is further constructed on the basis of the first-order trend, which can reflect the speed of changing of the first-order trend.

Figure 9.5 shows the stock prices of Google and Microsoft from 2006 to 2018. The abscissa is the date, and the ordinate is the share price. On the whole, the stock prices of both companies have witnessed a certain upward trend, but Google's is more obvious.

9.2.2   Periodicity

Periodicity refers to the repeated fluctuations in a period of time series, which is the result of various factors such as climatic conditions, production conditions, holidays, and people's customs. Many time series forecasting problems are cyclical; for example, temperature change forecasting will be affected by seasonal cycles, and subway traffic forecasting will be influenced by morning and evening rush hours or weekday and weekend cycles.

Periodicity can be expressed month on month or week on week, that is, the data values of the same period last month and those of the same period last week are used

Fig. 9.5 Trends in Stock Prices of Google and Microsoft

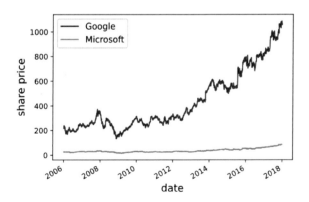

as characteristics respectively, because if there is periodicity, the label values of the same period may be more similar. Since there is periodicity, then the time features can also show periodicity, such as constructing the position of the current time in the period in which it is located, and the time difference between the current time and the peak in the period in which it is located.

As shown in Fig. 9.6, the horizontal axis is the time (date), and the vertical axis is the volume of viewing ads (traffic), illustrating the real mobile game data, which is used to investigate the advertising traffic viewed per hour and the daily in-game monetary expenditures. It is obvious that the data is repeated in a one-day cycle.

9.2.3 Correlation

Correlation in time series is also called autocorrelation. It describes that there is often a positive correlation or negative correlation in a certain period of sequence, and there will be a great correlation between the time points before and after. It is because of this correlation that the future becomes predictable.

The performance of correlation can be described by the label values of the neighboring moments, such as directly using the label values of the last moment and the last two moments as a feature. In addition, the statistical value of multiple moments in history can be used as a feature to reflect recent changes.

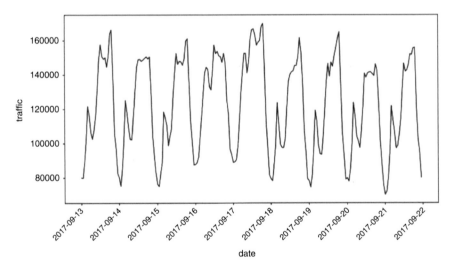

Fig. 9.6 Traffic of Ads Viewing

9.2.4   Randomness

Randomness describes random disturbances other than the above three modes. Due to the uncertainty of the time series, there will always be some accidents or noises, resulting in irregular fluctuations in the time series. For example, the stock market is a typical scenario with huge randomness, with complex historical dependence and nonlinear time series. Randomness is difficult to predict and is the place that brings the greatest errors. As shown in Fig. 9.7, which visualizes the demand for a product in Kaggle's M5 Forecasting—Accuracy competition, it is obvious that there are four anomalies that are difficult to predict compared to the overall demand.

For randomness, it can be solved by simple exception labeling, such as special dates, activities, etc. You can also preprocess to change the original data source by first removing these outliers due to randomness, and then correcting them.

9.3   Feature Extraction Methods

Competitions related with time series have different feature extraction methods, mainly focusing on the delay of time series (i.e. historical information data). The feature extraction methods here can be further divided into historical translation and window statistics. In addition, the sequence entropy feature and other additional features will be introduced.

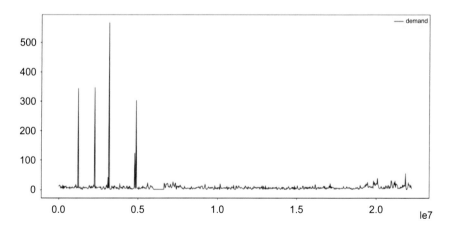

Fig. 9.7 Changes in Demand for Commodities

9.3.1 Historical Translation

Time series data have a context. For example, yesterday's sales volume is likely to affect today's sales volume, and tomorrow's weather temperature will be affected by today's temperature. In other words, the more similar labels in the time series, the higher the correlation. We can use this feature to construct historical translation features, that is, to directly use historical records as features. Specifically, if the current time is t, then the value of moments like $t - 1$, $t - 2$, ..., $t - n$ can be used as the feature. This value can be a label value, or a value related to the label value. For example, if the forecast target is passenger car sales, then passenger car production and GDP related to passenger car sales can be used as features.

   As shown in Fig. 9.8, for the unit of time d, by directly taking the value of the time $d - 1$ as a feature, you can use shift () to complete the translation operation in a direct way, where shift (1) means to translate 1 unit to the right, and shift (-1) means to translate 1 unit to the left. In this way, the value corresponding to each moment in the second row will all be used as the feature corresponding to the time in the first row.

9.3.2 Window Statistics

Unlike historical translation to extract features from a single sequence unit, window statistics is to extract features from multiple sequence units. Window statistics can reflect the status of sequence data within the interval, such as the maximum, minimum, mean, median, and variance within the window, etc.

   The window size is not fixed, and various attempts can be made. If it is a time series in days, it is a good decision to choose 3 days, 5 days, 7 days, 14 days as the window size for statistics. As shown in Fig. 9.9, if 3 is selected as the window size, then for the unit d time, the statistics are based on $d - 3$ to $d - 1$ time as the window.

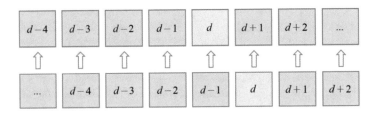

Fig. 9.8 Historical Translation

Fig. 9.9 Window Statistics

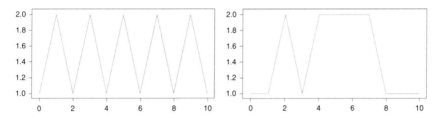

Fig. 9.10 Time Series Visualization

9.3.3   Sequence Entropy Characteristics

The concept of entropy was first applied to thermodynamics to measure the unavailability of energy in a system. The greater the entropy, the higher the unavailability of energy, and vice versa. Its physical meaning is a measure of the degree of confusion or complexity in a system. Similarly, in time series analysis, entropy can be used to describe the certainty and uncertainty of the series. As shown in Fig. 9.10, a visual display of two time series is given, namely (1, 2, 1, 2, 1, 2, 1, 2, 1, 2, 1) and (1, 1, 2, 1, 2, 2, 2, 2, 1, 1, 1).

If the mean, variance, median, and other results of the two sequences are counted separately, they will be found to be equal, so the statistical features have no distinction between the two time series, and it is difficult to mine the stability of the sequence. Just because of this, the concept of entropy is introduced to describe the sequence, and the calculation formula is as formula (9.4):

$$\text{entropy}(X) = -\sum_{i=1}^{N} P\{x = x_i\} \ln P\{x = x_i\} \tag{9.4}$$

9.3.4   Other Features

In addition to historical translation, window statistics, and sequence entropy features, there are many features that are often used. We summarize them into two categories: temporal features and statistical features.

Temporal features: such as hours, days, weeks, months, a certain period of the day (such as morning, noon), a few days before a certain day, whether it is a holiday, etc.

Statistical features: The primary statistical characteristics include maximum (max), minimum (min), mean, median, variance, standard variance, skewness, and kuriosis, etc. In addition, there are first-order difference, second-order difference, proportion correlation and other characteristics.

9.4   Model Diversity

There are still many models that can be tried for times series forecasting. This section will introduce traditional time series models, tree models, and deep learning models. You can try all these methods in the competition. By doing so, if there is no accident, you can not only find a good single model, but also prepare for the final integration.

9.4.1   Traditional Time Series Model

ARIMA (autoregressive integrated moving average model) is a common time series model. At the same time, ARIMA consists of three parts—AR, MA, and I, and contains three key parameters—p, q, and d. Here AR is an autoregressive model, p is the number of autoregressive terms, indicating the number of lagging observations included in the model; MA is a moving average model, q is the size of the moving average window, and d is the number of differences performed to become a stationary sequence; I (represents "integral") indicates that the data value has been replaced by the difference between the current value and the previous value (and the differential process may have been performed more than once).

The triple exponential smoothing mentioned earlier is based on the description of data trends and seasonality, while the ARIMA model mainly describes the interrelationship between data. The purpose of the three parts in ARIMA is to make the model easier to fit historical data and to obtain higher accuracy in a larger time series. The following code simply implements the ARIMA model:

```
from statsmodels.tsa.arima_model import ARIMA
model = ARIMA(train, order=(p,d,q) )
arima = model.fit()
pred = arima.predict(start= len(train), end= len(train)+L)
```

It can be noticed that the use of the ARIMA model is very convenient. The code is divided into three parts: creating the model, training the model, and making predictions. In addition to preparing the one-dimensional sequence tag value, it is also necessary to determine the parameters p, d, and q. Besides, the L in the prediction part means the length unit of the prediction.

Parameter Determination P, d, and q are all non-negative integers. The selection of p and q can be determined according to the ACF (autocorrelation coefficient) diagram and the PACF (partial autocorrelation coefficient) diagram. First, the d-order difference is performed to convert the time series data into a stationary time series, and then the ACF diagram and the PACF diagram of the stationary time series are obtained respectively. By analyzing these two diagrams, the best order p and order number q are obtained.

Extended Learning
SARIMA (seasonal autoregressive integrated moving average model) is an extension to ARIMA. This model can use univariate data including trend and seasonality for times series forecasting.

9.4.2   Tree Model

Tree models (XGBoost, LightGBM, etc.) are relatively general models that can show great power even in many competitions related to time series. Tree models are very suitable when trends and seasonality are relatively stable, and noise is low. Of course, some processing can be used to reduce the trend of time series and make it stable. In Chap. 10, the tree model will be used as a baseline scheme to complete basic predictions.

Extended Learning
The methods to convert time series into stationary series include logarithmic processing, first-order difference, seasonal difference, etc. In most situations, the stationarity adjustment is performed first, then training, and finally the result can be converted. These three methods can be used alone or in combination.

9.4.3   Deep Learning Model

The deep learning model can give more possibilities to times series forecasting, such as automatic learning of time dependence and automatic processing of time structures such as trends and seasonality. The deep learning model can also process large amounts of data, multiple complex variables, and multi-step operations, and extract sequence models from input data, which can provide great help to times series forecasting. This article mainly introduces how to use convolutional neural networks and long-term, short-term memory networks to solve the problem of times series forecasting and provides specific implementation codes. At the same time, this section will present more attempts and applications of deep learning in times series forecasting.

9.4.3.1   Convolutional Neural Network

The convolutional neural network (CNN) is a kind of neural networks designed to effectively process image data. It can automatically extract features from original input data. This ability can be applied to times series forecasting problems, that is, a series of observations are regarded as a one-dimensional image, and then the image is read and extracted as the most significant element. Convolutional neural networks also support multivariate inputs and multivariate outputs, and can learn complex functional relationships, but do not require models to learn directly from lagging observations. On the contrary, the model can learn feature representation from the input sequence most relevant to the prediction problem, i.e. automatically identify, extract, and refine salient features from the original input data, which are directly related to the prediction problem to be modeled.

As shown in Fig. 9.11, it is a structural diagram based on one-dimensional convolutional neural networks to solve the times series forecasting problem. First, an $n \times k$ matrix is initialized, where n represents the length of the time slice and k represents the number of features. Then, the first layer is a convolution masking layer, which defines multiple filters to extract features. Next, the maximum pooling layer is used to reduce the complexity of output data and prevent data overfitting. Finally, the output result is obtained through a fully connected layer.

The following is a reference code for solving times series forecasting problems with keras-based convolutional neural networks:

```
import numpy as np
import pandas as pd
from sklearn.model_selection import train_test_split
from keras import optimizers
from keras.models import Sequential, Model
from keras.layers.convolutional import Conv1D, MaxPooling1D
from keras.layers import Dense, LSTM, RepeatVector, TimeDistributed,
Flatten
# data preparation, divide training set and verification set in chronological order
# data means data set, features mean feature set, label represents tags
```

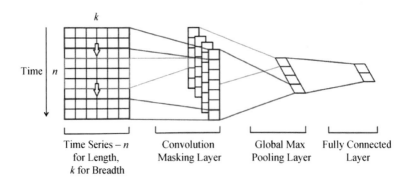

Fig. 9.11 Structure Diagram of Times Series Forecasting Based on One-dimensional Convolutional Neural Networks

```
X_train, X_valid, y_train, y_valid = train_test_split(data[features],
data[label],
   test_size=0.2, random_state=2020, shuffle=True)
# the format of input data is [sample, time step, feature]
X_train = X_train.values.reshape((X_train.shape[0], 1, X_train.shape
[1]))
X_valid = X_valid.values.reshape((X_valid.shape[0], 1, X_valid.shape
[1]))
# network design
# use a convolution masking layer and a maximum pooling layer
# the filter mapping is then smoothed before being interpreted by the fully connected layer and
prediction results output
model_cnn = Sequential()
model_cnn.add(Conv1D(filters=64, kernel_size=2, activation='relu',
   input_shape=(X_train.shape[1], X_train.shape[2])))
model_cnn.add(MaxPooling1D(pool_size=2))
model_cnn.add(Flatten())
model_cnn.add(Dense(50, activation='relu'))
model_cnn.add(Dense(1))
model_cnn.compile(loss='mean_squared_error', optimizer='adam')
# fitting network
model_cnn.fit(X_train, y_train, validation_data=(X_valid, y_valid),
epochs=20, verbose=1)
```

9.4.3.2   Long-term and Short-term Memory Network

Long-term & short-term memory network (LSTM) is a special kind of recurrent neural network (RNN), which is more suitable for longer time series. It is composed of a group of cells with the characteristics of memory data series. These cells have the function of storing data series. Long short-term memory network is suitable for processing time series data and can capture the dependency between current observations and historical observations.

As shown in Fig. 9.12, it is a schematic diagram of the long-term and short-term memory network, in which the output of each node will be used as the input of other

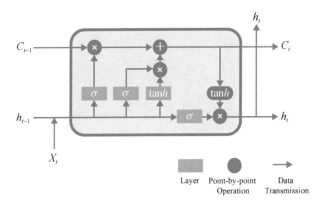

Fig. 9.12   Schematic Diagram of Long and Short-term Memory Network Structure

nodes; arrows indicate signal transmission (data transmission), blue circles indicate pointwise operation (i.e. point-by-point operation), such as node summation, green boxes indicate the network layer (neural networks layer) for learning, the two lines merged indicate connection, and the separated two lines indicate that information is copied into two duplicates and transmitted to different locations. Specifically, the input part consists of h_{t-1} (convolution masking layer at time $t - 1$) and x_t (feature vector at time t); the output part is h_t; and the main line part consists of C_{t-1} and C_t.

The following is a reference code for solving times series forecasting problems with keras-based long-term and short-term memory networks:

```
# network design
model_lstm = Sequential()
model_lstm.add(LSTM(50, activation='relu',
input_shape=(X_train.shape[1], X_train.shape[2])))
model_lstm.add(Dense(1))
model_lstm.compile(loss='mean_squared_error', optimizer='adam')
# fitting network
model_lstm.fit(X_train, y_train, validation_data=(X_valid,
y_valid), epochs=20, verbose=1)
```

9.5   Thinking Exercises

1. When making times series forecasting, under what circumstances is the effect of rule-based method better than that of a model?
2. How can the rule-based method be combined with the model in a better way?
3. Correlation feature extraction is mainly to extract the nearest label directly as a feature. So, how can the extracted time interval be determined?
4. Why can long-term and short-term memory networks solve the problem of long-time series dependence better than recurrent neural networks?

Chapter 10
Case Study: Global Urban Computing AI Challenge

This chapter will use a question in the 2019 Tianchi Competition (namely "Global Urban Computing AI Challenge: Metro Passenger Flow Forecast", as shown in Fig. 10.1) as a real-world case for issues related to time series analysis, mainly including question understanding, data exploration, feature engineering, and model training. In particular, in the part of the case study, besides providing a general idea for solving problems, what is more important is to guide everyone to learn the thinking process in different types of competition questions, and finally to sort out and further extend the knowledge, that is, the summary of competition questions.

10.1 Understanding the Competition Question

This section aims to let readers quickly understand the basic content of this actual case. In addition to the common background introduction, question data, and evaluation indicators, it also contains two unique parts: the question FAQ and the baseline plan. When facing a type of questions for the first time, asking questions and making assumptions helps to discover the core content and difficulties of the problem. The baseline plan allows us to quickly obtain offline and online feedback, and then continuously try and optimize.

10.1.1 Background Introduction

In 2019, the Hangzhou Municipal Public Security Bureau jointly launched the first Global Urban Computing AI Challenge with Alibaba Cloud Intelligence, and the title of this challenge was finally selected as "Metro Passenger Flow Forecast". At present, the subway is one of the main means of transportation for urban travel, and the sudden increase in passenger flow in subway stations is extremely easy to cause

W. He et al., *Machine Learning Contests: A Guidebook*,
https://doi.org/10.1007/978-981-99-3723-3_10

Fig. 10.1 Global urban computing AI challenge

congestion, triggering large passenger flow hedging, and causing potential safety hazards. Therefore, subway operation departments and public security organs urgently need to deploy corresponding security strategies in advance with the help of flow prediction technology to ensure the safe travel of passenger.

The competition was titled "Metro Passenger Flow Forecast". Participants needed to predict the future passenger flow changes of the station by analyzing the historical data of swiping card in the subway stations, and the prediction results could help passengers choose a more reasonable travel route, thus avoiding traffic jams. This might help subway operation departments and public security organs to deploy station security measures in advance, etc., and finally realize the use of big data and artificial intelligence technologies to empower future urban safe travel.

10.1.2   Competition Data

The competition opened 25 days of subway card swiping data from January 1, 2019 to January 25, 2019, involving 3 lines, 81 subway stations, and about 70 million pieces of data. These data were used as training data (Metro_train.zip) for players to build subway station passenger flow prediction models. After decompressing the training data, you could get 25 csv files, which stored the daily card swiping data, and the file name was prefixed with record. For example, the data of swiping card of all lines and all stations on January 1, 2019 were stored in the record_2019-01-01. csv file, and so on. The competition also provided road network maps, which were the connection relationship sheets between various subway stations, which were stored under Metro_roadMap.csv for players to use.

During the test phase, the competition would provide data records of card swiping of all stations on all routes on a certain day. Players needed to predict the number of people entering and leaving each station from 00:00 to 24:00 (in units of 10 min) in the following day.

In the qualifying round, the competition provided the card swiping data record on January 28, 2019 as testing set A (testA_record_2019-01-28.csv), and the contestants needed to predict the passenger flow of each subway station (in units of 10 min) throughout the day on January 29, 2019. In the knockout and final rounds, other batches of data would be updated as testing set B and testing set C respectively.

10.1.2.1   Users' Card Swiping Data Sheet (record_2019-01-xx.csv)

- In the record_2019-01-xx.csv file, except for the first line, each line contains a user's record of swiping card.
- For userID, the user identity cannot be uniquely identified when payType is 3, that is, this userID may be used by more than one person, but it can be regarded as that of the same user during each entry and exit. For payType with other values, the corresponding userID can uniquely identify a user.

Road Network Map (Metro_roadM ap.csv)

The competition provides a connection relationship sheet between subway stations. The corresponding adjacency matrix is stored in the roadM ap.csv, which contains an 81×81 matrix roadMap. In the roadM ap.csv file, the first row and the first column represent the subway station ID (stationID), with column values ranging from 0 to 80 and row values ranging from 0 to 80. Wherein, roadMap [i] [j] = 1 suggests the station with stationID of i is directly connected to the station with stationID of j; roadMap [i] [j] = 0 indicates the two stations with stationID of i and stationID of j are not connected to each other.

10.1.3   Evaluation Indicators

The evaluation indicator is used to judge whether the contestant's prediction is accurate. Here, the mean absolute error is used to evaluate the prediction results of the number of people entering and leaving the station separately, and then the final score is obtained by averaging the two.

10.1.4   Competition FAQ

◎The label of this competition needs to be built by contestants themselves. How can modeling enable them to achieve the greatest forecasting accuracy possible on a given data set?

Ⓐ Constructing the traffic label for entering and leaving the subway station is the first task of this competition. After observing the data for the first time, you will find that there are certain problems: for example, there are records where the traffic of entering and leaving the subway station before dawn is not 0, or there are certain differences in the data of different subway stations. In order to ensure the stability of the results, it may be possible to try to process the flow of entering and leaving the subway station in accordance with the subway stations (regularized, standardized, etc.).

Ⓠ **There are too many factors affecting the flow of subway stations, such as multiple subways arriving at the same station at the same time, emergencies, grand events, etc. So, how can outliers be dealt with to ensure the stability of the model?**

Ⓐ There are indeed many effects brought about by special factors. For such problems, common methods include exception removal, exception marking, and exception smoothing. We can try these methods one by one and compare the advantages and disadvantages.

10.1.5 Baseline Scheme

After understanding the content of the previous sections, we can start the basic modeling. The baseline scheme does not need to be too complicated, as long as it can give a correct result. This can also be seen as establishing a simple framework first, and then filling and optimizing in the following steps. Here, the card swiping data on January 29 is used as the testing set, and the card swiping data from January 1 to January 25 and that of January 28 are used as the training set for modeling. The data on January 26 and January 27 (weekend) are not officially provided.

10.1.5.1 Data Preparation

The following is the specific code for data reading and time unit conversion:

```
import numpy as np
import pandas as pd.
from tqdm import tqdm
Import lightgbm as lgb
# read data
path ='./input/'
for i in tqdm(range(1,26)):
if i < 10:
   train_tmp = pd.read_csv(path + 'Metro_train/record_2019-01-0' +
str(i) + '.csv')
```

```
    else:
    train_tmp = pd.read_csv(path + 'Metro_train/record_2019-01-' + str
(i) + '.csv')
    if i == 1:
        data = train_tmp
    else:
        data = pd.concat([data, train_tmp], axis=0, ignore_index=True)

  Metro_roadMap = pd.read_csv(path + 'Metro_roadMap.csv')
  test_A_record = pd.read_csv(path + 'Metro_testA/testA_record_2019-
01-28.csv')
  test_A_submit = pd.read_csv(path + 'Metro_testA/testA_submit_2019-
01-29.csv')
  data = pd.concat([data, test_A_record], axis=0, ignore_index=True)

  # convert data into that with units of 10 minutes
  def trans_time_10_minutes(x):
      x_split = x.split(':')
      x_part1 = x_split[0]
      x_part2 = int(x_split[1]) // 10
      if x_part2 == 0:
          x_part2 = '00'
      else:
          x_part2 = str(x_part2 * 10)
      return x_part1 + ':' + x_part2 + ':00'

  data['time'] = pd.to_datetime(data['time'])
  data['time_10_minutes'] = data['time'].astype(str).apply(lambda x: t
rans_time_10_minutes(x))
```

Next, construct the flow of entering stations (inNums) and flow of exiting stations (outNums) and aggregate them directly:

```
data_inNums = data[data.status == 1].groupby
(['stationID', 'time_10_minutes']).
   size().to_frame('inNums').reset_index()
data_outNums = data[data.status == 0].groupby
(['stationID', 'time_10_minutes']).
   size().to_frame('outNums').reset_index()
```

10.1.5.2 Key Part: Building a Training Set

The training set is still very troublesome to construct. Careful observation of the data shows that if a station does not have traffic counting for a certain period of time, then the data for this period of time is missing and needs to be filled by the players. The construction process is as follows:

```
  stationIDs = test_A_submit['stationID'].unique()
  times = []
  days = [i for i in range(1,26)] + [28, 29]
  for day in days:
```

```
        if day < 10:
            day_str = '0' + str(day)
        else:
            day_str = str(day)
        for hour in range(24):
                if hour < 10:
                    hour_str = '0' + str(hour)
                else:
                    hour_str = str(hour)
            for minutes in range(6):
                if minutes == 0:
                    minutes_str = '0' + str(minutes)
                else:
                    minutes_str = str(minutes * 10)
                times.append('2019-01-' + day_str + ' ' + hour_str +':' +
minutes_str + ':00')

    # compute the Cartesian product
    from itertools import product
    stationids_by_times = list(product(stationIDs, times))
    # construct new data set
    df_data = pd.DataFrame()
    df_data['stationID'] = np.array(stationids_by_times)[:,0]
    df_data['startTime'] = np.array(stationids_by_times)[:,1]
    df_data = df_data.sort_values(['stationID','startTime'])
    df_data['endTime'] = df_data.groupby('stationID')['startTime'].
shift(-1).values

def filltime(x):
  x_split = x.split(' ')[0].split('-')
  x_part1_1 = x_split[0] +'-'+x_split[1]+'-'
  x_part1_2 = int(x_split[2]) + 1
  if x_part1_2 < 10:
    x_part1_2 = '0' + str(x_part1_2)
  else:
    x_part1_2 = str(x_part1_2)

  x_part2 = ' 00:00:00'
  return x_part1_1 + x_part1_2 + x_part2
  # fill in missing values
  df_data.loc[df_data.endTime.isnull(), 'endTime'] =
  df_data.loc[df_data.endTime.isnull(), 'startTime'].apply(lambda
x: filltime(x))
  df_data['stationID'] = df_data['stationID'].astype(int)
```

After the above operation, the data has become very neat, in units of 10 min. This is also helpful for subsequent feature extraction. The complete training set is shown in Fig. 10.2.

Next, combine the data for flow of entering and exiting stations:

```
data_inNums.rename(columns={'time_10_minutes':'startTime'},
inplace=True)
data_outNums.rename( columns={'time_10_minutes':'startTime'},
inplace=True)
```

Fig. 10.2 The complete training set

	stationID	startTime	endTime
0	0	2019-01-01 00:00:00	2019-01-01 00:10:00
1	0	2019-01-01 00:10:00	2019-01-01 00:20:00
2	0	2019-01-01 00:20:00	2019-01-01 00:30:00
3	0	2019-01-01 00:30:00	2019-01-01 00:40:00
4	0	2019-01-01 00:40:00	2019-01-01 00:50:00
...	...	...	...
38875	9	2019-01-29 23:10:00	2019-01-29 23:20:00
38876	9	2019-01-29 23:20:00	2019-01-29 23:30:00
38877	9	2019-01-29 23:30:00	2019-01-29 23:40:00
38878	9	2019-01-29 23:40:00	2019-01-29 23:50:00
38879	9	2019-01-29 23:50:00	2019-01-30 00:00:00

```
df_data = df_data.merge(data_inNums, on=['stationID', 'startTime'],
how='left')
df_data = df_data.merge(data_outNums, on=['stationID', 'startTime'],
how='left')
df_data['inNums'] = df_data['inNums'].fillna(0)
df_data['outNums'] = df_data['outNums'].fillna(0)
```

10.1.5.3   Feature Extraction

In the baseline section, only some basic features can be extracted. For time series forecasting problems, simple time features and historical translation features are mainly extracted. The following is the specific code for extracting time-related features:

```
# time-correlated features
df_data['time'] = pd.to_datetime(df_data['startTime'])
df_data['days'] = df_data['time'].dt.day
df_data['hours_in_day'] = df_data['time'].dt.hour
df_data['day_of_week'] = df_data['time'].dt.dayofweek
df_data['ten_minutes_in_day'] = df_data['hours_in_day'] * 6 + df_data
['time'].dt.minute // 10
del df_data['time']
```

The features used to describe the position information of the current time in the cycle are very routine and their functions are also very large. For example, the week feature (day_of_week) helps to find similarities with the same number of weeks, similar to periodicity and correlation descriptions. The following is the specific code for extracting historical translation features:

```
# historical translation features
df_data['bf_inNums'] = 0
df_data['bf_outNums'] = 0
for i, d in enumerate(days):
    If d == 1:
    continue
    df_data.loc[df_data.day==d, bf_inNums] = df_data.loc[df_data.
day==days[i-1], inNums]
    df_data.loc[df_data.day==d, bf_outNums] = df_data.loc[df_data.
day==days[i-1], outNums]
```

10.1.5.4   Model Training

In order to quickly generate a reliable and stable result, we choose to use the LightGBM model; the offline verification method adopts a time series verification strategy, and the card swiping data of January 28 is used as the verification set. The code for model training is as follows:

```
# preparation of training set and verification set
cols = [f for f in df_data.columns if f not in
['startTime','endTime','inNums','outNums']]
    df_train = df_data[df_data.day<28]
    df_valid = df_data[df_data.day==28]

    X_train = df_train[cols].values
    X_valid = df_valid[cols].values

    y_train_inNums = df_train['inNums'].values
    y_valid_inNums = df_valid['inNums'].values
    y_train_outNums = df_train['outNums'].values
    y_valid_outNums = df_valid['outNums'].values
    # start training
    params = {'num_leaves': 63,'objective':
'regression_l1','max_depth': 5,
                'learning_rate': 0.01,'boosting': 'gbdt','metric':
'mae','lambda_l1': 0.1}
    model = lgb.LGBMRegressor(**params, n_estimators = 20000, nthread =
4, n_jobs = -1)
    model.fit(X_train, y_train_inNums,
                eval_set=[(X_train, y_train_inNums), (X_valid,
y_valid_inNums)],
                eval_metric='mae',
                verbose=100, early_stopping_rounds=200)
```

Only the flow of entering station is trained here, and the same operation could be used for training the flow of leaving. In this way, the basic score result can be obtained (the mean absolute error scores corresponding to the flow of entering and exiting stations are 19.6167 and 19.0041). There are still many points that can be optimized. In the following work, these optimization points will be gradually discovered, the structure and scores of the baseline scheme will be updated, and finally the scores will be ranked better with breakthroughs.

10.2 Data Exploration

In the data exploration part, there are obvious differences in the analysis methods for different business problems. In the time series forecasting problem, the key to data analytics lies in the analysis of time series patterns (trend, periodicity, correlation, and randomness), and the discovery of data features in multiple patterns.

10.2.1 Preliminary Research on Data

10.2.1.1 Traffic Flow Data

Figure 10.3 shows the basic traffic data, which is clean data that can be directly used for training and can construct time-related features from multiple dimensions.

Since it is a time series forecasting problem, the data naturally has a strong correlation with time. Figures 10.4 and 10.5 show the changes in the traffic flow of entering and exiting subway stations with stationID from 0 to 9 in time series, respectively. The data for January 1, 2019 (Tuesday) is selected.

In fact, there are some particularities in the data on January 1, 2019. Although it is a weekday, it is New Year's Day and belongs to a holiday. In the later analysis, we can observe the difference between it and other data.

10.2.1.2 Road Network Map

The competition provides a connection relationship table between subway stations, as shown in Fig. 10.6; it is an 81×81 matrix, with the column "Unnamed: 0" excluded.

Some preliminary assumptions can be made here. For subway transfer stations, especially stations with more adjacent stations (such as three adjacent stations and

	stationID	startTime	endTime	inNums	outNums	day	hours_in_day	day_of_week	ten_minutes_in_day
0	0	2019-01-01 00:00:00	2019-01-01 00:10:00	0.0	0.0	1	0	1	0
1	0	2019-01-01 00:10:00	2019-01-01 00:20:00	0.0	0.0	1	0	1	1
2	0	2019-01-01 00:20:00	2019-01-01 00:30:00	0.0	0.0	1	0	1	2
3	0	2019-01-01 00:30:00	2019-01-01 00:40:00	0.0	0.0	1	0	1	3
4	0	2019-01-01 00:40:00	2019-01-01 00:50:00	0.0	0.0	1	0	1	4
...	...	...	...	...	...	...	...	...	...
314923	9	2019-01-29 23:10:00	2019-01-29 23:20:00	0.0	0.0	29	23	1	139
314924	9	2019-01-29 23:20:00	2019-01-29 23:30:00	0.0	0.0	29	23	1	140
314925	9	2019-01-29 23:30:00	2019-01-29 23:40:00	0.0	0.0	29	23	1	141
314926	9	2019-01-29 23:40:00	2019-01-29 23:50:00	0.0	0.0	29	23	1	142
314927	9	2019-01-29 23:50:00	2019-01-30 00:00:00	0.0	0.0	29	23	1	143

314928 rows × 9 columns

Fig. 10.3 Basic traffic data

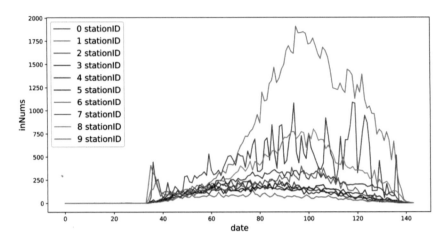

Fig. 10.4 Display of entering stations traffic on January 1, 2019. (see also the color illustration)

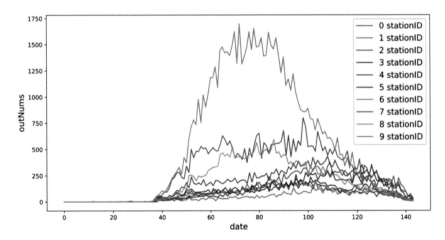

Fig. 10.5 Display of exiting stations traffic on January 1, 2019. (see also the color illustration)

four adjacent stations), assume that their traffic is relatively high; for transfer stations with only one adjacent station, it can be directly determined as the starting/terminal station, which is generally a relatively remote place with relatively low traffic.

10.2.2   Model Analysis

As mentioned in Chap. 9, to solve the time series problem, we first need to understand the key data patterns, and then express these patterns by extracting features. In addition, we also introduce 4 patterns, namely, trend, periodicity,

Unnamed: 0	0	1	2	3	4	5	6	7	8	...	71	72	73	74	75	76	77	78	79	80
0	0	1	0	0	0	0	0	0	0	...	0	0	0	0	0	0	0	0	0	0
1	1	0	1	0	0	0	0	0	0	...	0	0	0	0	0	0	0	0	0	0
2	0	1	0	1	0	0	0	0	0	...	0	0	0	0	0	0	0	0	0	0
3	0	0	1	0	1	0	0	0	0	...	0	0	0	0	0	0	0	0	0	0
4	0	0	0	1	0	1	0	0	0	...	0	0	0	0	0	0	0	0	0	0
...	...	...	...	...	...	...	...	...	...	...	...	...	...	...	...	...	...	...	...	...
76	0	0	0	0	0	0	0	0	0	...	0	0	0	0	1	0	1	0	0	0
77	0	0	0	0	0	0	0	0	0	...	0	0	0	0	0	1	0	0	0	0
78	0	0	0	0	0	0	0	0	0	...	0	0	0	0	0	0	0	0	1	0
79	0	0	0	0	0	0	0	0	0	...	0	0	0	0	0	0	0	1	0	1
80	0	0	0	0	0	0	0	0	0	...	0	0	0	0	0	0	0	0	1	0

81 rows × 82 columns

Fig. 10.6 Road network map data

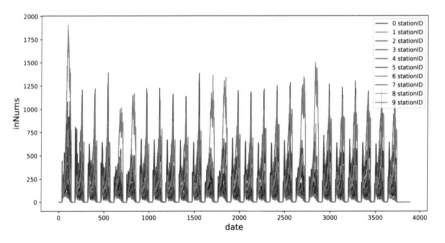

Fig. 10.7 Display of periodic visualization

correlation, and randomness. This section will also focus on these four aspects of data analytics.

- **Trend.** Trend is a common pattern in time series, and many things in real life include trend changes. Hangzhou subway traffic changes are no exception, such as the traffic from the start of the morning to the morning rush hour, or the traffic near the Christmas.
- **Periodicity.** We try to find the characteristics of periodicity from the data. As shown in Fig. 10.7, the abscissa and ordinate represent the time indicator (date, unit: second) and the flow of entering stations (inNums, unit: person),

respectively. It shows the traffic entering the subway stations with stationID from 0 to 9 in chronological order. The data selected here is from January 1, 2019 to January 28, 2019.

A closer look at the purple lines and blue lines (stationID 4 and stationID 9 respectively) in Fig. 10.7 shows that the time interval between each of these lines reaching the peak entering traffic is exactly 1 day, which means that 1 day is the most obvious cycle. It should be noted that January 1, at the beginning, is a holiday, and then the following 3 days are workdays. This further leads to the conclusion: the distribution of traffic data on weekdays and weekends is different, and special attention should be paid to this point when modeling.

Since Fig. 10.7 is a bit dense, the next step is to conduct a more detailed analysis. Select a subway station with stationID 4 and compare its traffic flow of entering station on weekdays and weekends in detail. The following code is used to generate a traffic comparison chart at different times on Fridays and Saturdays:

```
tmp = df_data.loc[(df_data.day.isin([4,5]))]
tmp.loc[tmp.stationID == 4].pivot_table(index='hours_in_day',
    columns='day',values='inNums').plot(style='o-')
```

The generated results are shown in Fig. 10.8. The abscissa and ordinate respectively represent time (unit: hour) and traffic flow of entering station. After comparing this traffic of each hour on Friday (day = 4) and Saturday (day = 5), it is easy to find that the time periods that see significant difference on Friday and Saturday are from 7 to 8 o'clock and from 17 to 19 o'clock, which happen to be the peak periods of entering the station. Therefore, the difference is mainly caused by the early and late peaks of working days. On the whole, it is also a cyclical change.

- **Correlation**. Generally speaking, correlation is prominent in two neighboring time units. For example, without cyclical influence, when the time interval is shorter, the traffic entering and leaving the station will be more similar. As shown in Fig. 10.9, the abscissa is time (unit: hour) and the ordinate is the traffic of

Fig. 10.8 Comparison of traffic at different times on Friday and Saturday

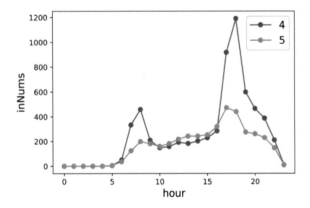

Fig. 10.9 Comparison of traffic at different times on Thursday and Friday

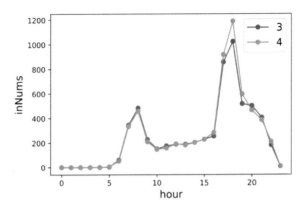

Fig. 10.10 The particularity of traffic on new year's day

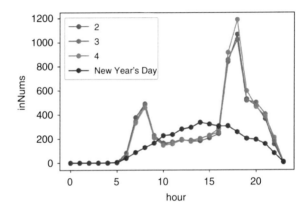

entering the station. This shows the incoming traffic of the subway on Thursday and Friday. The coincidence degree between the two lines is very high, which is very consistent with the concept of short-term correlation. In addition, comparing two similar moments in a day, it will be found that if special factors (morning and evening rush hours) are not considered, the more adjacent the moments, the more similar the traffic.

- **Randomness.** Random data changes are not easy to determine. Emergencies and special dates can lead to randomness. For example, it is difficult to predict the subway traffic on New Year's Day. As can be seen from Fig. 10.10, the subway traffic on New Year's Day is very different from that on other dates, which will bring difficulties to modeling and require special approach.

10.3 Feature Engineering

The content of this section is very important, all of which relies on my actual operation in the competition. The structure is clear, and it is easy to be optimized.

10.3.1   Data Preprocessing

Before formally extracting features, the data with large differences in the data distribution of the testing set (i.e. data of weekends and New Year's Day) is removed to ensure the consistency of the overall data distribution, and this part of the data is also removed first in the final scheme. The specific code is as follows:

```
# remove data of weekends and New Year's Day
df_data = df_data.loc[((df_data.day_of_week < 5) & (df_data.day !=
1))].copy()
# keep the date
retain_days = list(df_data.day.unique())
# re-compute rank to facilitate subsequent feature extraction
days_relative = {}
for i,d in enumerate(retain_days):
    days_relative[d] = i + 1
df_data['days_relative'] = df_data['day'].map(days_relative)
#### visualization code ####
dt = [r for r in range(df_data.loc[df_data.stationID==0,
'ten_minutes_in_day'].shape[0])]
fig = plt.fig.(1,figsize=[12,6])
plt.ylabel('inNums',fontsize=14)
plt.xlabel('date',fontsize=14)
for i in range(0,10):
    plt.plot(dt, df_data.loc[df_data.stationID==i, 'inNums'],
label = str(i)+'stationID' )
plt.legend()
# generate vector diagram
plt.savefig("inNums_of_stationID.svg", format="svg")
```

Run the above code and generate a visual diagram as shown in Fig. 10.11, specifically excluding the traffic data of weekends, so that the rest are data for workdays with similar traffic distribution.

10.3.2   Strong Correlation Features

The strong correlation information was mainly generated in the same time period of different days, so we constructed the features of flow for entering and exiting stations with a 10-min granularity and a 1-h granularity respectively. Considering the fluctuation of the flow before and after corresponding time periods, we added the flow characteristics before and after a certain period, or a certain two periods before and a certain two periods after. In addition, we also constructed the flow in the corresponding period of the previous n days. Further, taking into account the strong correlation between adjacent stations, we added the traffic characteristics of the corresponding time period of the two adjacent stations. The correlation code is as follows:

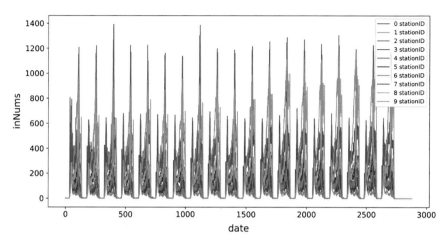

Fig. 10.11 A visualization display with only workday data retained

```
def time_before_trans(x,dic_):
if x in dic_.keys():
    return dic_[x]
else:
    return np.nan
```

```
  df_feature_y['tmp_10_minutes'] = df_feature_y['stationID'].values
* 1000 +
      df_feature_y['ten_minutes_in_day'].values
  df_feature_y['tmp_hours'] = df_feature_y['stationID'].values * 1000 +
      df_feature_y['hours_in_day'].values
```

```
  for i in range(1, n): # through the last n days
      d = day - i
      df_d = df.loc[df.days_relative == d].copy() # data of the day
```

```
      # feature 1: traffic flow of entering and exiting the station in this time period in the past (same
time period, 10-minute granularity)
      df_d['tmp_10_minutes'] = df['stationID'] * 1000 + df
['ten_minutes_in_day']
      df_d['tmp_hours'] = df['stationID'] * 1000 + df['hours_in_day']
      # Here, sum is used as a statistic, and mean, median, Max, min, STD and other statistics can be
further considered
      dic_innums = df_d.groupby(['tmp_10_minutes'])['inNums'].sum().
to_dict()
      dic_outnums = df_d.groupby(['tmp_10_minutes'])['outNums'].sum().
to_dict()
      df_feature_y['_bf_' + str(day-d) + '_innum_10minutes'] =
          df_feature_y['tmp_10_minutes'].map(dic_innums).values
      df_feature_y['_bf_' + str(day-d) + '_outnum_10minutes'] =
          df_feature_y['tmp_10_minutes'].map(dic_outnums).values
```

```
      # feature 2: traffic of entering and exiting stations during this period in the past (1-hour
granularity)
      dic_innums = df_d.groupby(['tmp_hours'])['inNums'].sum().to_dict()
```

```
    dic_outnums = df_d.groupby(['tmp_hours'])['outNums'].sum().
to_dict()
      df_feature_y['_bf_' + str(day-d) + '_innum_hour'] =
        df_feature_y['tmp_hours'].map(dic_innums).values
      df_feature_y['_bf_' + str(day-d) + '_outnum_hour'] =
        df_feature_y['tmp_hours'].map(dic_outnums).values

    # feature 3: traffic of entering and exiting stations in the first 10 minutes
      df_d['tmp_10_minutes_bf'] = df['stationID'] * 1000 + df
['ten_minutes_in_day'] - 1
      df_d['tmp_hours_bf'] = df['stationID'] * 1000 + df['hours_in_day'] - 1
      # sum statistic
      dic_innums = df_d.groupby(['tmp_10_minutes_bf'])['inNums'].sum
().to_dict()
      dic_outnums = df_d.groupby(['tmp_10_minutes_bf'])['outNums'].sum
().to_dict()
      df_feature_y['_bf1_' + str(day-d) + '_innum_10minutes'] =
        df_feature_y['tmp_10_minutes'].agg(lambda x:
        time_before_trans(x,dic_innums)).values
      df_feature_y['_bf1_' + str(day-d) + '_outnum_10minutes'] =
        df_feature_y['tmp_10_minutes'].agg(lambda x:
        time_before_trans(x,dic_outnums)).values

    # feature 4: traffic of entering and exiting stations in the first hour
      dic_innums = df_d.groupby(['tmp_hours_bf'])['inNums'].sum().
to_dict()
      dic_outnums = df_d.groupby(['tmp_hours_bf'])['outNums'].sum().
to_dict()
      df_feature_y['_bf1_' + str(day-d) + '_innum_hour'] =
        df_feature_y['tmp_hours'].map(dic_innums).values
      df_feature_y['_bf1_' + str(day-d) + '_outnum_hour'] =
        df_feature_y['tmp_hours'].map(dic_outnums).values

for col in ['tmp_10_minutes','tmp_hours']:
del df_feature_y[col]
return df_feature_y
```

Please observe the code carefully. When the characteristics of a certain day in history is constructed, it already contains periodically related features, such as the traffic entering and leaving the station at the corresponding time in the previous few weeks, and the statistical characteristics of the granularity of 10 min/1 h at the corresponding time in the previous few weeks.

Now please consider a problem. Section 10.1.5 does not deal with the problem of distribution differences caused by periodicity, which will have a great impact on the construction of features. The data of two adjacent days are correlated and affected by weekdays and weekends, so many noise features will be extracted. Confronted with this problem, we have a variety of ideas for modeling, such as removing weekend data to ensure consistency, retaining weekend data to enhance feature-related descriptions, and also considering a variety of modeling results.

> **Extended Thinking**
> As long as the data exists, it is reasonable. No modeling method is necessarily good. We are more about balancing the impact of the data on the modeling results. Although there are differences in data distribution, it is just because of the relationship between features. The new date will bring new feature combinations and lead to new modeling results.

10.3.3 Trend Features

Finding out trend is also the key for us to extract features. The definition of trend features we mainly construct is as follows:

$$A\_diff(n + 1) = A(n + 1) - A(n), A = in \text{ or } out$$

That is the difference between the data before and after the period. The data here can be either flow of entering the station or that of exiting the station. Similarly, we have taken into account the present period corresponding to each day, the corresponding prior period, etc. Of course, we can also consider the ratio, which is defined as follows:

$$A\_ratio\ (n + 1) = A\ (n + 1)/A(n), A = in \text{ or } out$$

This type of features is also very useful in actual competitions. These features are mainly used to assist the model to learn the change of trends. Generally speaking, the first-order trend feature and the second-order trend feature are constructed, wherein the first-order trend feature is the difference or ratio of data for adjacent time units, reflecting the changing trend; the second-order trend feature is the difference for the first-order trend feature, reflecting the speed of trend changing.

10.3.4 Station-Related Features

Since the objective is to predict the traffic flow of a subway station (stationID), more information from the station itself can be collected. What is mainly gathered is the heat of different stations and that of the combination of stations with other features. This type of feature is mainly used to describe entity information and is indispensable in real competitions. The following is the specific implementation code for constructing station-related features, mainly constructing frequency features (count) and category number features (nunique):

```
def get_stationID_fea(df):
  df_station = pd.DataFrame()
  df_station['stationID'] = df['stationID'].unique()
  df_station = df_station.sort_values('stationID')
  # related to nunique
  tmp1 = df.groupby(['stationID'])['deviceID'].nunique().
      to_frame('stationID_deviceID_nunique').reset_index()
  tmp2 = df.groupby(['stationID'])['userID'].nunique().
      to_frame('stationID_userID_nunique').reset_index()

  df_station = df_station.merge(tmp1,on ='stationID', how='left')
  df_station = df_station.merge(tmp2,on ='stationID', how='left')

# combine with stationID; get the feature of count
for pivot_cols in tqdm_notebook(['payType','hour',
'days_relative','ten_minutes_in_day']):
  tmp = df.groupby(['stationID',pivot_cols])['deviceID'].count().
  to_frame('stationID_'+pivot_cols+'_cnt').reset_index()
  df_tmp = tmp.pivot(index = 'stationID', columns=pivot_cols,
  values='stationID_'+pivot_cols+'_cnt')
  cols = ['stationID_'+pivot_cols+'_cnt' + str(col) for col in df_tmp.
columns]
  df_tmp.columns = cols
  df_tmp.reset_index(inplace = True)
  df_station = df_station.merge(df_tmp, on ='stationID', how='left')
return df_station
```

10.3.5   Feature Enhancement

It is a very important job to strengthen the constructed features. For example, in the 2019 Tencent Advertising Algorithm Competition, the new statistical features were required to be further expanded. Imagine if you treat the newly constructed feature as an unreal value and then cross it with the real value, then theoretically you can get a value close to the real value. This is just a direction. We can also perform cross-combination or aggregate statistics on the newly constructed features to get a more profound feature description.

The following will be specific feature enhancement. On the basis of correlation features, windows of different sizes are selected for summation and mean statistics, and differential features are extracted from window statistical features to obtain trend-related features:

```
columns = ['_innum_10minutes','_outnum_10minutes','_innum_hour','_
outnum_hour']
# compute the sum and mean for the flow in the past n days
for i in range(2,left):
    for f in columns:
        colname1 = '_bf_'+str(i)+'_'+'days'+f+'_sum'
        df_feature_y[colname1] = 0
```

```
        for d in range(1,i+1):
            df_feature_y[colname1] = df_feature_y[colname1] +
df_feature_y['_bf_'+
                str(d)+f]
        colname2 = '_bf_'+str(d)+'_'+'days'+f+'_mean'
        df_feature_y[colname2] = df_feature_y[colname1] / i

# differential features for the mean value of flow in the past n days
for i in range(2,left):
    for f in columns:
        colname1 = '_bf_'+str(d)+'_'+'days'+f+'_mean'
        colname2 = '_bf_'+str(d)+'_'+'days'+f+'_mean_diff'
        df_feature_y[colname2] = df_feature_y[colname1].diff(1)
        # process the first hour in the first day
        df_feature_y.loc[(df_feature_y.hours_in_day==0)&
                    (df_feature_y.ten_minutes_in_day==0),colname2] = 0
```

10.4   Model Training

This section will be the optimize solutions from handling the models. As a time series forecasting problem, there are still quite a lot of models to choose from, such as traditional time series models, tree models, and deep learning models. Parameter adjustment is also part of model selection. This section will optimize the parameters of the LightGBM model in order to further improve the score.

10.4.1   LightGBM

The LightGBM model same as the baseline scheme is used here to facilitate comparison of effects. The main differences before and after optimization lie in multi-angle feature extraction and adjustment of individual parameters of the model (learning_rate and feature_fraction).

Compared with the mean absolute error scores corresponding to entering and exiting stations in the baseline scheme, which are 19.6167 and 19.0041 respectively, the present mean absolute error scores for entering and leaving stations in the optimized scheme are 12.6477 and 13.1619 respectively, both having been greatly improved. There are three main reasons, namely, data preprocessing, feature extraction, and model parameter adjustment. If, in terms of importance, feature extraction has the greatest impact on the results, model parameter adjustment is only the icing on the cake.

10.4.1.1   Feature Importance Feedback

We know that the tree model can feed the importance score of features back, so let's take a look at the importance score of the LightGBM model. Execute the following code to visually display the importance score of features:

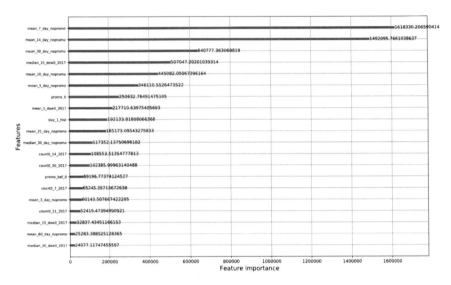

Fig. 10.12 LightGBM feature importance score

```
import matplotlib.pyplot as plt
import seaborn as sns
import warnings
warnings.simplefilter(action='ignore', category=FutureWarning)

feature_imp = pd.DataFrame(sorted(zip(model.feature_importances_,
cols)),
    columns=['Value','Feature'])

    plt.figure(figsize=(20, 10))
    sns.barplot(x="Value", y="Feature", data=feature_imp.sort_values
(by="Value",
                    ascending=False)[:20])
    plt.title('LightGBM Features Importance')
    plt.tight_layout()
    plt.show()
```

The resulting results are shown in Fig. 10.12, sorted from high to low in importance.

10.4.1.2 Feature Selection Based on Importance Score

In addition to measuring the importance of features, the feature importance score can also be used for feature selection, and features with high importance scores can be retained. Next, compare the effects of this type of feature selection method. Here, the features with importance of top100 are extracted, and then the model is retrained to compare the mean absolute error score of traffic entering and leaving stations:

```
new_cols = feature_imp.sort_values(by="Value",
    ascending=False)[:100]['Feature'].values.tolist()
```

After experimental feedback, the mean absolute error scores of the scheme after feature selection are 12.6248 and 13.1509. Compared with the previous ones, there is not much improvement, but the number of features has been reduced from 262 to 100, eliminating a large number of redundant features, and the overall performance has been greatly improved.

10.4.2 Time Series Model

In time series problems, it is a perfect choice to use time series models such as recurrent neural networks, LSTM, and GRU. Such models can automatically extract information related to time series and reduce the time-consuming work of manually constructing a large number of time series features.

Here we choose to use the LSTM model and then go through multiple layers of fully connected layers. In order to make the model more generalized, Batch Normalization and Dropout are also added. Generally, these two parts will be added while performing modeling related to deep learning, to make the model more robust. For Batch Normalization, we can use normalization to pull the more and more skewed distribution back to the normalized distribution, making the gradient larger, thereby accelerating the convergence speed of model learning and avoiding the problem of gradient disappearing. The modeling code is as follows:

```
from keras.models import Sequential
from keras.layers.core import Dense, Dropout, Activation
from keras.layers.normalization import BatchNormalization
from keras.layers import LSTM
from keras import callbacks
from keras import optimizers
from keras.callbacks import ModelCheckpoint, EarlyStopping,
ReduceLROnPlateau

def build_model():
    model = Sequential()
    model.add(LSTM(512, input_shape=(X_train.shape[1],X_train.
shape[2])))
    model.add(BatchNormalization())
    model.add(Dropout(0.2))

    model.add(Dense(256))
    model.add(Activation(activation="relu"))
    model.add(BatchNormalization())
    model.add(Dropout(0.2))

    model.add(Dense(64))
    model.add(Activation(activation="relu"))
```

```
model.add(BatchNormalization())
model.add(Dropout(0.2))
model.add(Dense(16))
model.add(Activation(activation="relu"))
model.add(BatchNormalization())
model.add(Dropout(0.2))
model.add(Dense(1))
return model
```

The above code is also very versatile and can show good results in general time series forecasting problems. Next, let's look at how to compile and train the model. Here, early stopping can also be performed like in a tree model:

```
# the compiling section
model = build_model()
model.compile(loss='mae', optimizer=optimizers.Adam(lr=0.001),
metrics=['mae'])
# callback function
reduce_lr = ReduceLROnPlateau(
    monitor='val_loss', factor=0.5, patience=3, min_lr=0.0001,
verbose=1)
earlystopping = EarlyStopping(
    monitor='val_loss', min_delta=0.0001, patience=5, verbose=1,
mode='min')
callbacks = [reduce_lr, earlystopping]
# the training section
model.fit(X_train, y_train_inNums, batch_size = 256, epochs = 200,
verbose=1,
    validation_data=(X_valid,y_valid_inNums), callbacks=callbacks)
```

The above simple and clear code can result in a good score. As shown in Fig. 10.13, the mean absolute error score of offline traffic flow of entering is 13.2967. There are also many areas that can be optimized here, such as adjusting the network structure and parameters. Let's simply adjust a parameter. Note that in the training section, there is a particularly important parameter—shuffle, which is used to set whether to break up the data before each round (epoch) of iteration. False

```
Epoch 00013: ReduceLROnPlateau reducing learning rate to 0.0005000000237487257.
Epoch 14/200
151632/151632 [==============================] - 19s 127us/step - loss: 19.9452 - mean_absolute_error: 19.9452 - val_loss: 14.0227 - val_mean_absolute_error: 14.0227
Epoch 15/200
151632/151632 [==============================] - 19s 126us/step - loss: 19.8370 - mean_absolute_error: 19.8370 - val_loss: 13.6583 - val_mean_absolute_error: 13.6583
Epoch 16/200
151632/151632 [==============================] - 19s 126us/step - loss: 19.7433 - mean_absolute_error: 19.7433 - val_loss: 13.4510 - val_mean_absolute_error: 13.4510
Epoch 17/200
151632/151632 [==============================] - 19s 128us/step - loss: 19.5663 - mean_absolute_error: 19.5663 - val_loss: 13.3501 - val_mean_absolute_error: 13.3501
Epoch 18/200
151632/151632 [==============================] - 19s 127us/step - loss: 19.5381 - mean_absolute_error: 19.5381 - val_loss: 13.2505 - val_mean_absolute_error: 13.2505
Epoch 19/200
151632/151632 [==============================] - 19s 126us/step - loss: 19.4834 - mean_absolute_error: 19.4834 - val_loss: 13.7082 - val_mean_absolute_error: 13.7082
Epoch 20/200
151632/151632 [==============================] - 19s 126us/step - loss: 19.3861 - mean_absolute_error: 19.3861 - val_loss: 14.1614 - val_mean_absolute_error: 14.1614
Epoch 21/200
151632/151632 [==============================] - 19s 126us/step - loss: 19.5545 - mean_absolute_error: 19.5545 - val_loss: 13.3941 - val_mean_absolute_error: 13.3941

Epoch 00021: ReduceLROnPlateau reducing learning rate to 0.000250000011874362Β.
Epoch 22/200
151632/151632 [==============================] - 19s 125us/step - loss: 19.2886 - mean_absolute_error: 19.2886 - val_loss: 13.2967 - val_mean_absolute_error: 13.2967
Epoch 23/200
151632/151632 [==============================] - 19s 125us/step - loss: 19.1401 - mean_absolute_error: 19.1401 - val_loss: 13.3935 - val_mean_absolute_error: 13.3935
Epoch 00023: early stopping
```

Fig. 10.13 Demonstration of the training process of the LSTM model

is the default. If it is set to True, the mean absolute error score of offline traffic flow of entering can reach 12.9747.

10.5 Reinforcement Learning

Endless learning should be reflected in this book. Simply solving problems is not the ultimate goal. The primary objective of this book is to be able to consider different solutions for optimization, learn different problem-solving ideas, and be able to "draw inferences from one instance" in different competition questions.

10.5.1 Sequential Stacking

Although this kind of modeling method is rarely used in time series prediction problems, its power is very obvious, and the rationality of the modeling method is indispensable. Specifically, because there are some unknown singular values in historical data. For example, some large-scale activities will cause a sudden increase in traffic at certain sites at certain times, and the impact of these data is very significant. Another example is that there are differences between the distribution of data far from the current time and the current data.

In order to reduce the impact of singular value data, we use sequential stacking to solve the problem. If there is a big gap between the predicted results of the model and the real results, then such data is abnormal. Next, let's learn the modeling ideas of sequential stacking and discover its value. First, the structure of the scheme is given, as shown in Fig. 10.14. Through the following operations, what is offline and online can both be steadily improved.

Figure 10.14 is the part of timing stacking that constructs intermediate features. In order to ensure that the verification set does not have the problem of data crossing,

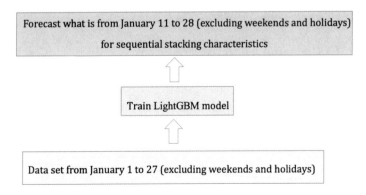

Fig. 10.14 Constructing intermediate features in sequential stacking

First use the data before January 28 for model training; then use the complete data for prediction to obtain intermediate features (temporal stacking features); finally splice the intermediate features with the rest of the features and select the data closer to the testing set as the training set (January 10 to January 28). The specific implementation code is as follows:

```
# training set preparation
y_inNums = df_data[df_data.day<28]['inNums'].values
y_outNums = df_data['outNums'].values
df_train = df_data[df_data.day<28][cols].values
df_data = df_data[cols].values
# model training
params = {'num_leaves': 63,'objective': 'regression_l1','max_depth': 5,
          'feature_fraction': 0.9, 'learning_rate': 0.05,'boosting':
'gbdt','metric':
          'mae','lambda_l1': 0.1}
model = lgb.LGBMRegressor(**params, n_estimators = 1500, nthread = 4)
model.fit(df_train, y_inNums)
# intermediate features
inNums_stacking = model.predict(df_data)
```

This code is divided into two parts, namely, intermediate feature acquisition and merging intermediate features and training.

Without adjusting any parameters, the average absolute mean of traffic entering and exiting stations after merging intermediate features can lead to a score of 12.5814. The time interval of this question is not too large. I believe that with the expansion of the time interval of the data set, the final effect of the optimization strategy of sequential stacking will be better and better.

> **Tips**
> The process of constructing intermediate features here is a simplified version. You can also use K-fold cross-validation to obtain intermediate features, and then splice the remaining features to determine the final training set interval for final training.

10.5.2   Top Scheme Analysis

After all, the content that this chapter can introduce in detail is limited. I hope readers can learn more excellent ideas and programs in the future. This section will introduce more top programs to expand their ideas.

10.5.2.1 Top 1 Scheme

In addition to the traditional LightGBM model, the champion also used the NMF model, which is mainly used to take into account the relationship between passenger inflow and outflow volume at each station. Note that if the interval is 10 min, the passenger inflow matrix and passenger outflow matrix for the same period are different. However, if the interval is 1 day, the passenger inflow matrix and passenger outflow matrix for the whole day are basically the same. Then the NMF model is transformed, and the two matrices W and H are obtained by non-negative matrix factorization. The hidden variables can be understood as the representations of the respective stations learned by the algorithm. At the same time, the matrices A and B of the learning trend are added, and the loss function is optimized to train the parameters of W, H, A, and B.

The extracted features are mainly divided into three categories: conventional features, location determination features, and subway network features. Among them, location determination features are station location features obtained from subway passenger flow changes during working hours, rest hours, rush hours, and weekend leisure hours, such as whether a station belongs to a working area or a residential area, etc. The characteristics of the subway network mainly refer to some ideas in the SNA (social network analysis), and transform centrality, betweenness, closeness, etc., not only to measure the location and physical relationship between subway stations, but also to measure the flow of passengers between stations. Due to time constraints and not very familiar with Hangzhou Metro, there is no road_map station given by the organizer corresponding to the real Hangzhou Metro station. It is recommended to refer to the strategies of other teams, as well as the real stations of Hangzhou Metro to extract features or introduce peripheral POI data, relevant POI corresponding event data, etc., in order to improve the final accuracy of forecasting.

10.5.2.2 Top 2 Scheme

The runner-up team only used the LightGBM model in the aspect of model. However, considering the strong periodicity of the data in this topic, two models were constructed, namely the full data model and the model of previous 2 days data, and finally the results were integrated. I also designed a time series weighted regression model, introducing similarity calculations of different dates to determine the weight when weighting. The final result is weighted by the results of the LightGBM model and the time series weighting regression model.

In terms of features, we have also made a more detailed attempt, mainly divided into original features, statistical features, and derived features. In the derived characterizing portion, the time difference between the peak time of entering and leaving the station, the change trend of the flow of people in the first 2 days, and the fixed number of people entering and leaving the station within a fixed time range every day are obtained through the user ID (this feature can effectively exclude the influence of the fixed number of people on the model, so that the training model pays more attention to the number of random trips and reduces errors), and the rest of the features are conventional.

10.5.3 Relevant Questions Recommendation

There are still quite a lot of competition questions related to time series forecasting, and there are fixed methods, such as the pattern analysis of time series and the way to extract features from multiple angles, but often some different operations are brought about due to different businesses and data. Next, we will recommend a few classic competition questions. I believe that after in-depth comparison and learning of multiple competition questions, we can easily deal with such problems.

10.5.3.1 2019 Tencent Advertising Algorithm Competition: Advertising Exposure Estimation

The title of this algorithm competition originates from the real business product of Tencent's advertising business for advertisers—advertisement exposure estimation. The purpose of advertisement exposure estimation is to provide advertisers with future reference to the advertisement exposure effect when advertisers create new advertisements and modify advertisement settings. Through this estimation reference, advertisers can avoid aimless attempts for optimization, effectively shorten the optimization cycle of advertisements, reduce trial and error costs, and make the advertisement effect reach their expected range as soon as possible.

This competition provides n-day historical exposure data of ads (sampled on a specific traffic), including the traffic characteristics corresponding to each exposure (temporal and spatial information such as user features and advertising place), as well as the settings and competitiveness scores of exposure of advertisements; the testing set is a new batch of advertisement settings (some are brand-new advertisement IDs, and some have modified the settings for old advertisement IDs), requiring an estimate of the daily exposure of this batch of advertisements.

Basic ideas: There are still quite a lot of schemes for this question, and each type of scheme can get the previous ranking. Here are three kinds, namely, traditional tree model, deep learning model, and rule-based strategy. The feature aspect mainly focuses on time series correlation for prediction, and the extraction idea is mainly considered from two parts: historical information and overall information, and what is more detailed is the statistical features of the day before the present day, the latest 5 days, five-fold cross-validation statistics, and all days except that day. In specific competitions, we will find a large number of new ad IDs in the testing set. New ads have no historical information, so how to construct the characteristics of new ads, and how to describe the history and integrity of new ads becomes the key to scoring.

We do fuzzy feature construction here. Although we do not know the historical information of the new advertisement, we know the historical information of the old advertisement contained under the advertisement account ID. Therefore, by combining the advertisement account ID with the advertisement winning rate of the old advertisement, we can construct the average and median of the advertisement winning rate under the advertisement account ID. In this way, we get the statistical

Fig. 10.15   Home page of web traffic time series forecasting competition

value of the advertisement winning rate of the new advertisement under the advertisement account ID.

10.5.3.2   Kaggle Competition: Web Traffic Time Series Forecasting

This question (the home page of the question is shown in Fig. 10.15) belongs to the multi-step time series forecasting problem, and the time span is very large. This problem has always been one of the most challenging problems in the field of time series. Specifically, this question provides the daily traffic of approximately 145,000 Wikipedia articles in the past year or so and requires the contestants to predict the traffic of these articles in the following 3 months.

The competition is divided into two stages and will include predictions of actual future events. In the first stage, the rank list will be scored based on historical data; in the second stage, participants' submissions will be scored based on real future events.

The training set data consists of about 145,000 time series, including data from July 1, 2015 to December 31, 2016. Each of the time series represents the number of daily views of different Wikipedia articles. The first stage of the ranking is based on the traffic from January 1, 2017 to March 1, 2017. The second stage will use training data as of September 1, 2017. The final ranking of the competition will be predicted based on the daily article views of each article in the data set during the period from September 13, 2017 to October 13, 2017.

Basic ideas: Simply using the rule-based method in this competition enables contestants to get a silver medal. Of course, the rule-based strategy should be considered very carefully (periodicity, trend, and similarity representation). For most time series forecasting problems, this method can play a certain role. Even if it is not the final scheme, it can also be used as one of the ideas for extracting features. The model scheme can be roughly divided into three categories: RNN seq2seq, convolutional neural networks, and median method prediction (turning the prediction target into forecasting the median).

Chapter 11
Case Study: Corporación Favorita Grocery Sales Forecasting

This chapter is based on a classic competition question—commodity sales forecasting—on the 2018 Kaggle competition platform, that is, the Corporación Favorita Grocery Sales Forecasting shown in Fig. 11.1. This is also the second actual practice for issues related to series analysis. Similarly, the content also mainly includes problem understanding, data exploration, feature engineering, and model training. As an international competition topic, there are many contents worth digging deeply. In addition to giving general ideas for solving problems, this chapter is more important to guide everyone to make different attempts, avoid fixed mindset, and finally sort out knowledge points and make further extension, which means to summarize competition questions.

11.1 Understanding the Competition Question

11.1.1 Background Introduction

In brick-and-mortar grocery stores, the relationship between sales volume forecasting and customer purchase volume is always subtle. If sales volume forecasts are higher and customers purchase less, then the grocery store will have an excessive backlog of goods, especially having a greater impact on perishable goods; if sales volume forecast is small and customers purchase more, then the goods will be sold out quickly and the customer experience will deteriorate in a short period of time.

The problem becomes more complicated as retailers continue to add new locations and new products, and the variety of seasonal tastes and the unpredictability of product marketing keep increasing. Corporación Favorita, a large grocery retailer in Ecuador, also knows this very well. It operates hundreds of supermarkets and sells more than 200,000 kinds of goods.

So Corporación Favorita challenged the Kaggle community to build a model that could accurately predict the sales volume of goods. Corporación Favorita currently

W. He et al., *Machine Learning Contests: A Guidebook*,
https://doi.org/10.1007/978-981-99-3723-3_11

Fig. 11.1 Home page of corporación favorita grocery sales forecasting

relies on subjective predictions to back up data and rarely executes plans through automated tools. They are very looking forward to providing enough correct goods at the right time through machine learning to better satisfy customers.

11.1.2   Competition Data

This competition requires predicting unit sales of thousands of items sold in different stores of the Ecuadorian retailer Corporación Favorita. The training data provided by the competition questions include date, store (store_nbr), item (item_nbr), whether an item participates in promotion (onpromotion, about 16% of the values in the file are missing) and unit sales (unit_sales); other documents include supplementary information, which may be useful for modeling. The competition questions contain more documents, and the following is a brief introduction to each document. Stores. csv: details of stores, such as location and type.

- items.csv: products information, such as category, whether the product is perishable, etc. Special attention should be paid to the fact that perishable products have a higher rating than other products.
- Transactions.csv: the transaction volume of each store on different dates (only dates within the time range of training data are included).
- Oil.csv: Daily oil price. This data is related to sales volume, because Ecuador is an oil-dependent country, and its economic health is extremely vulnerable to oil prices.
- Holidays_events.csv: data on holidays in Ecuador. Some of these holidays may be transferred to another day (from a weekend to a working day), similar to compensatory holidays.

11.1.3   Evaluation Indicators

The submitted content is evaluated according to the normalized weighted root mean square logarithmic error (NWRMSLE), and the calculation method is as formula (11.1):

$$\text{NWRMSLE} = \sqrt{\frac{\sum_{i=1}^{n} w_i (\ln(\hat{y}_i + 1) - \ln(y_i + 1))^2}{\sum_{i=1}^{n} w_i}} \quad (11.1)$$

For the line i of the testing set, $\hat{y}_i$ is the predicted unit sales volume of the goods, yi is the real unit sales volume, and n is the total number of lines of the testing set, and the weight w_i can be found in the file items.csv. The weight of perishable goods is 1.25, and the weight of other items is 1.00.

This regression evaluation index is different from the common mean absolute error (MAE) and mean square error (MSE). Eq. (11.1) converts the sales volume and gives different weights to different commodities. Finally, the square root of the result is calculated.

11.1.4 Competition FAQ

How can the phenomenon of data crossing in time series forecasting be prevented?

In time-related modeling problems, the most important thing to pay attention to is data crossing. There are many times when we will add future information as a feature in the modeling process if we pay little attention, which will lead to serious overfitting, making the difference between online and offline scores become larger, and often there will be inconsistent online and offline evaluation results. Here are two most typical crossing cases:

1. Suppose we need to predict whether the user watched a video, the testing set needs to predict the probability of the user watching video B at 10:10 on April 16, but through the data in the training set, it is found that the user was watching video A at 10:09 on April 16, and also watching video A at 10:11. Then it is obvious that at 10:10, there was a high probability that the user did not watch video B. Through future information, it is easy to judge that at 10:10 on April 16, the user did not watch video B.
2. Assuming that we need to predict the amount of money the user spent on a bank card on August 17, but the training set has given the user's bank card balance on August 16 and August 18, then we can easily know what the user spent on August 17. The above two examples are obvious data traversal situations. At this time, we should filter out future data information and train the model with only historical data.

How can the noise problem in the training set and testing set be solved or anticipated?

In the training set of time series problems, there will be noise more or less, which will have a great impact on modeling. Here we divide the noise into three categories:

random noise, local obvious noise/local singular value, and noise caused by long time. The noise in the testing set is similar to that in the training set. The testing set data for this question covers a relatively large time range of more than 10 days, and includes some special times, such as the end of the month (the payday), which may cause certain difficulties in prediction.

◨How do you understand the variable onpromotion and what impact might it have?

ⒶWe have found a strange variable onpromotion in this question. First, this is a crossing variable because it contains future promotion information; second, this variable has 16% missing data, and all rows where unit_sales is 0 in the training set are omitted. That is to say, a large part of onpromotion information is also lost, but as long as onpromotion information appears, our model will think that unit_sales is not 0 under many circumstances, so the model will be biased.

11.1.5   Baseline Scheme

With the above understanding, you can start the basic modeling. The baseline scheme does not need to be too complicated and can give a correct result. In fact, you can also think of this process as building a simple framework first, and then filling and optimizing it.

11.1.5.1   Data Reading

The relevant code for reading the data set is as follows:

```
import pandas as pd
import numpy as np
from sklearn.metrics import mean_squared_error
from sklearn.preprocessing import LabelEncoder
import lightgbm as lgb
from datetime import date, timedelta

  path = './input/'
df_train = pd.read_csv(path+'train.csv',
  converters={'unit_sales': lambda u: np.log1p(float(u)) if float(u) >
0 else 0},
  parse_dates=["date"])
df_test = pd.read_csv(path + "test.csv",parse_dates=["date"])
items = pd.read_csv(path+'items.csv')
stores = pd.read_csv(path+'stores.csv')
# type conversion
df_train['onpromotion'] = df_train['onpromotion'].astype(bool)
df_test['onpromotion'] = df_test['onpromotion'].astype(bool)
```

In the above code, first log1p () preprocessing is performed on the unit_sales. The advantage of this is that data with large skewness can be converted and compressed into a smaller interval, and finally log1p () preprocessing can play a role in smoothing the data. In addition, the same processing is performed on the unit_sales in the evaluation index part, and this part of the operation is also preprocessing.

Another operation is to process the date, converting the time string in the table file into a date format. Early processing not only facilitates subsequent operations, but also reduces the amount of code.

11.1.5.2   Data Preparation

The data set contains data from 2013 to 2017. The time span is very large, and there will be a lot of uncertainty in the 4 year development process. When using too long-term data to predict the future, there will be a certain amount of noise and there will be differences in distribution, which can also be found in Sect. 11.2. In addition, due to performance considerations, only data for 2017 will be used as the training set. Execute the following code to filter the data before 2017:

```
df_2017 = df_train.loc[df_train.date> = pd.datetime(2017,1,1)]
del df_train.
```

Next, the basic data format conversion is carried out, and finally the store, goods and time are used as indexes to construct a data table of whether the promotion is promoted or not, so as to carry out statistics related to promotion or not. This construction method is conducive to feature extraction afterwards. The relevant codes are as follows:

```
promo_2017_train = df_2017.set_index(["store_nbr", "item_nbr",
  "date"])[["onpromotion"]].unstack(level=-1).fillna(False)
promo_2017_train.columns = promo_2017_train.columns.
get_level_values(1)

promo_2017_test = df_test.set_index(["store_nbr", "item_nbr",
  "date"])[["onpromotion"]].unstack(level=-1).fillna(False)
promo_2017_test.columns = promo_2017_test.columns.
get_level_values(1)

promo_2017 = pd.concat([promo_2017_train, promo_2017_test], axis=1)
df_2017 = df_2017.set_index(["store_nbr", "item_nbr",
  "date"])[["unit_sales"]].unstack(level=-1).fillna(0)
df_2017.columns = df_2017.columns.get_level_values(1)
```

11.1.5.3   Feature Extraction

The historical translation feature and the window statistical feature are the core features of the time series forecasting problem. Here, only the historical translation feature (one unit) and the window statistical feature of different window sizes are simply used as the basic features. What is implemented below is a general code for extracting features:

```
def get_date_range(df, dt, forward_steps, periods, freq='D'):
    return df[pd.date_range(start=dt-timedelta
(days=forward_steps), periods=periods,
    freq=freq)]
```

The following feature extraction is mainly carried out around the get_date_range function just implemented, which is very general. Its entry parameters df, dt, forward_steps, periods, and freq are the data source method, start time, historical span, period, and frequency extracted by the window respectively. The specific code for feature extraction is as follows:

```
def prepare_dataset(t2017, is_train=True):
X = pd.DataFrame({
    # the historical translation feature; the sales of the day 1, day 2, and day 3
    "day_1_hist": get_date_range(df_2017, t2017, 1, 1).values.ravel(),
    "day_2_hist": get_date_range(df_2017, t2017, 2, 1).values.ravel(),
    "day_3_hist": get_date_range(df_2017, t2017, 3, 1).values.ravel(),
    })
for i in [7, 14, 21, 30]:
# the window statistical feature; sales diff/mean/meidan/max/min/std
X['diff_{}_day_mean'.format(i)] = get_date_range(df_2017, t2017, i,
    i).diff(axis=1).mean(axis=1).values
X['mean_{}_day'.format(i)] = get_date_range(df_2017, t2017, i,
    i).mean(axis=1).values
X['median_{}_day'.format(i)] = get_date_range(df_2017, t2017, i,
    i).mean(axis=1).values
X['max_{}_day'.format(i)] = get_date_range(df_2017, t2017, i,
    i).max(axis=1).values
X['min_{}_day'.format(i)] = get_date_range(df_2017, t2017, i,
    i).min(axis=1).values
X['std_{}_day'.format(i)] = get_date_range(df_2017, t2017, i,
    i).min(axis=1).values

for i in range(7):
# average weekly sales for the first 4 weeks and first 10 weeks
X['mean_4_dow{}_2017'.format(i)] = get_date_range(df_2017, t2017,
28-i, 4,
    freq='7D').mean(axis=1).values
X['mean_10_dow{}_2017'.format(i)] = get_date_range(df_2017, t2017,
70-i, 10,
    freq='7D').mean(axis=1).values

for i in range(16):
# whether the following 16 days are with promotion activities
X["promo_{}".format(i)] = promo_2017[str(t2017 +
    timedelta(days=i))].values.astype(np.uint8)
    if is_train:
        y = df_2017[pd.date_range(t2017, periods=16)].values
        return X, y
    return X
```

As can be seen from the above code, the baseline scheme only extracted historical translation, window statistics, statistical features of the first N weeks, whether the next 16 days are promotional days and type features, etc. The overall structure is very simple.

Among them, special attention should be paid to the statistical characteristics of the first N weeks, especially the parameter part of the get_date_range function where freq = '7D' indicates that the extraction interval (frequency) is 7 days, the periods 4 indicates that the extraction is done in 4 cycles, and 28-i (the value of i is 0, 1, 2, 3, 4, 5, 6) is the historical span. When i takes 1, it represents the calculation 4 day average sales for dates 2017-06-08, 2017-06-15, 2017-06-22, and 2017-06-29.

Next, determine the interval for extracting features, and introduce how the training set, verification set, and testing set extract features respectively:

```
X_1, y_1 = [], []
t2017 = date(2017, 7, 5)
n_range = 14
for i in tqdm(range(n_range)):
  delta = timedelta(days=7 * i)
  X_tmp, y_tmp = prepare_dataset(t2017 - delta)
  X_1.append(X_tmp)
  y_1.append(y_tmp)

X_train = pd.concat(X_1, axis=0)
y_train = np.concatenate(y_1, axis=0)
del X_1, y_1

# The training set takes the data from July 26th to August 10th
X_val, y_val = prepare_dataset(date(2017, 7, 26))
# The data of the testing set is taken from August 16th to August 31st
X_test = prepare_dataset(date(2017, 8, 16), is_train=False)
```

When dealing with the problems of time series forecasting, how to choose the verification set is very important. The testing set contains the data from August 16 to August 31, 2017, and the starting time is Wednesday. Because the verification set must not only be close to the testing set in time, but also conform to the periodic distribution, it is most appropriate to choose the data from July 26 to August 10, 2017 as the verification set; that is, the starting time is Wednesday and the ending time is Thursday. In addition, considering the stability of the verification set, multiple rounds of rolling verification can be carried out, i.e. one verification set can be selected 1 week or several weeks apart, such as the data from July 19 to August 3.

Another question worth considering is why you choose to use 7 days as a cycle to construct training data and extract features. Will this waste a lot of data? In order to clarify the rationality of using 7 days as a cycle, you can choose different cycles for experimental comparison. For example, changing the cycle to 1 day is to extract 7×16 days of data. It will be found that this not only greatly increases the training time of the model, but also the score is not as good as before. To solve this problem, a reasonable explanation is that taking 7 days as a cycle can well ensure the periodicity of the data set, and the data set has the same distribution as the verification set and the testing set.

11.1.5.4   Model Training

Here, LightGBM is used as the base model to predict the unit sales volume of goods in the next 16 days. There are many modeling methods for the multi-step prediction in problems related to time series forecasting: first, based on single-step prediction, the predicted value is added to the training set as the real value to predict the next unit, but this will lead to error accumulation. If there is a large error at the beginning, the effect will become worse and worse; second, directly predict the results of all testing sets, that is, as a multiple-output regression problem, so that although the problem of error accumulation can be avoided, it will increase the difficulty of model learning, because the model needs to learn a many-to-many system, which will raise the difficulty of training. We temporarily choose the first modeling method to build a baseline scheme:

```
params = {
'num_leaves': 2**5 - 1,
'objective': 'regression_l2',
'max_depth': 8,
'min_data_in_leaf': 50,
'learning_rate': 0.05,
'feature_fraction': 0.75,
'bagging_fraction': 0.75,
'bagging_freq': 1,
'metric': 'l2',
'num_threads': 4
}
MAX_ROUNDS = 500
val_pred = []
test_pred = []

for i in range(16):
print("====== Step %d ======" % (i+1))
dtrain = lgb.Dataset(X_train, label=y_train[:, i])
dval = lgb.Dataset(X_val, label=y_val[:, i], reference=dtrain)
bst = lgb.train(
   params, dtrain, num_boost_round=MAX_ROUNDS,
   valid_sets=[dtrain, dval], verbose_eval=100)

val_pred.append(bst.predict(X_val, num_iteration=bst.
best_iteration or MAX_ROUNDS))
test_pred.append(bst.predict(X_test, num_iteration=bst.
best_iteration or MAX_ROUNDS))
```

Until now, the basic baseline scheme has been set up. In terms of feature extraction and model training, there is no consideration of too complicated operations. The final score is 0.51837 (721/1624) for the public score and 0.52798

(695/1624) for the private score. In real business scenarios, such a scheme can already achieve good prediction results, and the rest of the work is continuous data analytics, feature extraction, and model tuning.

11.2 Data Exploration

11.2.1 Preliminary Research on Data

There are still a lot of data tables in this competition. This section mainly takes you to understand the structure and basic situation of each table one by one. Analyzing the basic table is the initial work of the whole data analytics, which helps to clarify the relationship between tables and tables.

11.2.1.1 Data Sheet Train

The following code will show the basic information of the training set, including the number of attributes (nunique) of each feature, the proportion of missing values, the proportion of maximum attributes, and the feature type.

```
stats = []
for col in train.columns:
    stats.append((col, train[col].nunique(),
        round(train[col].isnull().sum() * 100 / train.shape[0], 3),
        round(train[col].value_counts(normalize=True, dropna=False).
        values[0] * 100,3), train[col].dtype))

stats_df = pd.DataFrame(stats, columns=['features ', 'the number of attributes
', 'the proportion of missing values ', 'the proportion of maximum attributes ', 'feature type
'])
stats_df.sort_values('the proportion of missing values ', ascending=False)
[:10]
```

As shown in Fig. 11.2, you can have a general understanding of the basic information in the train.csv. The number of attributes refers to the number of categories included in the feature, and the maximum proportion of attributes refers to the proportion of the most frequently occurring attributes to the total data volume. In addition, the size of the train.csv file is larger than 5.6 GB, which contains 115,497,040 pieces of data.

11.2.1.2 Test.csv

As shown in Fig. 11.3, there are no missing values in the test.csv, which is larger than 106.1 MB and contains 3,370,464 pieces of data.

Features	Attributes No.	Proportion of Missing Values	Proportion of Maximum Attributes	Feature Types
onpromotion	2	17.257	76.519	object
id	125497040	0.000	0.000	int64
date	1684	0.000	0.094	object
store_nbr	54	0.000	2.799	int64
item_nbr	4036	0.000	0.067	int64
unit_sales	258474	0.000	18.682	float64

Fig. 11.2 Basic Information in train.csv

Features	Attributes No.	Proportion of Missing Values	Proportion of Maximum Attributes	Feature Types
id	3370464	0.0	0.000	int64
date	16	0.0	6.250	object
store_nbr	54	0.0	1.852	int64
item_nbr	3901	0.0	0.026	int64
onpromotion	2	0.0	94.108	bool

Fig. 11.3 Basic Information in test.csv

Features	Attributes No.	Proportion of Missing Values	Proportion of Maximum Attributes	Feature Types
date	1682	0.0	0.065	object
store_nbr	54	0.0	2.010	int64
transactions	4993	0.0	0.108	int64

Fig. 11.4 Basic Information in transactions.csv

11.2.1.3   Transactions.csv

As shown in Fig. 11.4, there are no missing values in the transactions.csv, which is larger than 1.9 MB and contains 83,488 pieces of data.

11.2.1.4   Items.csv

As shown in Fig. 11.5, there are no missing values in the items.csv, which is larger than 118.2 KB and contains 4100 pieces of data.

11.2.1.5   Stores.csv

As shown in Fig. 11.6, there are no missing values in the stores.csv, which is larger than 2.2 KB and contains 54 pieces of data.

11.2.1.6   Oil.csv

The information in this data sheet is daily oil prices, including data from January 1, 2013 to August 31, 2017.

Features	Attributes No.	Proportion of Missing Values	Proportion of Maximum Attributes	Feature Types
item_nbr	4100	0.0	0.024	int64
family	33	0.0	32.537	object
class	337	0.0	3.244	int64
perishable	2	0.0	75.951	int64

Fig. 11.5 Basic Information in items.csv

Features	Attributes No.	Proportion of Missing Values	Proportion of Maximum	Feature Types
store_nbr	54	0.0	1.852	int64
city	22	0.0	33.333	object
state	16	0.0	35.185	object
type	5	0.0	33.333	object
cluster	17	0.0	12.963	int64

Fig. 11.6 Basic Information in stores.csv

11.2.1.7   holidays_events.csv

The information in this data sheet is vacation data for Ecuador, which contains 350 records. As shown in Fig. 11.7, where locale indicates the region involved in the vacation, description indicates the description related to the vacation, and transferred indicates whether the vacation is transferred.

11.2.2   Univariate Analysis

11.2.2.1   Train.csv:Date

Figure 11.8 shows the daily sales volume, which is divided into three parts: total unit sales), promotional sales (On Promotion), non-promotional sales volume (Not On Promotion). It can be seen that the total unit sales volume has increased significantly every year, which may be because the company is growing. In addition, it can be found that there are basically daily sales records for promotion and non-promotion, and some records are empty, approximately before the second quarter of 2014.

There are also some data that show cliff-like decline or significant growth, which may be affected by other special factors, such as oil prices, holidays, or natural disasters, etc. These factors should also be considered in specific structural characteristics.

11.2.2.2   Train.csv:store_nbr

Next, analyze the store_nbr (store number) in the training set file. As shown in Fig. 11.9, the store number with the highest transaction frequency is 44, which is

Features	Attributes No.	Proportion of Missing Values	Proportion of Maximum Attributes	Feature Types
date	312	0.0	1.143	object
type	6	0.0	63.143	object
locale	3	0.0	49.714	object
locale_name	24	0.0	49.714	object
description	103	0.0	2.857	object
transferred	2	0.0	96.571	bool

Fig. 11.7   Basic Information in holidays_events.csv

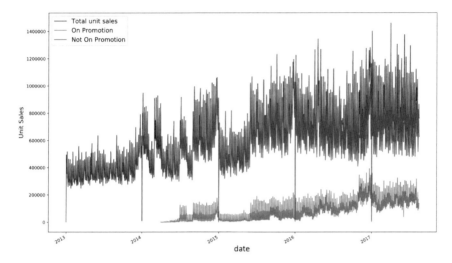

Fig. 11.8 Change of sales volume of different dates

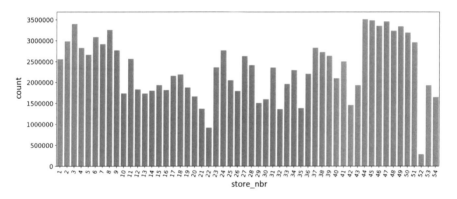

Fig. 11.9 Transaction frequency of stores in the training set

close to 3.5 million, and the store number 52 has the lowest transaction frequency. These may be caused by store type, business location, business hours, or promotion intensity.

11.2.2.3 Train.csv:item_nbr

Figure 11.10 shows the distribution of transaction frequency of different commodities through line graphs. It can be found that the number of transactions of commodities varies greatly, with the maximum number of transactions being more than 80,000 and the minimum number being only a few. In fact, this is also in line with intuition. For example, the sales volume of fast-moving consumer goods is

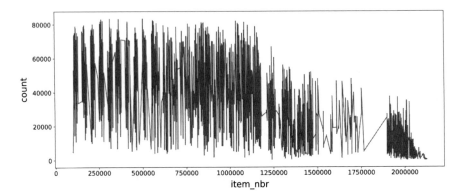

Fig. 11.10 Trading frequency of commodities in training set

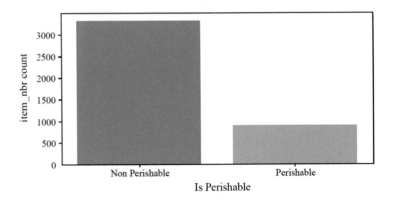

Fig. 11.11 Distribution of perishable or non-perishable commodities

generally better, while the sales volume of less commonly used and expensive commodities will be much worse.

11.2.2.4 Items.csv:Perishable

Now look at the balance between Perishable and Non-perishable goods, as shown in Fig. 11.11. At present, we only have a simple understanding of the single variable, and then we will analyze the distribution of perishable goods in different households or different stores.

11.2.2.5 Oil.csv: Dcoilwtico

This is a very interesting data set, which contains daily oil prices. Since Ecuador is an oil-dependent country, we can try to understand the relationship between commodity sales and oil prices. The knowledge contained here is largely related to economics.

Figure 11.12 shows the changes in oil prices from 2013 to 2017. In addition to the lack of records of oil price on a few dates, the overall oil price has changed significantly in several stages, such as the trough in early 2015, the trough in early 2016, the oil price from 2013 to the first half of 2014 being between 80 and 100 yuan, and the oil price from the beginning of 2015 to the second half of 2017 being mostly within 50 yuan. In addition, it can be clearly seen that the oil price changes in 2017 are basically stable, so consider selecting only the data of 2017 as training sets int the phase of modeling.

11.2.2.6 Stores.csv:State

Figure 11.13 shows the distribution of the number of stores in each state through a vertical graph, including a total of 16 states. Among them, Guayas and Pichincha have the most stores—Pichincha has 19 stores (the largest number), Guayas has 11 stores, and other states have no more than 3 stores.

11.2.2.7 Stores.csv:City

Feature city is also a very important. It has its own unique entity information. As shown in Fig. 11.14, the stores.csv file contains a total of 22 different cities, of which Guayaquil and Quito have more stores. As can be seen from the graph, the number of

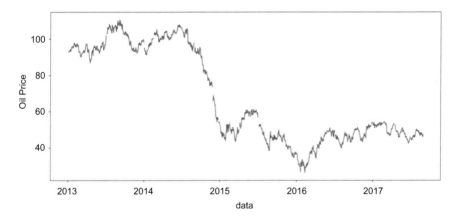

Fig. 11.12 Change in oil prices

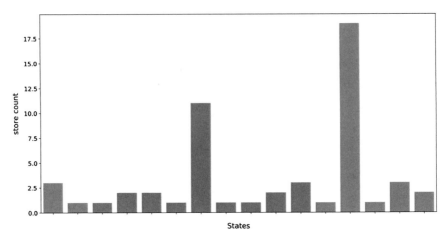

Fig. 11.13 Distribution of the number of stores in each state

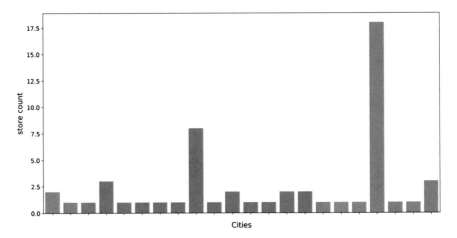

Fig. 11.14 Distribution of the number of stores in each city

stores (store count) in different cities is quite different, probably due to the different levels of economic development in different regions.

11.2.3   *Multivariate Analysis*

This section mainly analyzes the relationship between variables and variables, and that between variables and labels. On the one hand, it explores the distribution relationship between variables; on the other hand, it explores whether variables and labels are distinguishable, or whether there is some characteristics different from what we get through our intuition.

11.2.3.1 Variables and Variables

First analyze the holidays_events.csv file. Figure 11.15 shows the regional distribution of different holiday types. It is illustrated that the Holiday type occurs most frequently, and most Holidays occur in Local. Most of the other holiday types occur in National.

11.2.3.2 Variables and Labels (the Label of this Competition Question is Sales Volume)

Let's first look at the relationship between store (stroe_nbr) and sales volume. As can be seen from Fig. 11.16, the total sales volume corresponding to different stores is different. For example, the sales volume of Store 44 and Store 45 is very high, while the sales volume of Store 22 and Store 52 is very low. Referring to Fig. 11.9, it can be found that the sales volume of most stores with high transaction frequency is also relatively high. Of course, it may also be affected by the price of the product itself, promotional activities, and economic factors.

Since there are too many categories of commodities (item_nbr), the relationship between commodities and sales volume is shown in the box chart here, as shown in Fig. 11.17. The figure ignores specific commodities and only considers the distribution of commodity sales volume after aggregation. The sales volume of different

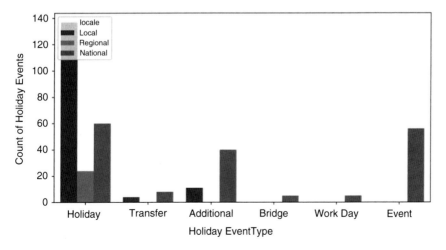

Fig. 11.15 Regional distribution of different holiday types

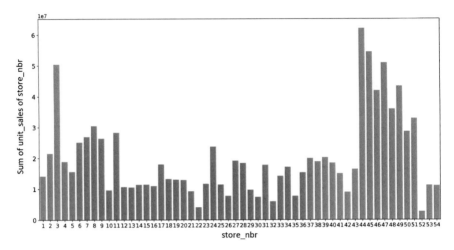

Fig. 11.16 Distribution of sum of sales volume corresponding to different stores

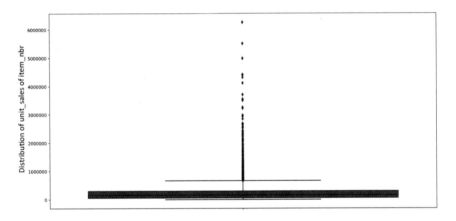

Fig. 11.17 Distribution of sales volume for different commodities

commodities varies greatly, with more than 80% below 1,000,000 and only a small portion above 1,000,000.

An important parameter in the evaluation index affects the importance of the product in the scoring stage, that is, whether it is perishable (perishable). The weight of perishable products is 1.25, and the weight of other items is 1.00. Next, let's see how the sales volume of the two types of products will vary over time.

From Fig. 11.18, it can be concluded that the sales volume of non-perishable goods is higher, and perishable goods have better stability and lower jitter. The overall growth or decline trends of the two are basically the same. However, perishable goods have higher weights, so a single perishable commodity has a greater impact on the score than a single non-perishable commodity in the evaluation

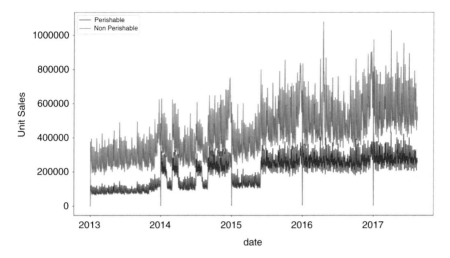

Fig. 11.18   Distribution of sales for perishable and non-perishable commodities

stage. You can consider adding perishable features or adding sample weights in the model training stage.

> **Extended Learning**
> This section of data analytics does not show the visual analysis of all variables, combinations of variables and variables, and combinations of variables and labels. However, in general, we can find that many of the analyses are still in line with our prior judgment, and there is also some information that can only be learned after performing visual analysis. For the time series forecasting problem, the most important thing is the change of the target over time, so in the extended learning section, I hope you will continue to explore the changes of different variables over time and analyze the change tendency of unit sales to see if there will be unexplained phenomena.

11.3   Feature Engineering

The content of this section is very important and is the key to getting high scores in this competition. It will start with the idea of feature extraction, and finally explain the feature enhancement method, which is clear and easy to optimize. In addition, it will also take readers to practice efficient feature selection methods to improve the performance of the overall training.

Figure 11.19 is a visual display of the daily sales volume of goods with an item_nbr of 502,331, of which the part to be predicted is that after August 15, 2017. From this line graph, we can also see the three core patterns of the time series prediction problem, namely periodicity, trend, and similarity. The cycle of

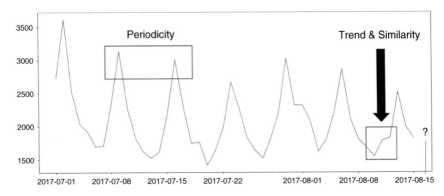

Fig. 11.19 The core idea of feature extraction

sales volume change is 1 week, which is still very obvious in the figure, so it is meaningful to extract more features related to multiples of seven, such as sales volume last week, sales volume last week, and average sales volume on the Nth day in the last 3 weeks. Of course, it is necessary to predict and consider changes in trends and similarities more accurately, periodically determine the approximate sales volume at this moment, adjust the increase or decrease of future sales volume through trends, and ensure the recent sales volume through similarities.

In the problem of time series forecasting, micro and macro changes should also be considered. If item_nbr sales volume is a sales volume description from a micro perspective, then the sales volume aggregation corresponding to store_nbr, class, and city can be regarded as a sales volume description from a macro perspective. If a store experiences a general depression this year, then such condition is related to every commodity it sells. The high sales volume of first-tier cities indicates that the sales volume of commodities is also very high. Macro changes in sales volume will not be affected by local commodity anomalies and can well reflect the overall change trend. Therefore, more expansion will be carried out in the specific feature extraction stage, describing the sales volume step by step from micro to macro.

11.3.1   Historical Translation Features

The historical translation feature is to extract information with similarity, and the basic historical translation is to use the sales volume of t - 1 time unit as the feature of the t-th time unit. Generally speaking, the time interval for feature extraction is within 1 month, because in the specific construction of features, data that is too far apart in time not only does not have similarity, but also brings noise.

```
for i in range(1,31):
# the historical translation feature, sales volume of the first N days
X["day_{}_hist".format(i)] = get_date_range(df_2017, t2017, i, 1).
values.ravel().
```

11.3.2 *Window Statistical Features*

There are two types of window statistics constructed here. The first is the window statistics N days before the structural feature day, and the second is the window statistics of the i-th day of the week in the N weeks before the structural feature day. The second one is more difficult to understand; why is it in weeks? Why only the i-th day of the week is selected? This can be explained from the perspective of periodicity. A complete periodicity can ensure that the statistics have practical significance, and a high degree of similarity in time units in a cycle can be guaranteed for selecting the same i-th day every week.

In addition to counting the traditional mean, median, extreme values, and variance in the window, the difference calculation before and after the sequence in the window can also be performed and the mean can be counted. Another thing that needs to be introduced is the power function attenuation weighting, which is similar to the exponential weighted average. The weight gradually decays over time, and these methods are often used as rule-based strategies.

11.3.2.1 Window Statistics N Days before the Structural Feature Day

The basic code is given below:

```
# The mean of the difference between the forward and backward values
X['before_diff_{}_day_mean'.format(i)] = get_date_range(df_2017,
  t2017-timedelta(days=d), i, i).diff(1,axis=1).mean(axis=1).values
X['after_diff_{}_day_mean'.format(i)] = get_date_range(df_2017,
  t2017-timedelta(days=d), i, i).diff(-1,axis=1).mean(axis=1).
values
# exponential decay summation
X['mean_%s_decay_1' % i] = (get_date_range(df_2017, t2017-timedelta
(days=d),
  i, i) * np.power(0.9, np.arange(i)[::-1])).sum(axis=1).values
# mean/meidan/max/min/std
X['mean_{}_day'.format(i)] = get_date_range(df_2017, t2017-
timedelta(days=d),
  i, i).mean(axis=1).values
X['median_{}_day'.format(i)] = get_date_range(df_2017, t2017-
timedelta(days=d),
  i, i).median(axis=1).values
X['max_{}_day'.format(i)] = get_date_range(df_2017, t2017-timedelta
(days=d),
  i, i).max(axis=1).values
X['min_{}_day'.format(i)] = get_date_range(df_2017, t2017-timedelta
(days=d),
  i, i).min(axis=1).values
X['std_{}_day'.format(i)] = get_date_range(df_2017, t2017-timedelta
(days=d),
  i, i).std(axis=1).values
```

Among them, we need to pay attention to the two parameters i and d. The parameter i means the window size, and d means crossing d days in the historical direction. It can be explained as: first cross d days in the historical direction to reach a certain time unit, and then count the eigenvalues of this time unit on day i in the past. In order to ensure periodicity in actual operation, the value of d is 0, 7, 14, and the value of i is 3, 4, 5, 6, 7, 10, 14, 21, 30, 90, 110, 140, 356.

11.3.2.2 Window Statistics of Day i of Each Week in the First N Weeks Before the Structural Day

The basic code is given below:

```
for i in range(7):
# sales volume of day i of each week in the first N weeks
for periods in [5,10,15,20]:
  steps = periods * 7
  X['before_diff_{}_dow{}_2017'.format(periods,i)] = get_date_range
(df_2017,
    t2017, steps-i, periods, freq='7D').diff(1,axis=1).mean(axis=1).
values
  X['after_diff_{}_dow{}_2017'.format(periods,i)] = get_date_range
(df_2017,
    t2017, steps-i, periods, freq='7D').diff(-1,axis=1).mean
(axis=1).values
  X['mean_{}_dow{}_2017'.format(periods,i)] = get_date_range
(df_2017, t2017,
    steps-i, periods, freq='7D').mean(axis=1).values
  X['median_{}_dow{}_2017'.format(periods,i)] = get_date_range
(df_2017, t2017,
    steps-i, periods, freq='7D').median(axis=1).values
  X['max_{}_dow{}_2017'.format(periods,i)] = get_date_range
(df_2017, t2017,
    steps-i, periods, freq='7D').max(axis=1).values
  X['min_{}_dow{}_2017'.format(periods,i)] = get_date_range
(df_2017, t2017,
    steps-i, periods, freq='7D').min(axis=1).values
  X['std_{}_dow{}_2017'.format(periods,i)] = get_date_range
(df_2017, t2017,
    steps-i, periods, freq='7D').std(axis=1).values
```

It is necessary to pay attention to the three parameters i, periods, and steps. The parameter i indicates the day i of the week, periods indicates the first N cycles, and steps-i indicates the start date.

Fig. 11.20 Hierarchical relationship

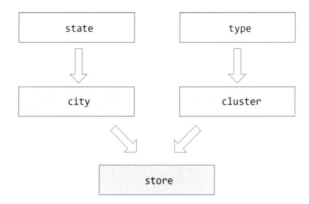

11.3.3   *Structural Granularity Diversity*

However, in the actual feature extraction, besides constructing the combination of stores and commodities, the combination of first-order and second-order features such as item_nbr, store_nbr, city, class, store_class, city_class, etc. are also considered.

So how can we determine whether the two features can be combined? The main basis is the degree of sparsity and hierarchical relationship. If each attribute is unique after the combination of the two features, then this feature combination is very sparse, and this feature has no structural significance; as shown in Fig. 11.20, it is a hierarchical diagram of each field in the stores.csv. The state and city fields have a hierarchical relationship, and the number of attributes after the combination of the two is consistent with the number of city attributes, so the combination of state and city is meaningless.

11.3.4   *Efficient Feature Selection*

A large number of features can be constructed in the time series forecasting problem. For historical translation features, you can consider translating in any feasible unit; for window statistical features, you can consider counting in different window sizes. If constructed in this way, the overall feature may reach thousands. So many features will lead to feature redundancy, and features containing noise may be constructed. In order to solve this problem, we choose to use the tree model to generate feature selection methods of feature importance and perform online and offline score verification.

Generating features through the tree model has certain interpretability, which is also a common method in the competition world and in real world practice. This part will also explore the effect of selecting the number of features according to the importance of features through experiments. Here only the training data of 6 training

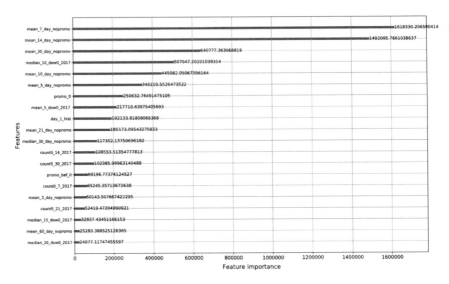

Fig. 11.21 Feature importance score

windows are extracted, and the offline score comparison of top500, top1000, top2000, tail2000 features and complete features are compared respectively. Also, only the offline score of the first day of the verification set needs to be noticed. Execute the following code to generate a score visualization diagram of feature importance:

```
import matplotlib.pyplot as plt
fig, ax = plt.subplots(figsize=(10,10))
lgb.plot_importance(bst, max_num_features=20, ax=ax,
importance_type='gain')
plt.show()
```

The generated result is shown in Fig. 11.21, which helps us quickly understand the importance of features in model training.

The experimental code is as follows:

```
# the order of feature importance
imps = sorted(zip(X_train.columns, bst.feature_importance("gain")),
        key=lambda x: x[1], reverse=True)
# extract top500 features
top_500 = [items[0] for items in imps[:500]]
# the offline score of the first day of the verification set
dtrain = lgb.Dataset(X_train[top_500], label=y_train[:, 0],
weight=train_weight)
dval = lgb.Dataset(X_val[top_500], label=y_val[:, 0],
reference=dtrain,
  weight=val_weight)
bst = lgb.train(params, dtrain, num_boost_round=MAX_ROUNDS,
        valid_sets=[dtrain, dval], verbose_eval=100)
```

Fig. 11.22 Scores display
of extracting feature sets of
different parts

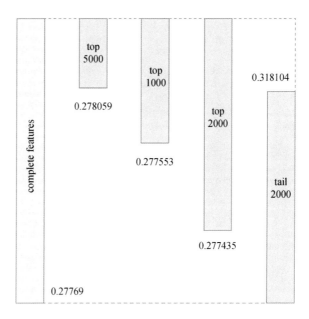

After comparing several rounds of experiments, as shown in Fig. 11.22, it can be
noticed that the effect of extracting top2000 features is the best (the score is
0.277435), and the effect of extracting tail2000 features is the worst (the score is
0.318104), which clearly shows that obtaining feature importance scores through the
tree model is effective.

11.4   Model Training

This section will optimize the scheme from the model side. For time series prediction
problems, there are still many models to choose from, such as traditional time series
models, tree models, and deep learning models. This section selects LightGBM,
LSTM, and Wavenet as the final models of the scheme, and will also learn model
integration to make the final ranking of a higher level.

11.4.1   LightGBM

Although the model used here is the same as that used in the baseline scheme, it is
optimized here. First, a weight is set for each sample because the evaluation index
will be perishable. Then, the historical translation features, window statistical features, and granularity diversity features mentioned in Sect. 11.3 are added, which

basically contain most of the features that can be extracted from the time series forecasting problem.

```
# the sample weight construction part
item_perishable_dict = dict(zip(items['item_nbr'],items
['perishable'].values))
train_weight = [] # weight of the training set
val_weight = [] # weight of the verification set

items_ = df_2017.reset_index()['item_nbr'].tolist() * n_range
for item in items_:
    train_weight.append(item_perishable_dict[item] * 0.25 + 1)
items_ = df_2017.reset_index()['item_nbr'].values
for item in items_:
    val_weight.append(item_perishable_dict[item] * 0.25 + 1)

# add the sample weight and specify the category feature
dtrain = lgb.Dataset(X_train, label=y_train[:, i],
weight=train_weight)
dval = lgb.Dataset(X_val, label=y_val[:, i], reference=dtrain,
weight=val_weight)
bst = lgb.train(params, dtrain, num_boost_round=MAX_ROUNDS,
    valid_sets=[dtrain, dval], verbose_eval=100)
```

Compared with the baseline scheme, the public score is 0.51837 (721/1624) and the private score is 0.52798 (695/1624). After more detailed feature extraction, modeling strategy adjustment, adding sample weights, and specifying category features, the score of the model has been greatly improved. Specifically, the public Score is 0.51319 (497/1624), and the private score is 0.51571 (13/1624).

11.4.2   LSTM

In the problem of time series forecasting, the author still chooses the most popular LSTM model. Compared with the LightGBM model, the number and granularity of features extracted by the LSTM model will be greatly reduced, mainly because it has the capability to extract sequence information from historical data. By comparison, the characteristics of tree models such as LightGBM cannot extract historical information by themselves, requiring a large number of artificial features.

This section first gives the basic network structure. BatchNormalization and Dropout methods will be used in the training process to help improve the generalization of the model. The following are the specific codes implemented to build the LSTM model:

```
def build_model():
model = Sequential()
model.add(LSTM(118, input_shape=(X_train.shape[1],X_train.shape
[2])))
```

```
model.add(BatchNormalization())
model.add(Dropout(0.2))

model.add(Dense(64))
model.add(Activation(activation="relu"))
model.add(BatchNormalization())
model.add(Dropout(0.2))

model.add(Dense(16))
model.add(Activation(activation="relu"))
model.add(BatchNormalization())
model.add(Dropout(0.2))
model.add(Dense(1))
model.compile(loss='mse', optimizer=optimizers.Adam(lr=0.001),
metrics=['mse'])
return model
```

The modeling method of LSTM is basically the same as the previous one, which is also the result of 16 training sessions to obtain 16 units in the testing set, and only the weight of the training sample is added during the training. In addition, it should be noted that the label is converted by subtracting the label mean, mainly for scaling processing and reducing the jitter of the prediction result. The following is the implementation code of 16 times of training and 16 times of forecasting:

```
for i in range(16):
y_mean = y_train[:, i].mean()
 # compilation part
  model = build_model()
  # callback function
  reduce_lr = ReduceLROnPlateau(
    monitor='val_loss', factor=0.5, patience=3, min_lr=0.0001,
verbose=1)
  earlystopping = EarlyStopping(
    monitor='val_loss', min_delta=0.0001, patience=3, verbose=1,
mode='min')
  callbacks = [reduce_lr, earlystopping]
  # the training part
  model.fit(X_train, y_train[:, i]-y_mean, batch_size =4096, epochs =
50, verbose=1,
    sample_weight=np.array(train_weight),
    validation_data=(X_val, y_val[:, i]-y_mean),
    callbacks=callbacks, shuffle=True)

  val_pred.append(model.predict(X_val)+y_mean)
  test_pred.append(model.predict(X_test)+y_mean)
```

The score of the LSTM model is also very good, with a public score of 0.51431 (557/1624) and a private score of 0.52067 (116/1624), which can be used for model integration.

In fact, there are many optimization directions in the model part, and the most basic one is to determine the number of hidden layers of deep neural networks. Theoretically, the deeper the number of layers, the stronger the ability of the fitting function, and the better the prediction effect of the model, but in fact, the deeper number of layers may lead to overfitting, while increasing the training difficulty, making it difficult for the model to converge. Of course, in order to better determine the number of layers, a simple experimental comparison was carried out next.

The experiment compared the deep neural network structures of 8 layers and 4 layers (in order to obtain the experimental results quickly, only the scores of the first day of the verification set are compared). Among them, the offline evaluation score of 8 layers is 0.2790 and that of 4 layers is 0.2795, which are basically the same. However, in terms of running time, 8 layers are more than three times that of 4 layers. In view of this, using a 4-layer deep neural network structure can not only achieve the same effect for the score as using 8-layer, but also have very efficient running time.

> **Extended Learning**
> In addition to determining the number of hidden layers, you can also select the number of nerve cells. Using too few nerve cells in the hidden layer will lead to underfitting. Using too many nerve cells may lead to overfitting. When the number of nodes in the neural network is too large (the information processing capacity is too high), the limited information contained in the training set is not enough to train all nerve cells in the hidden layer, which will lead to overfitting. In addition, even if the training set contains enough information, too many nerve cells in the hidden layer will increase the training time, making it difficult to achieve the desired effect. Obviously, it is very important to choose the right number of hidden layer nerve cells.

11.4.3   Wavenet

Although the Wavenet model basically did not appear in this competition, the runner-up contestant achieved a good second place by using this model. Since it has such great power, we might as well get to know this model.

In 2016, Google DeepMind published "WaveNet: A generative model for raw audio" on ISCA. The Wavenet model is a sequential generative model originally used for speech generation modeling. Compared with traditional ARIMA, Prophet, LightGBM, or LSTM, Wavenet model has unique advantages in solving time series forecasting problems. It is a time series model based on convolutional neural networks, and its core is the extended Dilated Casual Convolutions, as shown in Fig. 11.23.

The Wavenet model can correctly handle the time sequence and can handle long-term dependencies, thus avoiding model explosion. The specific implementation code is as follows:

```
def build_model(shape_):

def wave_block(x, filters, kernel_size, n):
  dilation_rates = [2**i for i in range(n)]
```

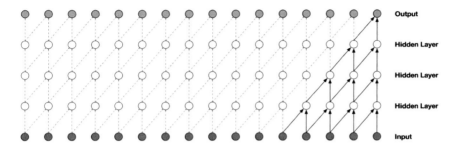

Fig. 11.23 Dilated casual convolutions

```
  x = Conv1D(filters = filters, kernel_size = 1, padding = 'same')(x)
  res_x = x
for dilation_rate in dilation_rates:
    tanh_out = Conv1D(filters = filters,
      kernel_size = kernel_size, padding = 'same',
      activation = 'tanh', dilation_rate = dilation_rate)(x)
    sigm_out = Conv1D(filters = filters,
      kernel_size = kernel_size, padding = 'same',
      activation = 'sigmoid', dilation_rate = dilation_rate)(x)
    x = Multiply()([tanh_out, sigm_out])
    x = Conv1D(filters = filters,
      kernel_size = 1, padding = 'same')(x)
    res_x = Add()([res_x, x])
return res_x

inp = Input(shape = (shape_))
x = wave_block(inp, 32, 3, 8)
x = wave_block(x, 64, 3, 4)
x = wave_block(x, 118, 3, 1)

out = Dense(1, name = 'out')(x)

model = models.Model(inputs = inp, outputs = out)

return model
```

The training part of the model is basically the same as before, but the format of the label needs to be adjusted. At the same time, in order to shorten the training time, the batch_size and epochs have also been adjusted. The code is as follows:

```
model.fit(X_train, y_train[:, i].reshape((y_train.shape[0], 1, 1)),
  batch_size = 4096, epochs = 5, verbose=1,
  sample_weight=np.array(train_weight),
  validation_data=(X_val, y_val[:, i].reshape((y_val.shape[0], 1, 1))),
  callbacks=callbacks, shuffle=True)
```

Compared with the previous LSTM model, if you do not use GPU for training, it will take a lot of time. However, the final result is also good. If you want to get better scores, you have to continue to optimize.

11.4.4   Model Integration

This part can use a simple weighted average, which is also the fusion method used by the vast majority of the top-ranked teams in this competition, which is not only effective, but also simple and intuitive. Specifically, the optimal score of the LightGBM model and the optimal score of the LSTM model are fused, and the weight is roughly determined according to the online score, that is, the final result is $0.7 \times$ the optimal score of the LightGBM model $+0.3 \times$ the optimal score of the LSTM model. Finally, the public score is 0.51089 (119/1624) and the private score is 0.51456 (6/1624), which shows that the score of the model after integration has been significantly improved. If the results of the Wavenet model are added and the characterizing portions of the three models are refined, the score will be improved even more greatly.

11.5   A Summary of the Competition Question

11.5.1   More Schemes

11.5.1.1   Top 1 Scheme

In terms of models, the champion team used LightGBM and NN models, and they had constructed many of them. These models had differences in features or sample selection. In terms of data, the champion team only used data for 2017 to extract features and build samples. Specifically, the training set used data from May 31, 2017 to July 19, 2017 or that from June 14, 2017 to July 19, 2017 (different models used different data sets), and the verification set used data from July 26, 2017 to August 10, 2017.

The characteristics of the structure of champion team were generally divided into basic features and statistical features, of which the basic features included category features, promotion features, and cycle-related features. Statistical features were taken as the main scoring features, and some methods (mean, maximum, standard deviation, difference, etc.) were used to count some target values (sales volume, promotion, etc.) of different keys (item_nbr, store_nbr, store_nbr_class, etc.) in different time windows.

11.5.1.2   Top 2 Scheme

The runner-up team used the Wavenet model and introduced the division of the training set and verification set. The randomly sampled sequence with a length of 128 was fed into the model in small batches, and then the start of the target date was randomly selected. As a result, it could be said that the model would see different data in each training iteration. Because the total data set was about 170,000 (seq) × 365 days, we believed that the Wavenet model trained in this way could handle the overfitting problem well. The verification method they adopted was step by step and kept the verification data of the last 16 days.

The runner-up team also found that on July 1, 2015, September 1, 2015, October 1, 2015, and November 27, 2016, many items began to have unit_sales records, but the difficulty was not knowing when most new items in the test data began to have this record. By examining the promotional information, it was found that the promotional information with new items was only in the data on August 30 and August 31. By looking at the data on October 1, 2015 or November 27, 2016, those "old new items" would be displayed on the day of the promotion (not before). Therefore, the runner-up team believed that most new items would have unit_sales records starting on August 30 and August 31. Before that, only some new commodities might have been sold. In this regard, models were not used to predict to prevent overfitting, but only some regular values were used to verify the loss calculation for those "old new commodities".

11.5.1.3 Top 3 Scheme

The third-place team built three models, namely LightGBM, CNN, and GRU models, which had almost equal weights when the models were merged. If you only look at a single model, the GRU model has a better prediction effect than other models.

In the above scheme sharing, the third-place team highlighted that in the time series forecasting problem, two important things are validation and bagging. Through correct verification, that was, to simulate the split of the training set and testing set, future information disclosure problems caused by data breakdown would be avoided.

In view of this, in the specific split method, only the history of 80 corresponding sales days for each "item and store pair" is recorded for the training set, and the next 16 sales days are used as the verification set. The division of the training set and the verification set is always from Tuesday to Wednesday, which is guaranteed to be the same as the division of the original testing set and is designed to capture weekly dynamics.

In the above partitioning mode, the model is trained and the optimal number of iterations of the model is estimated, and then the model is retrained by splicing verification sets to use the latest information.

Regarding bagging, the third-place team will train each model 10 times, each time initializing different weights, so the results are different. Averaging them can help improve the solution, especially when dealing with an uncertain future. Another bagging method is to predict the target after each training period (including the initial period), and then average these prediction results, which will also greatly improve the final result.

11.5.2   Knowledge Points

Sorting out knowledge points is also an important task after the competition, which can be roughly divided into sorting out key schemes and sorting out core codes. The core of this competition lies in feature engineering. If this part of the effort is done enough, then the results will be very good; the part of sorting out codes is mainly to optimize the code, improve its readability and modularization, so as to facilitate the reuse of it in the competition afterwards. This section mainly summarizes the feature extraction methods in feature engineering and tries to show the best and complete time series feature extraction ideas.

11.5.2.1   Time Characteristic

Year, quarter, month, week, day, hour, etc. are the basic time characteristics. Of course, the day can also be divided into morning, noon, afternoon, evening, late night, and before dawn.

There is also a class of time characteristics, which is the record of a certain time span, such as a certain interval of time, a few days from a certain day, the time difference between performing a certain action the last N times and the next N times.

11.5.2.2   Time Series Features

Time series features can be divided into historical translation and window statistics, which are also the core part of time series forecasting.

Historical translation requires only simple translation, such as taking the sales volume on the 1st, 2nd, 3rd, 7th, and 14th days in history as the characteristics of the day.

Window statistics is first to determine the window size, and then aggregate statistics; the specific statistical methods are mean, median, extreme values, percentiles, deviation, skewness, and kurtosis, etc. First-order difference or second-order difference can also be performed within the time window, and then the difference values will be aggregated.

11.5.2.3   Cross Features

Cross features are generally divided into three categories, namely, combinations of category features and category features, combinations of category features and continuous features, and combinations of continuous features and continuous features. The combination of category features and category features is equivalent to Cartesian product, such as combining days and hours to obtain a certain hour of a specific day; the combination of category features and continuous features is generally a polymerization operation; the combination of continuous features and continuous features includes those of year-over-year and month-on-month, first-order difference, and second-order difference, etc.

11.5.2.4   Advanced Features

Such features are generally obtained by traditional time series models, such as AR, ARMA, ARIMA, Prophet, etc. Such models only fit the prediction results based on historical target variables, and the prediction results can be regarded as high-level features combined with the final feature set.

11.5.3   Extended Learning

This section will recommend some competitions related to commodity sales as an extension of the learning content, in order to deepen the understanding of this type of competition questions, and to understand and find out which pits need to be paid attention to and which problem-solving routines need to be mastered in different commodity sales problems.

11.5.3.1   Kaggle's M5 Forecasting: Accuracy—Estimate the Unit Sales of Walmart Retail Goods

This competition (shown in Fig. 11.24) is to predict the unit sales of Walmart retail goods. The competition provides hierarchical information on the data (state, store, dept., and item) and from January 29, 2011 to June 19, 2016. The goal is to predict the sales volume of different commodities in the next 28 days.

Basic ideas: The modeling scheme of the contest question is basically divided into two types: recursive and non-recursive. The first type is to cycle and predict the sales volume of each day in the next 28 days. The predicted sales volume (day $t + 1$) is also merged into the training set to continue to predict the next sales volume (day $t + 2$), or the predicted sales volume can be used as a feature; the second type

Fig. 11.24 Home page of m5 forecasting—accuracy—estimate the unit sales of walmart retail goods

Fig. 11.25 Home page of predict future sales competition

will not expand the training set. When predicting the sales volume on day $t + 1$, only the sales volume from day of start to day t is used as training data for training and forecasting.

11.5.3.2 Predict Future Sales on Kaggle Platform

The data of the Predict Future Sales competition (shown in Fig. 11.25) is a time series data set composed of daily sales data, which is provided by a Russian company. It includes stores, goods, prices, and daily sales for 34 consecutive months. It is required to predict the sales volume of each product in each store in the 35th month. The evaluation index is the root mean square error.

Basic ideas: This competition is entitled the conventional commodity sales prediction problem, involving a total of 60 stores which are located in 31 cities. The main work focuses on data exploration and feature engineering. When constructing features, it is necessary to consider the feature combinations of different fine grains, which is very similar to the competition questions mentioned in this chapter.

11.5.3.3 IJCAI-17 Customer Flow Forecast of WOM Merchants

With the popularity of mobile positioning services, Alibaba and Ant Financial Services Group have gradually accumulated a large amount of online and offline transaction data from users and merchants. The O2O platform "Word of Mouth" of Ant Financial uses these data to provide merchants with back-end business intelligence services including transaction statistics, sales analysis, and sales recommendations. For example, Word of Mouth is committed to providing sales forecasts for each merchant.

Based on the forecast results, merchants can optimize operations, reduce costs, and improve the user experience. In this competition, the data provided is payment data of Ant Financial, which specifically provides the user's browsing and payment history, as well as relevant information of the merchant. By giving the past daily traffic counting of the store (brick and mortar store), daily traffic flow of the store in the following 14 days is predicted.

Basic ideas: If this competition question is used today, it will be a more conventional question. However, in 2017, it would be a very novel competition question. There were not many similar schemes for reference. The top players also gave a variety of ideas for solving the problem. The champion used a weighted integration of the results of the time series weight model and the tree model, in which the time series weight model was a rule method that took into account the combination of constant factors, time attenuation factors, week factors, and weather factors. Even in today it is a very detailed rule-based method; the tree model uses XGBoost and random forest. It is particularly important to note that the contestants perform a 1.1 time amplification for the prediction results customer traffic counting of the Black Friday.

Part IV
Precise Delivery, Optimized Experience

Chapter 12
Computational Advertising

In the early days of China's Internet boom, there were mysterious BAT giants famous in all corners of the country, namely Baidu (B), Alibaba (A), and Tencent (T), which occupied the top of search engines, e-commerce, and social software. Recently, TDJ (TikTok, DiDi, JD.com) appeared. In the 1980s and 1990s, what was happening in the entertainment circle of Hong Kong (China) dominated the entertainment industry in mainland China, and all kinds of kings, queens, and superstars surged, all thanks to media such as television, radio, and posters. From 2010 onwards, with the emergence of 4G mobile Internet, smart end points, and the development of the film and television industry, the way people chase stars has also changed, and it is not known since when famous actors and singers were called stars, and were divided into A-list, B-list ... D-list. With this rapid change, traffic flow has gradually become a standard to measure the popularity of stars, and stars with massive traffic flow are different from traditional stars and have formed a new group. Top traffic is the praise of A+ list stars. The reason why traffic is so valued is that it can be monetized and has great potential value.

12.1 What Is Computational Advertising

Imagine, on a morning when you are preparing to go to work, you will be surrounded by ads. After entering the elevator, you will be flanked by posters advertising programming education and the New Year's Shopping Festival; when you are about to ride a shared bicycle, you can find the sign of a certain App on the bike; the big screens flanking the subway corridor all show the flagship device just launched by a famous brand. In today's highly developed commercial civilization, there will be advertisements where there are people moving, because the existence of people indicates there is traffic flow. No matter whether there is a certain difference between the type of advertisement and the advertising material, they can both monetize the traffic flow. Even though there are more and more advertising patterns under the market economy, its essence is still based on traffic. For example, businesses ask for celebrity endorsement with the help of the traffic of stars, who themselves can achieve traffic monetization by charging advertising fees; the content and form of ads is also a kind of traffic, which is based on the public consensus of the target consumer or popular culture. In addition, the channel of advertising will affect the size of the traffic, which in turn affects making money from traffic. Mobile Internet has spawned many new nouns, of which a pair of nouns different from traditional sales methods are offline and online. Most of the contents introduced in this chapter are applicable to online scenarios. Advertising is realized by means of traffic, while computational advertising is to allow advertising to get access to more traffic.

Computational Advertising refers to the use of big data analysis and modeling, so that advertising can cover a wide range of areas and customers will be exposed to accurate, multi-span ads, enabling the same advertisement to reach as much effective traffic and more people interested in the corresponding advertising as possible. In this way, with the same cost, the advertising effect can be as good as possible; therefore, products and services can achieve more commercial success. With the gradual development of big data, artificial intelligence, and the Internet of Things in human society, the significance and possibility of computational advertising will become increasingly higher.

12.1.1 Main Issues

The purpose of computational advertising is to find as many traffic channels and target consumers as possible on the basis of controlling the cost within a certain range, so as to carry out commercial monetizing. The key is to use big data and artificial intelligence for targeted delivery, which involves computational advertising's three-factors (as shown in Fig. 12.1), that is, the interaction between

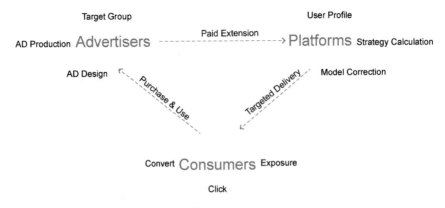

Fig. 12.1 Three factors for computational advertising

advertisers, platforms, and consumers. For advertisers, they want to rely on the platform to promote their products, by investing a certain cost to promote the product, so as to increase sales and turnovers; the platform charges advertisers promotion fees, in order to better build their platforms, serving consumers more excellently and effectively; consumers can filter out the types of advertising they are not interested in while enjoying other free services provided by the platforms. In fact, in essence, it is the consumers who feed the advertisers and the platforms.

Therefore, the main problem that computational advertising needs to solve is how to coordinate the interests of the three parties, that is, the interests of advertisers, platforms, and consumers. In response to this problem, many core technologies have emerged, such as bidding strategies between advertisers and advertisers, and those between advertisers and platforms, the technology to predict user click-through rates or conversion rates enabling the platform to deliver appropriate advertisements to users, and the technology to optimize advertising scheduling and control budgets to support operations related to advertisers and platforms.

12.1.2   Architecture of Computational Advertising System

Although there are great differences in details between computational advertising systems of different companies or different businesses, there are still general parts in terms of the architecture of these computational advertising systems. Here, three major parts are mainly introduced, namely, online delivery engines, distributed computing platforms (offline), and stream computing platforms (online), as shown in Fig. 12.2. The online delivery engine carries out advertisement retrieval, advertisement sorting, and revenue management according to the relevant information such as users and contexts corresponding to the advertisement request of the Web server, and finally transfers the relevant records to the distributed computing platform and stream computing platform; the distributed computing platform periodically processes the data in the past period of time in a batch processing manner to obtain offline user tags and CTR models and features, and then stores these in the database for use when making online delivery decisions; the stream computing platform is responsible for dealing with the data in a short period of time, obtaining real-time user tags and model parameters, and also storing these in a database for use in making online delivery decisions.

12.1.2.1   Online Delivery Engines

- Advertisement retrieval: when the Web server sends an advertisement request, the system searches for qualified advertisements from the advertisement index list according to the page label or user label of the advertisement position. The advertisement retrieval stage mainly uses the recall rate as the evaluation index.

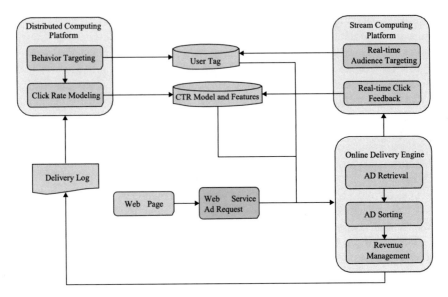

Fig. 12.2 Architecture of computational advertising system

A high recall rate means that the possibility of missing advertisements that may be clicked by users can be avoided.

- Advertisement sorting: when multiple advertisers snatch an advertising space, it is necessary to estimate the benefits that may be generated by placing each advertisement, i.e. calculating the eCPM value, and then sorting the advertisers from high to low according to the value.

12.1.2.2   Distributed Computing Platforms

- Behavior targeting: this module is used to find out the user behavior attributes in the advertisement delivery log, endow users with various tags and store them in a structured label library for subsequent advertisement delivery.
- Click-through rate modeling: the function of this module is to train and obtain the parameters and corresponding features of the click-through rate model on a distributed computing platform, and then load them into the cache to assist the advertisement delivery system in making decisions.

12.1.2.3   Stream Computing Platforms

- Real-time audience targeting: the function of this module is to process the user behavior and advertisement delivery logs that have occurred in a short period of time into real-time user tags in time to assist the advertisement retrieval module.

For the online computational advertising system, this part is of greater significance for improving the effect.

- Real-time click feedback: this module is also real-time feedback of changes in user behavior and ad placement logs, mainly generating real-time click-through rate-related features to assist the ad sorting module. In many cases, capturing short-term behavior records can better reflect the user's preference information, and the effect of ad placement is more significant.

12.2 Advertising Types

In order to maximize benefits and continuously meet a wide variety of needs, advertising types are constantly updated and iterated, and their development and evolution process is shown in Fig. 12.3. This section will introduce advertisement types according to the development of advertising business models, including CPT advertising, targeted advertising, bidding ads, and programmatic trade advertising.

12.2.1 Contract Advertising

CPT advertising and targeted advertising can be collectively referred to as contract advertising, and contract advertising can be specifically divided into advertisement with non-targeted contract transaction and coarse-grained targeted contract transaction. CPT advertising is billed at a time cost, and advertisers buy out advertising spots within a period at a fixed price to display their own advertising, such as splash ads, rich media advertising, or drop-down keywords in app stores; as for the targeted advertising, advertisers select the interest labels they want to deliver, and then the algorithm matches them with the corresponding audience and then advertises.

12.2.2 Bidding Advertising

After the generation of targeted advertising, the market develops in the direction of refinement; more and more advertisers participate, and targeted labels are becoming increasingly accurate. To improve revenue, media owners have introduced the

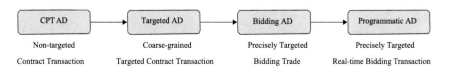

Fig. 12.3 Evolution of advertising types

bidding ads mode. In this mode, media owners no longer promise the amount of display to advertisers in the form of contracts but adopt the scheme of "the highest price gets" to decide which advertisement to display each time, so that media owners can compare prices on different advertisements in real time, to maximize revenue. This model also gives birth to advertising products such as ADN (advertising network) and ATD (advertising transaction terminal).

12.2.3   Programmatic Trade Advertising

The further development of bidding ads gave birth to the mode of real time bidding (RTB), which enabled advertisers to select their target audience in each advertisement display in real time and participate in bidding. Later, a series of ad exchanges with RTB as the core gradually evolve into a model that relies on programs to complete ad exchange decisions between machines. Therefore, this kind of advertising is called programmatic trade advertising, and the related advertising products that have been spawned include DSP, SSP, ADX, DMP, etc.

Computational advertising technology is the key to supporting the advertising application business. It can mainly coordinate the interest relationship between advertisers, platforms, and consumers. There are three core technologies in computational advertising, namely, advertisement recall, advertisement ranking and advertisement bidding. These technologies can not only ensure that advertisements are placed in the right crowd, but also ensure the interests of advertisers and platforms.

12.3   Advertising Recall

Advertising recall is advertising retrieval. The main work in this stage is to retrieve backup advertisements that meet the delivery conditions from the advertising index (Ad index) according to user or commodity attributes and page context attributes. The recall (retrieval) methods that will be used are also varied. Next, let's look at the specific recall methods.

12.3.1   Advertising Recall Modules

The modules of advertisement recall are divided into the following three parts.

- **Boolean expression recall**: Boolean expressions are combined according to the targeted labels set by advertisers. In the huge, targeted label system, advertisers recall the targeted audience of advertisements according to the Boolean expressions composed of users' interests, age, and gender.

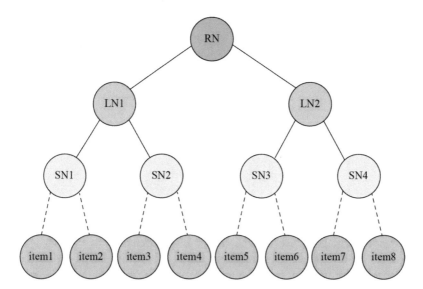

Fig. 12.4 TDM depth tree structure

- **Vector retrieval recall**: this technology can be divided into three types. The first is to obtain the vector representation of the advertisement through traditional Word2Vec, Item2Vec, or Node2Vec, and then recall the targeted group through similarity calculation. Its characteristics are simple to implement and are strong in expression. The second is to obtain the vector representation of the advertisement through the deep learning model. For example, YouTube DNN uses the deep learning model to map the advertisement, user, and other information to a vector, and then to recall the targeted group through the vector's nearest neighbor retrieval algorithm. There is also a classic DSSM double tower model (it will be described in detail in sect. 12.3.2).
- **Recall based on TDM (Tree-based Deep Match model)**: this is a large-scale (above 10 million) recommendation system algorithm framework based on deep learning, independently developed by the Precisely Targeted Advertising Algorithm team in Alibaba. This technology combines deep learning model and tree structure search to solve the balance between high-performance requirements in the recall problem and the use of complex models for content search. This technique can transform the recall problem into a process of classification and screening layer by layer. With the help of the hierarchical retrieval nature of the tree, the time complexity can be reduced to the logarithmic level. If the target recommendation number is K, the total number of goods is N, then the time complexity is $O(K \, logN)$.

As shown in Fig. 12.4, each leaf node of the depth tree corresponds to an item in the data, while non-leaf nodes represent a collection of items. Such a hierarchical structure intuitively reflects the item architecture from coarse to fine granularity.

At this time, the recommendation task is converted into how to retrieve a series of leaf nodes from the depth tree and return these leaf nodes as the items that the user is most interested in. It is worth mentioning that although the tree shown in Fig. 12.4 is a binary tree, there is no such limitation in practical applications.

Of course, in addition to the above three, there are many recall methods, such as classic system filtering, recall based on graph calculation, recall based on Knowledge Graph, etc. In addition, the current recall strategy is mostly a combination of multi-channel recall and weight retrieval, and in actual business, it is often a combination of recall methods of more than ten channels.

12.3.2 DSSM Semantic Recall

This section will introduce a semantic modeling method based on deep neural networks—DSSM (Deep Structured Semantic Model), which is proposed by a paper published by Microsoft on the similarity calculation model of Query and Doc. In the advertisement recall problem, this multi-tower structure constructs different towers for user-side features and advertisement-side features respectively. After multiple layers of full connection, the embedding vectors of the last output layer are spliced together and then input to the softmax function. Moreover, the output vector is in the same vector space, which can directly calculate the similarity between Query and Doc through point multiplication or cosine function and perform advertisement retrieval. Next, let's look at the network structure of DSSM, as shown in Fig. 12.5.

1. First, the input layer converts the Query (or User) vector and the Doc (or Item) vector (one-hot encoding) into embedding vectors. The original paper proposes a special embedding method called word hashing for English input to reduce the dictionary scale. When embedding for Chinese, we can use the Word2Vec class for routine operations.
2. Next is the presentation layer. After embedding, the word vector is mapped through multiple layers of full connection to obtain the semantic feature vector representation for Query and Doc.
3. Finally, the matching layer calculates the cosine similarity of the Query vector and the Doc vector to obtain the similarity, and then performs softmax normalization to obtain the final index posterior probability P. The training target fits the positive sample of the click to P as 1, and vice versa fits P as 0.

This method is also widely used in recall and ranking problems in search, recommendation, and other fields. The biggest feature of the two-tower model is that the user side and the advertising side are two independent sub-networks, the two towers can be cached separately, and only the vectors in the cache need to be taken out for similarity calculation during online recall. Retrieving and matching vectors is a very time-consuming job, and the retrieval efficiency can be improved through the nearest

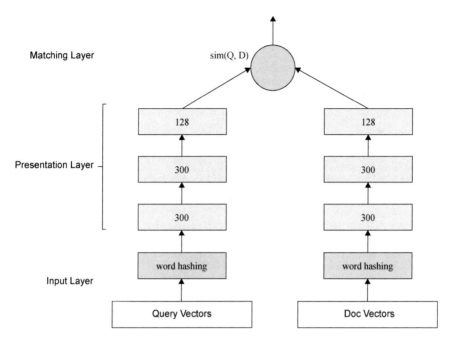

Fig. 12.5 Network structure of DSSM

search method. Using common python packages such as Annoy and Faiss can easily deal with such scenarios.

12.4 Advertising Sorting

Advertising sorting is the fundamental part of computational advertising. Its main function is to calculate the eCPM (effective cost per mile) of the advertisement backup set sent by the advertisement recall module, and sort it backwards according to the size of the obtained value. The calculation of eCPM depends on the click-through rate calculated offline by the audience targeting platform. Since the final ads are all from the results of sorting, this module is very important, and it is also a place for various algorithm models and strategies to play their role.

12.4.1 Click-Through Rate Prediction

Click-through rate (CTR) prediction is one of the most important algorithm modules to help advertising. At the same time, it is also crucial in industrial applications such as information retrieval, recommendation systems, and online advertising systems. To a certain extent, click-through rate represents the user's experience. For example, in the recommendation system of the e-commerce platform, a major sorting target GMV (gross merchandise value) can be disassembled into traffic × click rate × conversion rate × the guest unit price, which shows that click rate is an important factor to optimize the sorting target.

The click-through rate prediction task can be abstracted into a binary classification problem, that is, to put an advertisement to the user, and then predict the probability of the user clicking on the advertisement. When an advertisement is displayed in front of the consumer, the proportion of click behavior generated by the link of the advertisement can be reflected in two indicators according to the different delivery methods. One is click rate:

$$\text{Click} - \text{through rate} = \text{number of clicks (times)}/\text{number of exposures (times)}$$

Obviously, the higher the click rate, the better the effect of ad placement. The other is the conversion rate (CVR), which is a further extension of the click rate. It represents whether consumers have further completed the corresponding conversion behavior on the basis of clicking on the advertisement link; that is, the conversion rate is the proportion of customers who pay or sign up to customers who click ads, whose definition is similar to the click rate:

$$\text{Conversion rate} = \text{number of conversions (times)}/\text{number of exposures (times)}$$

Similarly, the higher the conversion rate, the better the effect of ad placement.

Problems related to click-through rate also often appear in competitions. For example, the IJCAI 2018 Alimama (A separate marketing platform that serves Alibaba) Advertising Prediction Algorithm Competition, Tencent Advertising Algorithm Competition, and the iFLYTEK AI Marketing Algorithm Competition are all competitions around click-through rate, and such competitions often face a large number of discrete features and feature combinations. The following will give common solutions to these problems and introduce common models.

12.4.2   Feature Processing

Feature engineering has always received much attention in the competition, which of course also includes the feature engineering of the click-through rate prediction problem. The data in the advertising business is not only rich, but also has very high dimensions, which requires extremely high accuracy. In addition to the model, there are also many skills to learn in feature processing.

12.4.2.1   Feature Crossing Combination

There are a lot of category features in the click-through rate problem, such as user tags, advertising tags, etc., so extracting fine-grained feature expression becomes the key, such as the combination of user occupation and advertising type: programmers _ anti alopecia advertising. Of course, you can also combine the three category features to construct more fine-grained features.

12.4.2.2   Processing of Continuous Features

Continuous features have practical statistical significance, such as the number of user behaviors, ad impressions, etc., though these features can be directly fed into the model for training. However, the importance of features in different intervals may be different. Continuous features default to the importance of features. There is a linear relationship between the degree and feature values, but in practice there is often a nonlinear relationship between the two; that is, the importance of feature values in different intervals is not the same.

Here a neural networks model will be introduced—key-value memory, which is used to realize the mapping from floating-point numbers to vectors. As shown in Fig. 12.6, the input of this model is a dense feature q, and the output is a feature vector v, realizing the feature space conversion from one-dimension to multi-dimensions.

The specific steps in Fig. 12.6 are as follows.

Key addressing section: addressing procedures, where the softmax function is used, as shown in formula (12.1). Calculate the probability value of each memory selected above.

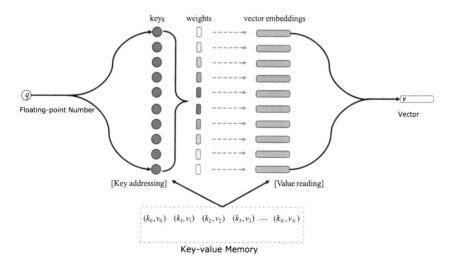

Fig. 12.6 Model structure of key-value memory

$$w_i = \text{softmax}\left(\frac{1}{|q - k_i + e^{-15}|}\right) \tag{12.1}$$

Attention Section: Different parts of key addressment have different feature importance, so Attention is used to give different weight probabilities.

Value-reading Section: Perform the weighted sum under the weight of the previous step to obtain the answer information.

$$\left(v = \sum_{i=1}^{N} w_i v_i\right)$$

12.4.2.3   Smooth Click Rate

When constructing features related with the click-through rate, the calculation is often biased due to sparse data. For example, if an advertisement is delivered 100-times and there are 2 clicks, the click-through rate is 2%. However, when the advertisement is put in 1000-times, and the click-through rate is only 10-times, then the click-through rate is 1% at this time, which is only a half of the former. Therefore, the calculation result will be corrected by smoothing, i.e. adding a relatively large constant to the numerator and denominator, which can alleviate low exposure data, highlight high exposure data, such as popular advertisements and commodities, and fill in cold start samples.

Bayesian smoothing is often used for processing. The basic idea is to select a prior distribution with a smooth distribution, and then use the prior distribution to find the final smooth distribution in some way, as shown in formula (12.2).

$$\text{SmoothCTR} = \frac{C + \alpha}{I + \alpha + \beta} \tag{12.2}$$

Here, C is the number of clicks, I is the exposure, α and β are calculated by Bayesian smoothing.

12.4.2.4   Vectorized Representation

Many traditional feature extraction methods have limited characterization ability; therefore, we will try to use some embedded representation methods (such as Word2Vec, DeepWalk, etc.) or through deep learning models to learn embedded vector representation, as shown in Fig. 12.7. For example, extract the user history click advertisement sequence, combine all the user sequences into a text input to Word2Vec for advertisement vector training, and finally a vectorized representation of the advertisement will be obtained. Of course, you can also get the vector representation of the user by exposing the user sequence through the advertisement.

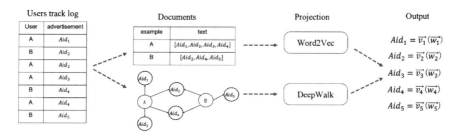

Fig. 12.7 Advertising embedding vector extraction process

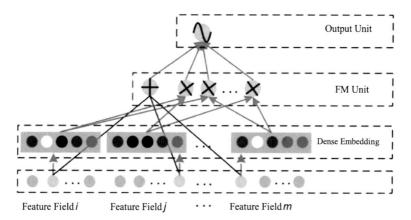

Fig. 12.8 FM model structure

12.4.3   Common Models

Click-through rate prediction is an extremely important part in the field of recommendation systems and Computational Advertising. The prediction effect will directly affect the user experience and advertising revenue. In order to continuously improve performance, the update iteration of related models is also very fast. Because this field has the characteristics of large amount of data and highly sparse features, most of the model improvements are optimized around these characteristics. Here we will introduce FM, Wide & Deep, DeepFM, and DIN, the four quite classic models that have evolved from different directions.

12.4.3.1   FM: Factorization Machines (2010) Implicit Vector Learning Improves Model Expression

FM (Factorization Machine) can be said to be a very classic algorithm in the field of recommendation systems and computational advertising. It is improved on the basis of the LR model. Its model structure is shown in Fig. 12.8. The LR model is an early model used for advertising click-through rate or conversion rate problems. The traditional LR model cannot learn the cross information between features, but it can only rely on a large amount of physical effort to construct features. Faced with this difficulty, FM came into being. It can not only directly introduce second-order feature combinations but can also calculate the weight of feature combinations.

FM solves the above problems by learning the interaction of pairs of features in the potential feature space. In the feature space, each feature has a hidden vector associated with it, and the interaction between the two features is the inner product of their respective hidden vectors. In FM, the model can be expressed as formula (12.3):

$$\varphi(w, x) = w_0 + \sum_{i=1}^{n} w_i x_i + \sum_{i=1}^{n} \sum_{j=i+1}^{n} \langle V_i, V_j \rangle x_i x_j \qquad (12.3)$$

Wherein,

$$\langle V_i, V_j \rangle = \sum_{f=1}^{k} v_{i,f} \cdot v_{j,f}$$

Since FM will learn all cross-combination features, which will definitely contain many useless combinations. These combinations will introduce noise and thus reduce the performance of the model. Therefore, in general, we cannot directly put all category features into the model. Certain feature selection should be carried out first to reduce the risk of introducing noise.

In addition, FM cannot learn the information of the feature field (Field) when performing feature combination, which means it cannot perceive the existence of the feature field, that is, the feature combination uses the same hidden vector. Obviously, this will appear very rough, and features belonging to the same feature field and different feature fields should use different hidden vectors when combining. In order to improve this, FFM (Field-aware Factorization Machine) was born.

The idea of FFM is to divide the original potential space into many smaller potential spaces and use one of them according to the feature domain. For example, in terms of "male" and "basketball", "male" and "cosmetics", the potential role of the combination of these two features is different, and it is very necessary to introduce the concept of feature field. In FFM, the model can be expressed as formula (12.4).

$$\varphi(w, x) = w_0 + \sum_{i=1}^{n} w_i x_i + \sum_{i=1}^{n} \sum_{j=i+1}^{n} \langle V_{i,f_2}, V_{j,f_1} \rangle x_i x_j \qquad (12.4)$$

Here, f_j is the feature field to which the feature j belongs. If the length of the hidden vector is k, then there are $n \times f \times k$ quadratic parameters of FFM, which is much more than $n \times k$ of the FM model. In addition, because the hidden vector is related to the feature field, it cannot simplify the quadratic term in the FFM expression, and its prediction complexity is $O(kn^2)$.

12.4.3.2   Wide & Deep: Wide and Deep Learning (2016)—Memory and Generalization Information Complementation

The core idea of the Wide & Deep model is to combine the memory ability of linear models and the generalization ability of deep neural networks models to improve the performance of the overall model. The feature combination that can learn high-frequency co-occurrence from historical data is the model's memory ability, and the model's generalization ability represents the model's ability to use correlation and transitivity to explore feature combinations that have never appeared in historical data. The Wide & Deep model combines memory ability and generalization ability and has successfully landed in the scene of Google Play store, becoming a classic model.

As shown in Fig. 12.9, the Wide & Deep model structure consists of Wide and Deep. The Wide part is mainly a generalized linear model (such as LR). The linear model usually inputs one-hot sparse representation features or continuous features for training. The Wide model can efficiently realize memory ability through cross features to achieve the purpose of accurate recommendation; the Deep part can be simply understood as a common structure where embedding vectors combine with multi-layer perceptron (MLP), and the generalization ability of the model can be realized through the learned low-dimensional dense embedding vectors. Even for commodities that have not appeared in history, you can get good recommendations.

The training of the Wide & Deep model adopts joint training, and its training error will be fed back to the linear model and the deep neural networks model at the same time for parameter update, so the weight update of a single model will be affected by the influence of both the Wide part and the Deep part on the training error of the model. The mathematical representation of the Wide & Deep model is as formula (12.5):

$$Y_{\text{wide\&deep}} = \text{sigmoid}\left(w_{\text{wide}}^{\text{T}} \cdot [x, \varphi(x)] + w_{\text{deep}}^{\text{T}} \cdot a^{(l)} + b\right) \qquad (12.5)$$

In the formula above, b represents bias; $a^{(l)}$ represents the last layer of output of the Deep model; x represents the original input feature. Please note that there is also a

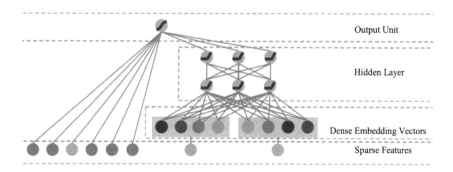

Fig. 12.9 Wide & deep model structure

$\phi(x)$, which is the feature intersection of the original feature. The outputs of the Wide and Deep parts are combined by weighting, and the final output is performed through the logistic loss function.

12.4.3.3   DeepFM: Deep Factorization Machines (2017)—Introducing Implicit High-Order Cross Information of Neural Networks Based on FM

The similarities between DeepFM and Wide & Deep are both considered from the Wide and Deep parts at the same time. The difference is to be able to better capture cross feature information. DeepFM uses FM as the model of the Wide part. Traditional linear models cannot extract high-order combined features, but the effect achieved by manually discovering feature cross combinations is very limited, so FM is used as the model of the Wide part to make full use of its ability to extract feature cross combinations.

As shown in Fig. 12.10, FM and Deep share input vectors and dense embedding vectors, which make training not only faster, but also more accurate. In contrast, in the Wide & Deep model, the input vector is very large, which contains a large number of artificially constructed and paired combination features, which will undoubtedly increase the computational complexity of the model.

The physical implication given by the DeepFM paper is that the FM part is responsible for the first-order features and second-order cross features, and the deep neural network part is responsible for the high-order cross features above the second order. Formula (12.6) gives the output formula of the FM model, and formula (12.7) gives the prediction result expression of the DeepFM model:

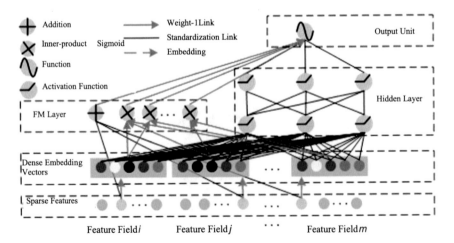

Fig. 12.10 DeepFM model structure

$$Y_{FM} = w_0 + \sum_{i=1}^{n} w_i x_i + \sum_{i=1}^{n} \sum_{j=i+1}^{n} \langle V_i, V_j \rangle x_i x_j \tag{12.6}$$

$$Y_{DeepFM} = \text{sigmoid}(Y_{FM} + Y_{DNN}) \tag{12.7}$$

12.4.3.4 DIN: Deep Interest Network (2018)—A Deep Learning Model Integrating the Attention Mechanism

DIN (Deep Interest Network) adaptively learns the representation of user interest from the historical behavior of a particular advertisement by designing a local activation unit, which greatly improves the expressiveness of the model.

The structure of the DIN model is shown in Fig. 12.11. One part is the basic model, and the other part is an improved model after adding Attention (attention mechanism). The basic model is an embedded vector combined with a multi-layer perceptron, which first converts different features into corresponding embedded vector representations, and then splices the embedding vectors of all features together, and finally inputs them into a multi-layer perceptron for calculation. In order to ensure the fixed-length input to the multi-layer perceptron, the basic model uses pooling, which is generally the vector sum and/or vector average, to perform sum and mean pooling operations on the embedding vectors of commodities in the user's historical behavior sequence.

However, there are great limitations. For each recommended product to be predicted, whether the product is clothes, cosmetics, electronic products, etc., the user's representation vector is determined to be unchanged, which will result in indiscriminate recommendation. We know that in the e-commerce scene, users' interests are various. With the migration of time or other situations, the change of users' interests and the existence of unrelated behaviors cannot help with the click rate prediction. Therefore, we should consider setting different weights, such as changing the weights according to the change of time, but this cannot completely solve the problem.

In order to solve the above-mentioned problem, Alimama's algorithm team proposed DIN. This model adjusts the processing of the user's click on the product sequence, and the other parts have not changed. The core idea is that in the pooling part, the weights of the products related to the recommendation are set to be larger, and the weights of the products not related to the recommendation are set to be smaller. This is an idea of Attention, which makes the products to be recommended interact with each product in the click sequence to calculate the attention mechanism score.

The key point of DIN is the design of the local activation unit. DIN will calculate the correlation weight of the product to be recommended and the product in the user's recent historical behavior sequence and use this weight as a weighting coefficient to carry out sum pooling for the embedded vectors of the product in these behavior sequences. The user's interest is represented by this weighted sum

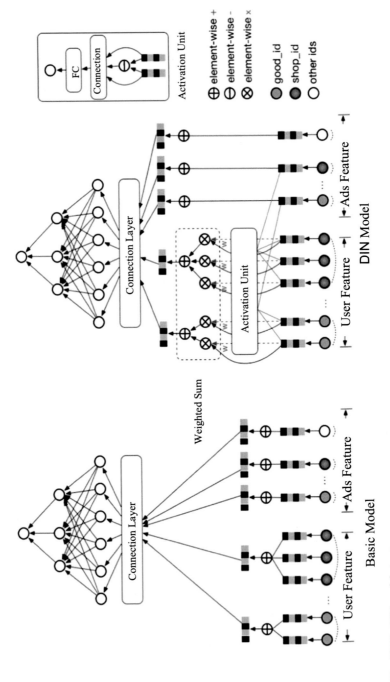

Fig. 12.11 DIN model structure

vector. The calculation formula for the representation of user interest is as formula (12.8):

$$V_u = \sum_{i=1}^{n} w_i \times V_i = \sum_{i=1}^{n} G(V_i, V_a) \times V_i \tag{12.8}$$

Here V_u is the user's embedding vector, V_a is the embedding vector of the product to be recommended, V_i is the embedding vector of the user's i-th behavior, and G (V_i, V_a) is Attention. You can see that the candidate embedding vector will be used in the user's behavior sequence. The embedding vector of each product calculates a weight, and the final weighted sum is the user's interest representation.

The seemingly perfect representation of user interest actually has some shortcomings. The user's interest is constantly evolving, while the user's interest extracted by DIN is independent and unrelated and does not capture the dynamic evolution of interest; in addition, the user's explicit behavior is used to express the user's implicit interest, and its accuracy cannot be guaranteed. In this regard, Alimama's algorithm team proposed DIEN (Deep Interest Evolution Network), which replaced the Attention and sum pooling parts of DIN with sequential model and attention. However, due to the complexity of the network structure, it will bring difficulties to engineering applications.

12.5 Advertising Bidding

There may be hundreds or even more companies that provide the same product and service; so, how can we determine the priority of them at this time? In response to this, the platform side has invented a bidding ranking with mixed reviews based on its own interests. Simply put, whoever pays more will be given priority to displaying their ads. This model of tying search and bidding has pros and cons. The simple logic is certainly efficient, but the effectiveness of this method still needs to be checked. In particular, the one-size-fits-all approach to all searches leads to higher risks in some specific industries. For example, a scandal of hospital ads published in a browser was revealed, which had an extremely bad impact on the browser. For advertisers, platforms, and consumers, consumers are relatively vulnerable groups. The strategy of search bidding ignores the quality and adaptability of advertisements. The search matching of platforms can only reach the granularity of keywords and cannot provide personalized services for thousands of people. Everyone is a unique individual. The adoption of group strategy will inevitably lead to poor consumer experience. At the same time, advertisers will push up costs in bidding and transfer them to consumers in disguised forms. Even if the platform can earn a lot of advertising expenses for a while, it is difficult to maintain it for a long time.

Usually in the advertising auction mechanism, the actual exposure of the advertisement depends on the size of the traffic coverage of the advertisement and the relative competitiveness level in the competitive ads. The former depends on the

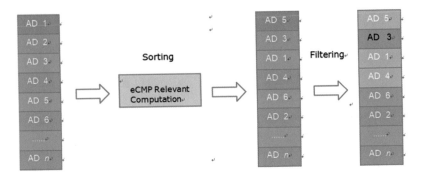

Fig. 12.12 Advertising auction process

crowd orientation of the advertisement (the number of users matching the corresponding characteristics), the size of the advertisement material (the advertising space matched), and the setting items such as delivery time and budget; the factors that affect the latter mainly include bidding, advertising quality (such as pCTR/pCVR, etc.), and control strategies for user experience. Generally speaking, the basic competitiveness can be expressed by $eCPM = 1000 \times cpc\_bid \times pCTR = 1000 \times cpa\_bid \times pCTR \times pCVR$ (cpc and cpa represent cost per conversion mode and cost per acquisition mode respectively). In summary, the size of advertising traffic coverage determines the number of times the advertisement can participate in the competition and the competition object, and the relative competitiveness level determines the probability of the advertisement winning in each competition, and the two jointly determine the daily exposure of the advertisement.

As shown in Fig. 12.12, the retrieved advertisements are sorted by eCMP computation, and then diversity filtering is performed to obtain the advertisements to be placed. Of course, eCMP only reflects the basic competitiveness, and will usually be combined with factors such as advertisement quality for final calculation.

Ad impression, as its name implies, is the number of advertisements exposed in front of consumers. It can be calculated either by the number of people or by person times, but it should be noted that the dimension of calculation should be consistent with the subsequent click-through rate and conversion rate. The relationship between ad impressions and the effect of ad placement is not absolutely proportional, as shown in Fig. 12.13. Generally, the effect of moderate exposure is the best, neither too little nor too much. In the 2019 Tencent Advertising Algorithm Contest, the task is to estimate the daily exposure of future ads, providing historical n-day ad impression data (sampling a specific traffic), including the traffic characteristics corresponding to each impression (user attributes, ad space, and other temporal information) as well as the settings and competitiveness scores of ads exposed. It is worth mentioning that Wang He, the co-author of this book, and his team won the championship of the contest. The competition will be analyzed and explained in depth in Chap. 13.

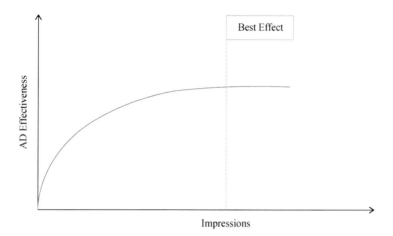

Fig. 12.13 Relationship between ad impressions and effect of ad placements

This chapter takes the actual computational advertising field as the background, and introduces the advertising system framework, advertising types, and advertising core technologies. From the perspective of competition, computational advertising can conduct in-depth study around the quantitative indicators of advertising, so as to clarify the costs and benefits of advertisers and platforms and force the model to be improved. Since this book focuses on the actual practical application related to machine learning competitions, it mainly considers the online placement and application of computational advertising, and there are some exclusive evaluation indicators in this field to evaluate the effect of placements. For a certain product and service, the process from the birth of the concept to the landing on the consumer is relatively long and involves a lot of consideration. The corresponding computational advertising is mainly divided into three stages. First, the advertisement should be displayed in front of the consumers, which is the so-called impressions, so that the consumers can see the advertisement; secondly, consumers who are interested in this advertisement will click on it and browse the advertisement content and its corresponding product and service; finally, consumers who have the desire to buy will make corresponding paid transactions or register for use. In order to facilitate advertisers and platforms to analyze and count the effect of advertising, the three stages are respectively provided with three indicators—exposure, click-through rate, and conversion rate. Although the definitions of the three are different, they are all rooted in the number of consumers who carry out the corresponding operations. Except for exposure, which consumers cannot choose on their own, click and conversion are both spontaneous behaviors of consumers.

12.6   Thinking Exercises

1. What are the significant similarities and differences between computational advertising and ordinary advertising?
2. Please briefly describe the advantages and disadvantages of computational advertising under the search bidding mode.
3. Please think about how the platform should negotiate with advertisers on the pricing strategy of advertising fees through various indicators of advertising.
4. Please carefully observe the advertising pages pushed to you by 10 apps you use most frequently. Can you find out what they have in common?
5. The combination of GBDT and LR is also a classic model in click-through rate prediction. How do they combine with each other then?
6. There is another important part of computational advertising that is not mentioned, that is, advertisement detection. This part is often accompanied by cheating and anti-cheating problems. What is the specific modeling method?

Chapter 13
Case Study: 2018 Tencent Advertising Algorithm Competition—Audience Lookalike Expansion

This chapter will take the second Tencent Advertising Algorithm Competition in 2018 (as shown in Fig. 13.1) as an example to analyze the practical cases related to computational advertising and will explain the complete process and precautions of the cases in detail. This chapter is mainly divided into six parts, namely, competition question understanding, data exploration, feature engineering, model training, model integration, and contest question summary. This is not only the organizational structure of all chapters of case examples in this book, but also an important process for a competition. I believe that under the guidance of this book, you can quickly become familiar with the competition process and apply what you have learnt into practice.

13.1 Understanding the Competition Question

As the saying goes, sharpening your axe will not delay your job of chopping wood. Before the competition, you should fully understand the information related to the competition questions, know the needs behind these questions, and then achieve the purpose of examining the questions correctly. This competition is based on Audience Lookalike Expansion of computational advertising problems. The contestants are lucky because the organizer of the contest have the largest social platform in China. Both the quality of the data they provide, and the professionalism of the competition are impeccable. The content of this section is also mostly from the official description of the competition questions given by Tencent.

Fig. 13.1   2018 tencent advertising algorithm competition

Seed Population   Lookalike Audience

Fig. 13.2   Audience lookalike expansion

13.1.1   Competition Background

Advertising based on social relationships (that is, social advertising) has become one of the fastest growing types of advertising in the Internet advertising industry. Tencent social ad platform is a commercial ad platform that relies on Tencent's rich social products, rooting in Tencent's massive social data, and using powerful data analytics, machine learning, and cloud computing capabilities to create a service for tens of millions of businesses and hundreds of millions of users. Tencent social ad platform has always been committed to providing accurate and efficient advertising solutions, and complex social scenarios, diverse advertising forms, and huge user data have brought considerable challenges to achieve this goal. In order to overcome these challenges, Tencent social advertising platform is also constantly trying to find better algorithms for data mining and machine learning.

The topic of this algorithm competition originates from a real advertising product in Tencent's social advertising business—Audience Lookalike Expansion (later referred to as Lookalike). The purpose of this product is to find other groups similar to the target group from a large number of people based on the target group provided by advertisers, in order to achieve the goal of audience expansion, as shown in Fig. 13.2.

In an actual advertising business application scenario, Lookalike can find potential consumers similar to these existing consumers from the target consumers based on the existing consumers of advertisers, so as to effectively help advertisers reach new customers and extend their business. At present, Lookalike is based on the first-party data provided by advertisers and the effect data of advertising (that is, the seed population mentioned later), combined with Tencent's rich data tags. Through the mining of deep neural networks, it has realized the function of expanding

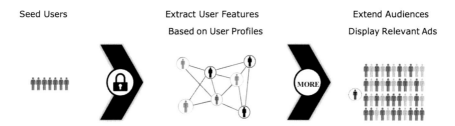

Seed Users   Extract User Features   Extend Audiences
Based on User Profiles   Display Relevant Ads

Fig. 13.3 How lookalike works

high-quality potential customers with similar characteristics for multiple advertisers simultaneously, online in real time. The working mechanism of Lookalike is shown in Fig. 13.3.

13.1.2 Competition Data

The question data (desensitized) extracted for this competition is that for certain 30 consecutive days. Usually, the data file can be divided into four parts: training set file, testing set file, user feature file, and advertisement feature file corresponding to the seed package. These four parts are introduced separately below.

- **Training set file train.csv**: Each row in this file represents a training sample, and the fields are separated by commas. The format is "aid, uid, label", where aid is used to uniquely identify an advertisement; uid is used to uniquely identify a user; label is the sample label with a value of +1 or −1; + 1 is the seed user, and −1 is the non-seed user. In order to simplify the question, a seed package corresponds to only one aid, and the two are one-to-one correspondence.
- **Testing set file test.csv**: Each line in this file represents a test sample, and the fields are separated by commas in the format "aid, uid". The meaning of two fields is the same as that of the training set file.
- **User feature file userFeature.data**: Each line in this file represents a user's feature data, with a vertical line "|" between each field fields; the format is "uid | features". Among them, features are composed of many feature_group, and each feature_group represents a feature group; the multiple feature groups are also separated by a vertical line "|" in the format "feature_group1 | feature_group2 | feature_group3 |...". If a feature group consists multiple values, it is separated by spaces as in the format "feature_group_name fea_name1 fea_name2 ...", and the fea_name in it adopts the format of data numbering.
- **The advertisement feature file adFeature.csv corresponding to the seed package**: the format of each line in this file is "aid, advertising serId, campaignId, creativeId, creativeSize, adCategoryId, productId, productType ". The first field aid is used to uniquely identify an advertisement, and the remaining fields are

advertisement features, separated by commas. For data security reasons, we encrypt uid, aid, user features, and advertisement features as follows.

- uid: Randomize the number of each uid from 1 to n to generate an unduplicated encrypted uid - n is the total number of users (Assuming that the number of users is one million, all users are randomly scattered and arranged, and the sequence number after the arrangement is used as the user's uid. The value range of the sequence number is [1, one million]).
- aid: Refer to uid's encryption method to generate encrypted aid.
- User features: refer to the encryption method of uid to generate encrypted fea_name.
- Ad features: refer to the encryption method of uid to generate encrypted fields.
  Next, the values of user features and advertising features will be explained.

- **Description of value selection of user features**

  - Age: segmentation representation; Each number represents an age group;
  - Gender: male, female;
  - Marital status: single, married, etc. (multiple states can coexist);
  - Education background: doctor, master, undergraduate, high school, junior high school, primary school;
  - Consumption ability: high and low;
  - Geographical location (LBS): Each number represents a geographical location;
  - Interest category: Mining different data sources to obtain 5 interest feature groups, which are represented by interest 1, interest 2, interest 3, interest 4, and interest 5. Each interest feature group contains several interest IDs;
  - Keywords: Mining different data sources to obtain three keyword feature groups which are represented by kw1, kw2 and kw3 respectively. Each keyword feature group contains several keywords that users are interested in which can be more than interest categories. Fine grain indicates user preferences;
  - Topics: Use LDA algorithm to mine user preference topics. Specifically, mine different data sources to obtain 3 topic feature groups, which are represented by topic1, topic2, and topic3;
  - Recent app installation behavior (appIdInstall): including apps installed in the last 63 days, where each app is represented by a unique ID;
  - appIdAction: The ID of the app with a high user engagement rate;
  - Internet connection type (ct): Wi-Fi, 2G, 3G, 4G;
  - Operating system (os): Android, iOS (version number is not distinguished);
  - Mobile telecommunication operator (carriers): mobile, Unicom, telecommunications, others;
  - Real estate (house): having real estate, not having real estate.

- **Description of value selection of advertisement features**
  - Advertisement ID (`aid`): Advertisement ID corresponds to specific advertisements. Advertising refers to the advertising creativity (or advertising materials) created by advertisers and the settings related to advertising display, including the basic information of the advertisement (advertising name, delivery time, etc.), promotion objectives, delivery platforms, advertising specifications, advertising creativity, advertising audience (that is, targeting settings of advertisements), advertising bids, and other information;
  - Advertiser ID: The account structure is divided into four levels: account, promotion plan, advertisement, and material. Advertisers and accounts are one-to-one correspondence;
  - Campaign ID: The promotion plan is a collection of advertisements (similar to the folder function of a computer). Advertisers can place advertisements with the same conditions such as the promotion platform, budget limit, and whether to deliver at a constant speed in the same promotion plan for management;
  - Material ID (creativeId): advertising content that is directly displayed to users. There can be multiple groups of materials under one advertisement;
  - Material size (creativeSize): The material size ID is used to identify the size of the advertising material;
  - Ad category (adCategoryId): Ad classification ID, using the ad classification system;
  - Product ID (productId): the ID of the product promoted;
  - Product type (productType): The product type corresponding to the advertising target (such as commodities corresponding to JD.com, corresponding download in Apps).

13.1.3 Competition Tasks

Lookalike automatically finds a similar audience, called the extended audience reach, among the candidate population provided by the advertiser through computation based on the seed population (also known as the seed package). This competition question will provide the contestants with hundreds of seed population, user characteristics corresponding to a large number of candidate audiences, and advertising characteristics corresponding to the seed population. For the sake of business data security, all data have been desensitized. The entire data sets are divided into training sets and testing sets. In the training set, users belonging to the seed group as well as those not belonging to the seed population are calibrated respectively (i.e. positive and negative samples) in the candidate group. The model prediction will detect whether the contestant's algorithm can accurately calibrate whether the users in the testing set belong to the corresponding seed package. The seed population corresponding to the training set and the testing set is exactly the same. What is shown in Fig. 13.4 is the distribution of the population groups.

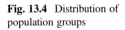

Fig. 13.4 Distribution of
population groups

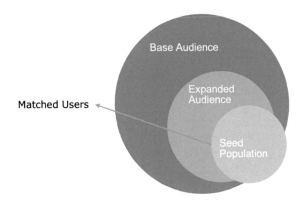

In order to test whether the contestant's algorithm can learn users and seed groups well, this competition requires participants to submit results that include the scores of candidate population for various seed groups belonging to the seed group (the higher the score, the more likely the candidate population is to be potential look-alike expansion users of this seed group). The seed groups provided by the preliminary and semi-final are the same except for the difference in order of magnitude.

13.1.4   Evaluation Indicators

If there is a relevant effect action (such as clicking or converting) after advertising to similar users extended is conducted, it is considered a positive example; if there is no effect behavior, it is considered negative. Each seed population to be evaluated will provide the following information: the advertisement ID (aid) corresponding to the seed population group, the ad characteristics, and the corresponding candidate group set (including the uid and user features of each candidate user). Contestants need to calculate the scores of users in the testing set for each seed group; the game will calculate the AUC index for each seed population group accordingly and use the average AUC value of all m seed groups to be evaluated as the final evaluation index; the formula is as in (13.1):

$$\frac{1}{m} \sum_{i=1}^{m} \mathrm{AUC}_i \tag{13.1}$$

wherein AUC_i represents the AUC value of the i-th seed population group.

13.1.5 Competition FAQ

💬What is the essential task of this competition?

🅰The basic task of the competition question is to carry out accurate user matching of the future advertisement push based on the previous advertisement push and the user click records, so as to improve the click-through rate by the users for the push ads, thus improving the conversion rate, bringing commercial value to advertisers, and charging ads marketing fees.

💬There are several evaluation indicators in Internet advertising. What is the correlation between these indicators?

🅰The first indicator is exposure, which refers to the number of times one advertisement is exposed to the users, that is, how many users have been the push ads; the second indicator is the number of clicks after the user sees the advertisement, that is, how many users click and come to the advertisement page; the third indicator is the conversion. If users see the advertisement and purchase corresponding products, the number of this part of users is the conversion. It can be seen that the exposure, click-through rate, and conversion form an inverted pyramid structure, that is, decreasing progressively. Of course, exposing ads can not only bring direct user conversion to advertisers, but also be a disguised form of marketing to advertisers' brands and popularity.

13.2 Data Exploration

This section will analyze and interpret the available information and data provided by the competition to explore possible modeling ideas. Generally speaking, if memory allows, contestants can generally use common third-party python open-source packages such as jupyter notebook, pandas, and numpy to explore data. Different functions can be used according to the analysis needs. The functions commonly used in pandas packages are read_csv (), head (), describe (), value_counts (), plot (), shape, etc.

13.2.1 Public Data Sets for the Competition

Take the data of the preliminary as an example, the data set files provided are train. csv (training set), test1.csv (testing set), test1_truth.csv (testing set label), adFeature. data (basic attribute of advertisements), and userFeature.csv (basic information of users).

13.2.2 *Training Sets and Testing Sets*

The training set and the testing set only list the ID column and the label column. For this part, the data set publicly provided by Tencent also gives the real label of the testing set. Participants need to make it clear that this is a problem of two primary keys for matching the users with advertisements. Therefore, you can properly view the overlap between aid and uid in the training set and the testing set to determine the difference between the training set distribution and the testing set distribution.

13.2.2.1 Distribution Differences

First of all, it is necessary to confirm that there are no missing values in the training set or in the testing set, and the proportion of positive samples in the training set is 4.8%, which is probability the data obtained after a certain sampling since it is hard for the click rate in actual business to reach this level. Then merge and count the unique values of aid and uid deduplicated in the training set and testing set respectively, as shown in Table 13.1.

It can be seen that less than 18% of the uid in the testing set appears in the training set, while the aid appears all the same in the testing set and in the training set. In fact, this is also in line with business logic—that is, in the case of a short period of time to maintain the same type of advertising, probability matching is predicted based on the click effect of the existing launch for users who have not been pushed the advertisement, thereby increasing the number of clicks, and bringing commercial benefits.

After the value difference of the single primary key is checked, it is necessary to confirm that the value of the two primary keys is also unique, that is, to confirm that the combination of aid and uid is unique. The unique representation here has only one definite label value. The code verification is as follows:

```
train_nunique = train[['uid', 'aid']].drop_duplicates().shape[0]
test1_nunique = test1[['uid', 'aid']].drop_duplicates().shape[0]
all_nunique = test1[['uid', 'aid']].append(train[['uid',
    'aid']]).drop_duplicates().shape[0]
assert train_nunique == train.shape[0]
assert test1_nunique == test1.shape[0]
assert train_nunique + test1_nunique == all_nunique
```

Finally, according to the above analysis, there is still a lack of logical closed loop, that is, whether the distribution of advertisement ID placed in the training set and the testing set is the same. The verification result is shown in Fig. 13.5.

Table 13.1 Distribution of uid and aid

	Train_nunique	Test_nunique	all_nunique	Duplicates	inbag_ratio (%)
uid	7883466	2195951	9686953	392464	18
aid	173	173	173	173	100

Fig. 13.5 Distribution of advertisement ids released in training set and testing set

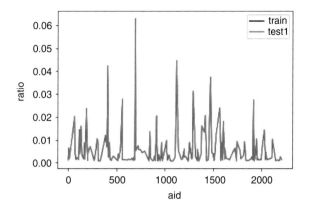

	aid	advertiserId	campaignId	creativeId	creativeSize	adCategoryId	productId	productType
0	2169	16770	38402	43877	35	89	9760	9
1	411	9106	163120	220179	79	21	0	4
2	894	452	38391	43862	35	10	12193	11
3	450	45705	352827	565415	42	67	0	4
4	313	243	531344	979528	22	27	113	9

Fig. 13.6 Advertising attributes data display

As can be seen from Fig. 13.5, the distribution of advertisements in the training set and the testing set is basically the same. Therefore, the focus is to examine the degree of interest of different users in the same advertisement, or it can be said that the participants need to find out the characteristics of the user group of the same advertisement, and then discover more users who may be interested in the advertisement by taking advantage of the existing click data, which is the theme of this competition—Audience Lookalike Expansion.

13.2.3   Advertising Attributes

Use pandas. DataFrame ().head () method to display the basic data. As shown in Fig. 13.6, although all the data have been desensitized, it does not prevent participants from understanding the meaning of each field. Section 13.1.2 has listed detailed descriptions of advertisement features. Participants can view the data by themselves under the help of the instructions.

13.2.4   User Information

Since the format of the user feature file is .data, which is not conducive to the direct analysis and statistics by the contestants, it is converted into a .csv format file first. The specific operation code is as follows:

```
# Check whether the file path already exists
if os.path.exists('data/preliminary_contest_data/userFeature.
csv'):
   user_feature=pd.read_csv('data/preliminary_contest_data/
   userFeature.csv')
else:
   userFeature_data = []
   with open('data/preliminary_contest_dataa/userFeature.data',
   'r') as f:
       for i, line in enumerate(f):
         line = line.strip().split('|')
               userFeature_dict = {}
           for each in line:
           each_list = each.split(' ')
           userFeature_dict[each_list[0]] = ' '.join(each_list[1:])
               userFeature_data.append(userFeature_dict)
         if i % 1000000 == 0:
             print(i)
       user_feature = pd.DataFrame(userFeature_data)
     user_feature.to_csv('data/preliminary_contest_data/
     userFeature.csv',
         index=False)
```

After converting the raw data source into pandas.dataframe format, the analysis becomes very convenient. Due to too many fields, only some fields are shown in screenshots here, as shown in Fig. 13.7. In addition to the user ID uid, other fields are user attributes. The user atrribues are divided into univariate attributes and multivariate attributes. Age, gender, marriageStatus, education, consumptionAbility, and LBS are univariate attributes with only one value per user. Interest 2, interest 5, and kw2 are multivariate attributes where each user will have multiple values. The processing of multivariate attributes will use algorithms related to natural language processing, which will be explained in Sect. 13.3.

13.2.5   Feature Splicing of Data Sets

After becoming familiar with the training set, testing set, advertising attributes, user information, participants are able to comprehend the relationship between these table files, i.e., using the ID columns of the training set and testing set as the basis to associate the advertising attributes with user information, and to form a wide table of features with ID columns and tags in the conventional sense; the remaining features can be directly used for modeling, but multivariate attribute features may require additional processing.

	uid	age	gender	marriageStatus	education	consumptionAbility	LBS	interest2	interest5	kw2
										15215
									77 53	80808
0	72068206	4	2	10	1	1	317.0	79 6	109 30 6	114283
									59	71854
										34525
									77 52	15571
									100 72	92783
1	44661871	5	1	11	7	1	458.0	NaN	131 37	34154
									116 4 79	33457
									71 109 8	31671
									69 41 6 …	
								47 22 58	100 72	
								24 79 73	80 131	11395
								9 46 32	37 116	79112
2	3036658	3	1	11	7	1	682.0	70 20 6	108 79	82720
								33 50 49	29 8 113	87384
								30 …	6 132 42	56195
									…	

Fig. 13.7 Display of basic user features

Because the original data source is relatively large, for some participants who have just started, they may not have enough computing resources readily available to use. Therefore, in order to facilitate participants to quickly understand and run a successful demo, this book conducts 1% random sampling of the training set and testing set in this round, so that the big data problem is converted into a small data problem, and participants can quickly carry out relevant data exploration, feature engineering, and model building. After the scheme is determined here, if there are enough resources, you can perform full data modeling. Here are the codes implemented for random sampling and data splicing:

```
train = train.sample(frac=0.01, random_state=2020).reset_index
(drop=True)
test1 = test1.sample(frac=0.01, random_state=2020).reset_index
(drop=True)
test1['label'] = -2
# Extract user information for the existing training sets and testing sets
user_feature = pd.merge(train.append(test1), user_feature,
    how='left', on='uid').reset_index(drop=True)

# Splicing advertising information
data = pd.merge(user_feature, ad, how='left', on='aid')

# Perform label conversion to facilitate the differentiation of training sets and testing sets
data['label'].replace(-1, 0, inplace=True)
data['label'].replace(-2, -1, inplace=True)
```

At the same time, in order to facilitate modeling, it is necessary to replace the -1 representing negative samples in the sample label with 0 and record the real labels of the testing set at the same time, so as to verify and compare the subsequent modeling. The display of label distribution is shown in Fig. 13.8.

Fig. 13.8 Display of label
distribution

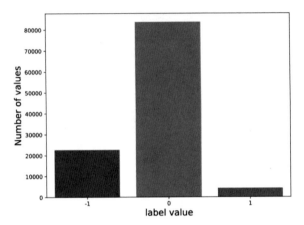

Then distinguish feature categories according to the univariate and multivariate attributes of features. The distinguishing methods are as follows:

```
cols = train.columns.tolist()
cols.sort()
se = train[cols].dtypes
# multivariate attributes
text_features = se[se=='object'].index.tolist()
#  univariate attributes
discrete_features = se[se!='object'].index.tolist()
discrete_features.remove('aid')
discrete_features.remove('uid')
discrete_features.remove('label')
```

Finally, it can be concluded that there are 16 multivariate features text_features in the data set, namely appIdAction, appIdInstall, ct, interest1, interest2, interest3, interest4, interest5, kw1, kw2, kw3, marriageStatus, os, topic1, topic2, topic3, and 14 discrete_features univariate features, which are LBS, adCategoryId, advertising Id, age, campaignId, carrier, consumptionAbility, creativeId, creativeSize, education, gender, house, productId, productType. Since the handling of different types of characteristics is very different, making simple distinctions makes it easier to work more efficiently later.

13.2.6 Basic Modeling Ideas

Through simple data exploration and splicing of table files, participants will be able to perceive that this data structure is very clear. In fact, there are two types of features, namely, multivariable text features text_features and univariate discrete features discrete_features. Therefore, this chapter will consider a novel modeling idea, which is to introduce a CatBoost model that can directly support text_features for modeling.

13.3 Feature Engineering

This section will perform some feature extraction on the basis of data exploration. The data of this competition is very representative. Except for the ID column and the tag column, the other columns are feature columns, and the feature columns here are all discrete columns, including multivariate features and univariate features. This data organization form and the organization form of Chap. 8 are two typical scenarios. The raw data in Chap. 8 is some user behavior records, which need to be designed and extracted before modeling. Of course, it does not mean that the actual cases in this chapter do not need feature design and extraction; rather it is just because the feature engineering here will be somewhat different from that in Chap. 8. This section will take the data of the 2018 Tencent Advertising Algorithm Competition as an example to illustrate another set of commonly used feature design and extraction schemes, in which classic features and business features are used to extract information from univariate fields, while text features are aimed at multivariate fields.

13.3.1 Classic Features

Intuitively speaking, ordinary models (such as LR, RF, GDBT, etc.) cannot distinguish and process univariate discrete features during training. Therefore, such features need to be transformed so that they can be characterized by continuous columns with large and small meanings, and then use models for quantitative differentiation and study. This section will introduce the meanings and extraction methods of three common statistical features.

13.3.1.1 Count Feature

This is a simple counting feature, which can measure the frequency of occurrence of a univariate discrete field and indicate whether a certain attribute of the sample is suitable to the majority or minority. The pandas.series value_counts () method is usually used for frequency statistics. The count coding feature corresponds to the countVectorizer used for multivariate fields introduced in Sect. 13.3.3, which is most obvious in the data value distribution such as the long tail distribution. Reflected in this competition, the count coding feature is called the exposure feature, which can be the exposure of a single field or a combination of multiple fields. The following

	uid	LBS	adCategoryId	advertiserId	age	campaignId
0	10971433	108.0	21	10055	1	86429
1	7657617	431.0	81	2509	2	141893
2	39031487	72.0	142	8203	1	37818
3	3671870	312.0	27	327	5	5616
4	81984219	399.0	142	5459	1	7527

Fig. 13.9 Training set raw input data

	exposure_LBS	exposure_adCategoryId	exposure_advertiserId	exposure_age	exposure_campaignId
0	150	18787	455	26029	92
1	885	105	105	25245	105
2	1815	2009	3781	26029	1271
3	147	13193	2448	26179	2210
4	254	2009	798	26029	144

Fig. 13.10 Exposure features output

part takes some univariate feature fields in this data set as examples to give the raw input data and the output exposure feature.

Figure 13.9 shows a portion of the input data:

Figure 13.10 shows a portion of the output data:

Take the exposure_age field as an example. The sample values of age are 1, 2, and 5, and the corresponding numbers are 26,029, 25,245, and 26,179 respectively. The other fields have similar meanings. The exposure_age_and_gender is the combinational counting of age and gender. Only one univariate feature field is needed to calculate the univariate exposure feature, and two univariate feature fields are needed to calculate the exposure feature above the second order. In this way, the value that does not have the size relationship is mapped into the quantity value, which can intuitively demonstrate whether the user belongs to the majority or minority, for example, in the dimension of age. Participants will realize that this can reflect the age difference of users to a certain extent. On this basis, you can even calculate third-order features and features of higher order, which is of course easy to cause dimension explosion. This is also the essence of the N-Gram algorithm for text feature extraction in the field of natural language processing.

13.3.1.2 Nunique Feature

The second type of feature is the number of attribute values feature nunique, which refers to the number of attribute values after the intersection of two univariate fields.

	nunique_adCategoryId_in_LBS	nunique_advertiserId_in_LBS	nunique_age_in_LBS	nunique_campaignId_in_LBS	nunique_carrier_in_LBS
0	25	44	5	59	4
1	36	70	5	116	4
2	36	72	6	123	4
3	18	32	5	49	4
4	27	50	5	81	4

Fig. 13.11   The nunique feature output

The two univariates can have an inclusion relationship or be independent of each other. Usually, when two univariates have an inclusion relationship and the number of attribute values between different branches varies greatly, the modeling effect is more obvious. For example, if the user's geographic location attribute, that is, the LBS, contains desensitized city ID and subway line information, it is possible to add additional information mined by the geographic location attribute to the user, since it is generally believed that cities with more subway lines have more prosperous economies and may have larger population.

The output nunique feature is shown in Fig. 13.11. Observing the nunique_adCategoryId_in_LBS feature column, it can be noticed that to some extent, this column can reflect the adCategoryId distribution range of different LBS. This is the representation of LBS at the adCategoryId level, which in turn can be characterized at the LBS level. However, this part of information is also only a first-order expression, and both the including items and the items being included for this type of feature can be extended at a higher order.

13.3.1.3   Ratio Feature

The ratio feature can be constructed by making use of the interaction between the two features during the construction of the second-order feature mentioned above in the count coding feature part. The calculation result of the count coding feature and the nunique feature are both integers. Unlike these two, the value obtained by calculating the ratio feature is a decimal between 0 and 1. If the nunique feature can reflect the influence of the distribution range of the feature, then the ratio feature can reflect the proportion, or preference level.

13.3.2   Business Features

Section 13.3.1 has introduced the three classic features of count coding, nunique, and ratio. This section will introduce another statistical feature that requires tags basing on this competition question, namely the business feature. This feature may be used in the classification model. In fact, it is the label distribution ratio of different values for each discrete field, which is reflected in this competition as the click-through rate.

13.3.2.1   Click-Through Rate Features

Before introducing click-through rate features, we must first clarify two concepts—over-fitting and leakage. Over-fitting refers to the model's over-learning of the training set during training, resulting in poor generalization performance, especially when the distribution of the training set and the testing set, especially the joint distribution of features and tags, is quite different. Leakage means that when the model is trained, the features are mixed with tag information, which leads to tagging as part of the features to some extent. Therefore, the model has excellent learning effect. However, the problem is that the tags of the testing set should have been unknown and there may be distribution differences, which will also lead to poor or even extremely poor generalization performance of the model, as well as over-fitting. Therefore, extreme care should be taken when using labels to extract and process related features. It is necessary to strengthen the expression of features on labels without over-expression, which will lead to over-fitting and leakage of label information.

In order to avoid label leakage to a certain degree, the idea of five-fold cross validation can be used for cross click rate statistics, so that the click-through rate characteristics of each sample obtained do not use the information of its label. The specific algorithm steps are as follows:

1. The training set is randomly divided into n equal parts;
2. The click-through rate feature corresponding to each training set obtained in step (1) is statistically mapped by the remaining n-1 training sets, and the click-through rate feature mapping result of one testing set is obtained at the same time;
3. After step (2) is completed, the click-through rate feature corresponding to the entire training set can be obtained, and the click-through rate feature mapping result of the testing set is averaged for n times of different n-1 training sets; then the click-through rate feature corresponding to the testing set can be obtained.

Next, the specific implementation code is given. Special attention should be paid to the fact that only the first-order click-through rate feature is given here, i.e. the original category feature is directly constructed, and the click-through rate feature after the cross combination of the category feature is not given.

```
# Step 1
n_parts = 5
train['part'] = (pd.Series(train.index)%n_parts).values

for co in cat_features:
    col_name = 'ctr_of_'+co
    ctr_train = pd.Series(dtype=float)
    ctr_test = pd.Series(0, index=test.index.tolist())
    # Step 2
    for i in range(n_parts):
        se = train[train['part']!=i].groupby(co)['label'].mean()
        ctr_train = ctr_train.append(train[train['part']==i][co].map
        (se))
        ctr_test += test[co].map(se)

    train_df[col_name] = ctr_train.sort_index().fillna(-1).values
    test_df[col_name] = (ctr_test/5).fillna(-1).values
```

13.3.3 *Text Features*

The above two sections have introduced some feature extraction methods for univariate discrete fields. However, there is another category of multivariate discrete fields in the competition introduced in this chapter, such as interests, keywords, and topics. How to process such fields and perform feature engineering are also worth discussing. This section will introduce relevant algorithms of natural language processing and process such fields as text features. Figure 13.12 shows the fields of interest.

The following is the basic preparations before extracting text features: mainly importing the library and initializing the data set:

```
from scipy import sparse
from sklearn.feature_extraction.text import CountVectorizer,
TfidfVectorizer
from sklearn.preprocessing import OneHotEncoder,LabelEncoder
from sklearn.decomposition import TruncatedSVD
train_sp = pd.DataFrame()
test_sp = pd.DataFrame()
```

Let's first look at the sparse matrix structure of the scipy library, which is a data storage method different from pandas.DataFrame(). The sparse matrix is characterized by its high total dimension, but each user only has a value in a small part of it, so it will not take up too much memory while maintaining ultra-high dimensions. Next, the sparse matrix features will be generated from three aspects.

13.3.3.1 OneHotEncoder

OneHotEncoder, also known as one-hot coding, refers to the encoding process of univariate discrete fields to form a sparse matrix structure. Simply put, it is to change a univariate discrete field with a unique value of n into an n-dimensional vectors of 0 and 1, and then store it as a sparse matrix structure. The DataFrame format is used here for codes implementation:

Fig. 13.12 Multi-valued feature interest1

	uid	interest1
0	10971433	70 86 109 76 45 28 29 49 5 18 72 36 11
1	7657617	70 100 47 76 28 33 106 29 59 49 27 7 9 17 56 3...
2	39031487	70 109 76 48 28 106 49 122 6 119 5 17 56 116 3...
3	3671870	70 76 59 49 36 11
4	81984219	109 59 49 89 111

```
ohe = OneHotEncoder()
for feature in cat_features:
    ohe.fit(train[feature].append(test[feature]).values.reshape(-1,
    1))
    arr = ohe.transform(train[feature].values.reshape(-1, 1))
    train_sp = sparse.hstack((train_sp, arr))
    arr = ohe.transform(test[feature].values.reshape(-1, 1))
    test_sp = sparse.hstack((test_sp, arr))
```

After one-hot encoding, the original single variable discrete field that does not
have a quantization size relationship is converted into multiple continuous fields
represented by 0 and 1, which can be directly used for logical regression models
(LR) and other models that do not directly support discrete fields.

13.3.3.2   CountVectorizer

Likewise, since univariate discrete fields can convert continuous features of 0 and
1 values, multivariate discrete fields also have corresponding conversion methods,
namely CountVectorizer. It makes perfect sense to count each field of multivariate
separately to represent the number of occurrences of samples on a certain value. Of
course, the data of this competition will not be repeated because of the multiple
values of a single user on features such as interest, so the converted value is still only
0 or 1. The specific implementation code is given below:

```
cntv=CountVectorizer()
for feature in text_features:
    cntv.fit(train[feature].append(test[feature]))
    train_sp = sparse.hstack((train_sp, cntv.transform(train
    [feature])))
    test_sp = sparse.hstack((test_sp, cntv.transform(test
    [feature])))
```

13.3.3.3   TfidfVectorizer

TfidfVectorizer is a statistical vector related to word frequency. Its similarity with
CountVectorizer is that their feature dimensions are the same. The difference
between them is that CountVectorizer calculates the number of values of an attribute
in different dimensions, while TfidfVectorizer calculates frequency. The importance
of an attribute increases in proportion to the number of times it appears in a sample,
but at the same time it will decrease inversely as appearing more frequently in the
entire data set. The specific code is as follows. Special attention should be paid to the
fact that TfidfVectorizer () contains parameters, but they are default parameters, that
is, no settings are made.

```
tfd = TfidfVectorizer()
for feature in text_features:
    tfd.fit(train[feature].append(test[feature]))
    train_sp = sparse.hstack((train_sp, tfd.transform(train
[feature]))))
test_sp = sparse.hstack((test_sp, tfd.transform(test[feature])))
```

So far, participants may have a question naturally: that is, the approach of this section will undoubtedly produce ultra-high dimensional features, which may cause performance problems. In view of this risk, in addition to using sparse matrix as the storage structure of data, there is also an auxiliary method to reduce dimension to a certain extent, remove redundant extremely sparse dimensions, or map features to low-dimensional space through feature transformation, thus realizing optimization of calculation speed and memory occupancy.

13.3.4 Feature Dimension Reduction

13.3.4.1 TruncatedSVD

The sklearn (scikit-learn) is a powerful machine learning Python open source package, which is consisted of various commonly used modules. The feature decomposition module contains multiple algorithms for feature dimension reduction to deal with different types and forms of features. In this book, in order to facilitate participants to quickly become familiar with the algorithm process and skills, the competition data (about 10 W data) was sampled in advance. However, participants who have undergone text feature processing will find that their feature dimensions explode to 25 W +, which will bring great performance challenges to modeling. Therefore, a certain degree of dimension reduction can be considered first. The decomposition module in the sklearn package has a TruncatedSVD arithmetic operator for dimension reduction of sparse matrix structures, which can specify the number of features of the principal component for matrix output. Its usage is similar to that of text feature processing operators. Here are the codes implemented for TruncatedSVD usage:

```
svd = TruncatedSVD(n_components=100, n_iter=50, random_state=2020)
svd.fit(sparse.vstack((train_sp, test_sp)))

cols = ['svd_'+str(k) for k in range(100)]

train_svd = pd.DataFrame(svd.transform(train_sp), columns = cols)
test_svd = pd.DataFrame(svd.transform(test_sp), columns = cols)
```

In addition to SVD, there are many dimension reduction methods that can be used, such as PCA (Principal Components Analysis), LDA (Linear Discriminant Analysis), and NMF (Non-negative Matrix Factorization), etc. These methods have great differences in the specific dimension reduction process, indicating that different dimension reduction methods have the possibility of common use.

13.3.5   Feature Storage

It should be noted that in order to achieve better results in the competition, the field information in the testing set is usually added to the computation and processing of features, but in actual business applications, this approach is impossible to achieve, and some competitions will explicitly require that the field information of the testing set should not be used for feature engineering. After the feature processing in the previous sections, in addition to the original data features, three other feature files are generated. As shown in Fig. 13.13, a description of all feature files is given.

13.4   Model Training

13.4.1   LightGBM

The LightGBM model is able to support category features during training, but the premise is that LabelEncoder coding processing needs to be performed first. Feature module includes LabelEncoder for univariate discrete fields, and SVD is the sparse matrix feature of multivariate discrete fields after dimension reduction processing. Combine the two with the LightGBM model and use a five-fold cross validation method to train the model. Finally, the verification set evaluation score of the model is 0.67922 (AUC index), and the testing set evaluation score is 0.61864.

It is obvious that the training set has over-fitting phenomenon; that is, the testing set evaluation score is much lower than the verification set evaluation score, which may be caused by the feature over-fitting in the feature module and the information loss after SVD dimension reduction.

Shortname	Training Seta	Training Seta	Feature Description
sample	train_sample.csv	test_sample.csv	Raw Data
feature	train_sample_feature.csv	test_sample_feature.csv	exposure+ratio+ nunique+ctr+ labelencoder
sparse	train_sample_sparse.csv	test_sample_sparse.csv	onehotencoder+ countvectorizer+ TfidfVectorizer
svd	train_sample_svd.csv	test_sample_svd.csv	sparse+TruncatedSVD

Fig. 13.13 Description of the feature file

13.4.2 CatBoost

Catboost is also one of the most commonly used models, because it directly supports the processing and modeling of text features, that is, multivariable field features, and can be trained and modeled using only the raw data source. It also uses the way of five-fold cross-validation to train the model, and the verification set score of the model is 0.64900 and the testing set score is 0.66501.

13.4.3 XGBoost

CatBoost is able to directly support text features (text_features) and category features (cat_features) because of the sparse processing of these fields within the model. Therefore, the relevant operators of the sklearn package can be used for processing in the outer layer first, and then XGBoost can be used for modeling. The verification set score of the model is 0.67905, and the testing set score is 0.67671.

13.5 Model Integration

13.5.1 Weighted Integration

A simple weighted integration is carried out according to the score of the testing set. The specific calculation method is: RandomForest result × 0.2 + LightGBM result × 0.3 + XGBoost result × 0.5. The verification set score of the model is 0.68147, and the testing set score is 0.68208. It can be seen that the effect of weighted integration is still relatively obvious, and there is no need for complicated operations.

13.5.2 Stacking Integration

Stacking structures have many alternatives. In this competition, we choose to use verification results and prediction results of LightGBM model and XGBoost model as the eigenvalues, and CatBoost model will be the final model, playing the role of training and prediction. This is because CatBoost model can obtain good prediction effect even in the case of only taking advantage of the original features with its relatively strong prediction ability. The following will specifically show the universal implementation code of Stacking integration which is often used:

```
def stack_model(oof_1, oof_2, oof_3, pred_1, pred_2, pred_3, y,
eval_type='regression'):
    # oof_1、 oof_2、 oof_3 are results of verification sets of the three models
    # pred_1、 pred_2、 pred_3 are results of testing sets of the three models
```

```
# y is the truth label of training sets, eval_type is the task type
train_stack = np.vstack([[oof_1, oof_2, oof_3]).transpose()
test_stack = np.vstack([[pred_1, pred_2, pred_3]).transpose()

from sklearn.model_selection import RepeatedKFold
folds = RepeatedKFold(n_splits=5, n_repeats=2, random_state=2020)

oof = np.zeros(train_stack.shape[0])
predictions = np.zeros(test_stack.shape[0])

for fold_, (trn_idx, val_idx) in enumerate(folds.split(train_stack, y)):
    print("fold n° {}".format(fold_+1))
    trn_data, trn_y = train_stack[trn_idx], y[trn_idx]
    val_data, val_y = train_stack[val_idx], y[val_idx]
    print("-" * 10 + "Stacking " + str(fold_) + "-" * 10)
    clf = BayesianRidge()
    clf.fit(trn_data, trn_y)

    oof[val_idx] = clf.predict(val_data)
    predictions += clf.predict(test_stack) / (5 * 2)

if eval_type == 'regression':
    print('mean: ',np.sqrt(mean_squared_error(y, oof)))
if eval_type == 'binary':
    print('mean: ',log_loss(y, oof))

return oof, predictions
```

Here, oof_1, oof_2, and oof_3 are the corresponding results of verification sets of the three models, and pred_1, pred_2, and pred_3 are the results of testing sets of the three models. As a general stacking framework, there are no specific constraints on the three models. The two parts are spliced separately to obtain one training set and one testing set with only three feature columns, and then the training set is fed into the BayesianRidge model for training; the final results are then stored in advance.

The final model verification set score is 0.70788, and the testing set score is 0.67445. It can be seen that the offline score of stacking integration is usually higher, but a consistent result that has been improved cannot be obtained on the testing set due to overfitting and other reasons.

13.6   A Summary of the Competition Question

13.6.1   More Schemes

13.6.1.1   GroupByMean

As what has been mentioned above, click-through rate features are extracted by combining univariate discrete fields with tags. From this point, it can be thought that the 0 and 1 columns similar to tags are obtained after sparse matrixing of multivariate discrete fields. Therefore, statistics can be done to compute the mean of the value of the univariate discrete field in a multivariate discrete field, namely groupby (cat_features) [text_features] .mean (). For example, when the value for age is 5, compute the proportion of people with an interest ID of 109 for interest1 in the group.

13.6.1.2 N-Gram

When extracting CountVectorizer features, the book uses the default parameter, that is, ngram = (1,1). It has not tried to take a higher-order N-Gram for statistics. The higher-order N-Gram essentially adds a layer of combination of features, so that information belonging to the same multivariable discrete field is tied together, such as identifying who likes running and cycling at the same time.

13.6.1.3 Graph Embedding

This kind of method is mainly used to extract vector representations of categories such as uid or aid, which can well mine user and advertisement information from the graph structure. Those uid or aid that have homogeneity or isomorphism in the graph can also be represented by the embedded vectors. Figure 13.14 shows two embedded vector extraction methods for DeepWalk.

13.6.2 Sorting Out Knowledge Points

13.6.2.1 Feature Engineering

As for feature engineering, this chapter introduces common feature extraction methods from three aspects: classical features, business features, and text features. Among them, the classical features are mainly interactive statistics between univariate discrete fields, including count coding, nunique and ratio features; the business features part introduces the click-through rate features combined with industry scenarios and domain knowledge, which are also features that need to be combined with tags; the text features portion introduces several different ways to generate

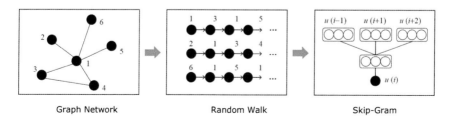

| Graph Network | Random Walk | Skip-Gram |

Fig. 13.14 Process of extracting embedding vectors

sparse matrices, which are especially useful when dealing with large-scale univariate and multivariate discrete fields.

13.6.2.2   Modeling Ideas

The competition question in this chapter represents a typical data organizing form and table data structure. For this kind of data, a relatively common feature engineering method can be abstracted. This is also one of the reasons why this competition question is used in this book to explain the case of computational advertising. The principle of Lookalike is to find potential users who are similar to the users who have clicked on the advertisement by taking advantage of the marketing results of previous advertisements, so as to achieve continuous exposure and clicks of advertisements. Therefore, the focus should be on finding similarities between users, especially the joint similarity in all dimensions. Unfortunately, machine learning is limited by feature engineering and cannot achieve the best results, while deep learning neural networks can perform nested combination and nonlinear function fitting on text fields, so the neural networks model had better performance in this competition.

13.6.3   Extended Learning

This competition requires participants to provide and submit the scores that show the candidate users of various seed population groups in the testing set belong to the corresponding seed groups (the higher the score, the more likely the candidate users are to be potential look-alike users of a certain seed group); then, can the probability of users clicking on an advertisement be regarded as a click-through rate prediction problem? This is very similar to 2017 Tencent Advertising Algorithm Contest. The basic feature construction method and model selection are the same. The difference is that there is no time-related information in the user behavior sequence of the 2018 Tencent Advertising Algorithm Competition, which lacks a lot of time-related features. Of course, this is also caused by business of Lookalike.

13.6.3.1   2017 Tencent Advertising Algorithm Contest: Conversion Rate Prediction of Mobile App Advertising

Computational advertising is one of the most important business models of the Internet. The effect of advertisement delivery is usually measured from three aspects: exposure, clicks, and conversion. Most advertising systems are limited by the function of returning advertising effect data and can only be optimized by using exposure or clicks to measure the effect of ad delivery. Tencent Social Ads makes the most of its unique capabilities in user identification and conversion tracking data to help advertisers track the conversion result after advertising, trains the predicted conversion rate model (pCVR) based on advertising conversion data, and introduces pCVR factors in advertising ranking to optimize the effect of ad delivery and improve ROI. This question takes mobile App ads as the research target and predicts the probability of activation of App ads after they are clicked: pCVR = P (conversion = 1 | ad, user, context); that is, given the advertisement, user, and context, predict the probability that the App ad will be activated after being clicked. The industry has always attached more importance to the research of advertisement click-through rate conversion (CTR), and the current application is relatively mature. Tencent's prediction of advertisement conversion rate (CVR) in this competition is unique. The competition has high research value both in academic research and industry application fields.

Basic ideas: 2017 Tencent Algorithm Contest is an early CTR contest, and many methods are worth learning from, including a lot of classic operations. In terms of models, most players chose the tree model and FFM model, and then combined various Stacking combinations to get the final result. At that time, the model used for predicting advertising clicks was relatively simple since the DeepFM, xDeepFM, AFM, etc. used today came out late.

In terms of feature construction, they are also similar, such as basic features, user category features, advertising category features, context features, interaction features, and other features. The focus here is on other features, which can be called trick features, specifically including the conversion of the user's repeated clicks on the day, the time difference between the first and last items of the repeated samples on the day (the feature variables are the same), and the repeated samples on the day are sorted by time.

$$p = f\left(\frac{f^{-1}(0.1) + f^{-1}(0.15) + f^{-1}(0.08)}{3}\right) = 0.1067 \qquad (13.2)$$

The champion's plan has great innovations in the model. In addition to the tree model, wide & deep, and PNN, it also uses an improved and innovative NFFM model, and the single model score is higher than that of the third place on the list. The final model integration method used is weighted average integration, but it is integrated after logit inversion. To be specific, first substitute the results of each model into the sigmoid inverse function, then get the mean, and finally use the sigmoid function for the mean value. Compared with the common weighted average, this method is more suitable for situations with small differences in results.

```
# sigmoid function
def f(x):
    res = 1 / ( 1 + np.e ** ( -x ) )
    return res
# sigmoid inverse function
def f_ver(x):
    res = np.log( x / ( 1 - x ) )
    return res
```

Chapter 14
Case Study: TalkingData AdTracking Fraud Detection Challenge

This chapter centers on the typical anti click fraud competition question of a contest held on the Kaggle competition platform in 2018, i.e., TalkingData AdTracking Fraud Detection Challenge (as shown in Fig. 14.1), which will also be used as the second practical case for issues related to computational advertising. The main content includes competition question understanding, data exploration, feature engineering, model training, and competition question summary. In fact, when selecting the competition question to be analyzed, the joint authors discussed many times, because the advertising field not only involved a lot of core technologies, but also had a variety of competition questions available for discussion. In the end, we took data quality, knowledge points that can be covered, and the popularity of competition questions as the main criteria for selection and finally decided to choose this competition question.

14.1 Understanding the Competition Question

14.1.1 Background Introduction

Fraud risks are everywhere. For companies that advertise online, enormous click fraud incidents may occur, resulting in a sea of abnormal click data appearing, wasting a lot of money. Being able to identify fraudulent clicks can greatly reduce costs. In China, more than 1 billion intelligent mobile devices are being used every month.

TalkingData is a relatively large independent big data service platform, with focus on more than 70% of China's mobile devices, handling 3 billion click events per day, 90% of which may be fraudulent. The current approach the platform provides to developers of an app to prevent click fraud is to evaluate the click process of users in their product portfolio and mark IP addresses that generate

Fig. 14.1 Home page of TalkingData AdTracking fraud detection challenge

numerous click events but never install the app in the end. Using this information, developers have established IP address blacklists and device blacklists.

Aiming to be able to foresee click fraud so that anti-fraud efforts could prevent it from happening, TalkingData launched algorithm challenges on the Kaggle community to further develop solutions. In the second contest cooperated with Kaggle, the contestants faced the challenge of constructing an algorithm that could predict whether users would download the app after clicking on the app's advertisement. To support the contestants in modeling, the organizers provided a data set containing approximately 200 million click events in 4 days.

14.1.2   Competition Data

This competition question provides the training data with a sample size of nearly 190 million, including data from November 6, 2017 to November 9, 2017. Each data record is an ad click event. The variables (features) involved in the training set are as follows.

- **ip**: the IP address where the click event occurs;
- **app**: app ID provided by the advertiser;
- **device**: the user's mobile device ID;
- **os**: the operating system version ID of the user's mobile device;
- **channel**: advertising delivery channel ID;
- **click_time**: click time (UTC time), the format is yyyy-mm-dd hh: mm: ss;
- **attributed_time**: If the user clicks and downloads the app, then this is the time of downloading the App;
- **is_attributed**: whether the user has downloaded the app after clicking, which is the target variable.

14.1.3   Evaluation Indicators

The competition requires participants to submit the probability that users will eventually download the app and calculate the AUC value based on this as a judging criterion.

14.1.4   Competition FAQ

Q The initial size of the raw data source set has reached 4 GB. What difficulties will this bring to the competition?

A In real competitions, a too large data set often restricts operations. If the memory configuration is not sufficient, not only may the data set fail to load, but also the memory must be used sparingly during feature construction. Therefore, when writing code, you should optimize the writing; otherwise the memory will explode if you are not careful, or the code will run for several hours before it is over; when training the model, there will also be great limitations; for example, the parameters of the model can no longer be adjusted so arbitrarily, and the loss of time must be taken into account in determining the learning rate, iteration number, and early stopping; further, as for verification method, five-fold cross validation becomes quite laborious, and it will be more efficient to use the leave-one-out method.

Q What if the distribution of positive and negative samples is extremely unbalanced?

A This is also a common problem in the field of recommendation advertising; the general solution to it is to carry out data sampling, and the most commonly used approach is random sampling; to be specific, when the sample distribution is not balanced, it is random negative sampling, which could be illustrated as randomly sampling a certain percentage of negative sample data, and then using this part of negative sample data and complete positive sample data to complete feature extraction and the final model training. Special attention should be paid to the fact that in the model integration stage, the model is trained and predicted with different proportions of negative sample data first, and then integrated, usually leading to unexpected results.

14.1.5   Baseline Scenario

With the preliminary properties described in Sect. 14.1.2, the basic modeling work can be started. The initial baseline scheme constructed does not need to be too complicated, as long as it can give a correct result.

14.1.5.1   Data Reading

The relevant code for reading data is as follows:

```
import gc
import time
import numpy as np
import pandas as pd
from sklearn.model_selection import train_test_split
import xgboost as xgb

path = './input/'
train_columns = ['ip', 'app', 'device', 'os', 'channel', 'click_time',
'is_attributed']
test_columns = ['ip', 'app', 'device', 'os', 'channel', 'click_time',
'click_id']
dtypes = {'ip' : 'uint32', 'app' : 'uint16', 'device' : 'uint16', 'os' :
'uint16',
'channel' : 'uint16', 'is_attributed' : 'uint8', 'click_id' :
'uint32'}
train = pd.read_csv(path+'train.csv', usecols=train_columns,
dtype=dtypes)
test   =   pd.read_csv(path+'test_supplement.csv',   usecols=test_
columns, dtype=dtypes)
```

There is a skillful operation when reading data, that is, optimizing memory. If you read table data directly, then the integer column will default to int64 and the floating-point column to float64, obviously a waste of space for those integer values between −128 and 127. This problem is solved by optimizing memory.

14.1.5.2 Preparing Data

First, merge the training set and the verification set, and delete the redundant variables before that to ensure that there will be no errors when merging. The specific operation code is as follows:

```
# training set
y_train = train['is_attributed'].values
# delete redundant variables
del train['is_attributed']
sub = test[['click_id']]
del test['click_id']
# merge the training set and the testing set
nrow_train = train.shape[0]
data = pd.concat([train, test], axis=0)
del train, test
gc.collect()
```

In particular, due to limited memory space, it is necessary to delete the original training set and testing set after the training set and the testing set are merged, and then use gc.collect () to release memory.

14.1.5.3 Feature Extraction

For the kind of problems discussed in this chapter, simple statistical features can play a big role, because these features have business significance; for example, the count coding feature can reflect the degree of clout or activity, the nunique feature can reflect the breadth of certain variables, and the ratio feature can describe the range ratio, etc. Then, simply construct the count coding feature as part of the basic feature, and the click time feature click_time, which can be converted into various types. The code is as follows:

```
for f in ['ip','app','device','os','channel']:
data[f+'_cnts'] = data.groupby([f])['click_time'].transform
('count')

  data['click_time'] = pd.to_datetime(data['click_time'])
data['days'] = data['click_time'].dt.day
data['hours_in_day'] = data['click_time'].dt.hour
data['day_of_week'] = data['click_time'].dt.dayofweek

train = data[:nrow_train]
test = data[nrow_train:]
del data
gc.collect()
```

14.1.5.4 Model Training

The relevant code for training model is as follows:

```
  params = { 'eta': 0.2,
      'max_leaves': 2**9-1,
      'max_depth': 9,
      'subsample': 0.7,
      'colsample_bytree': 0.9,
      'objective': 'binary:logistic',
      'scale_pos_weight':9,
      'eval_metric': 'auc',
      'random_state': 2020,
      'silent': True }
  trn_x, val_x, trn_y, val_y = train_test_split(train, y_train,
test_size=0.2,
     random_state=2020)
  dtrain = xgb.DMatrix(trn_x, trn_y)
  dvalid = xgb.DMatrix(val_x, val_y)
  del trn_x, val_x, trn_y, val_y
  gc.collect()
  watchlist = [(dtrain, 'train'), (dvalid, 'valid')]
  model = xgb.train(params, dtrain, 200, watchlist,
early_stopping_rounds = 20,
     verbose_eval=10)
```

```
[0]      train-auc:0.963463      valid-auc:0.962897
Multiple eval metrics have been passed: 'valid-auc' will be used for early stopping.

Will train until valid-auc hasn't improved in 20 rounds.
[10]     train-auc:0.969029      valid-auc:0.968335
[20]     train-auc:0.971793      valid-auc:0.970815
[30]     train-auc:0.974514      valid-auc:0.972974
[40]     train-auc:0.976252      valid-auc:0.973998
[50]     train-auc:0.977784      valid-auc:0.974714
[60]     train-auc:0.978459      valid-auc:0.975059
[70]     train-auc:0.979047      valid-auc:0.975191
[80]     train-auc:0.979595      valid-auc:0.975238
[90]     train-auc:0.980047      valid-auc:0.975299
[100]    train-auc:0.980454      valid-auc:0.975339
[110]    train-auc:0.980931      valid-auc:0.97536
[120]    train-auc:0.981351      valid-auc:0.975342
[130]    train-auc:0.981737      valid-auc:0.975344
Stopping. Best iteration:
[113]    train-auc:0.981055      valid-auc:0.975366
```

Fig. 14.2 Display of the XGBoost training process

As a baseline scheme, not too many features are constructed in it. In terms of model training, in order to get the result feedback quickly, the data is directly divided into two parts in a ratio of 8:2, and the learning rate is also adjusted to be relatively high, also in order to get the results in a short time. Figure 14.2 is the score feedback during model training, which needs to focus on the valid-auc score, and then use the trained model to predict the testing set results, and finally get the online score.

After training the model, it is necessary to predict the results of the testing set and submit the final results. The implementation code is as follows:

```
dtest = xgb.DMatrix(test[cols])
sub['is_attributed'] = None
sub['is_attributed'] = model.predict(dtest, ntree_limit=model.
best_ntree_limit)
sub.to_csv('talkingdata_baseline.csv', index=False)
```

The final score of the online private score is 0.96854, and the online public score is 0.96566. The current baseline plan ranks 1995/3946 in the private leaderboard, so it seems that there is still a lot of room for improvement; therefore, you can set a short-term goal: strive to get the silver medal. This chapter will also give more directions that can be tried, and we will advance towards the gold medal together.

14.2   Data Exploration

14.2.1   Preliminary Research on Data

14.2.1.1   Base Display

Figure 14.3 shows the training set data to help quickly grasp the internal structure of the data set.

	ip	app	device	os	channel	click_time	is_attributed
0	83230	3	1	13	379	2017-11-06 14:32:21	0
1	17357	3	1	19	379	2017-11-06 14:33:34	0
2	35810	3	1	13	379	2017-11-06 14:34:12	0
3	45745	14	1	13	478	2017-11-06 14:34:52	0
4	161007	3	1	13	379	2017-11-06 14:35:08	0

Fig. 14.3 Display of training set data

14.2.1.2 Label Distribution

The implementation code of building a bar chart visualization for label distribution:

```
plt.figure()
fig, ax = plt.subplots(figsize=(6,6))
x = train['is_attributed'].value_counts().index.values
y = train["is_attributed"].value_counts().values
sns.barplot(ax=ax, x=x, y=y)
plt.ylabel('Number of values', fontsize=12)
plt.xlabel('is_attributed value', fontsize=12)
plt.show()
```

First observe the label distribution. As shown in Fig. 14.4, the positive and negative sample distribution of the labels of the training set is extremely unbalanced, with the positive sample accounting for about 0.247%, even not reaching 1%. In Sect. 14.4, the problem of unbalanced label distribution will be optimized, usually undersampling negative samples and oversampling positive samples. In large-scale data sets, undersampling negative samples is the most common choice and can solve performance problems.

14.2.1.3 Variable Distribution

Next, observe the basic distribution of the variables of the remaining features, as shown in Fig. 14.5. Here we mainly show the number of unique values of the feature variables ip, app, device, os, and channel. The analysis for remaining univariates is in Sect. 14.2.2.

Fig. 14.4 Label
distribution map

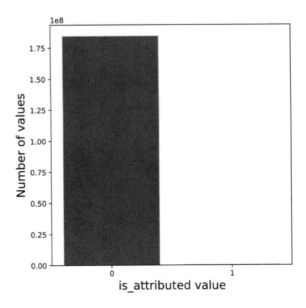

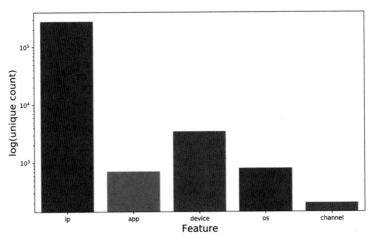

Fig. 14.5 Display of the number of unique values after taking the logarithm

14.2.2   Univariate Analysis

14.2.2.1   Attribute Distribution of Univariates

Understanding the distribution of univariate attributes helps us to establish a basic
understanding of features. This question mainly involves category features, and
generally we will observe the count distribution of each attribute.

First analyze the IP address. The following is the implementation code for generating visualization maps of the top 10 IP addresses with most downloads:

```
tmp = train.groupby('ip').is_attributed.sum()
data_plot = tmp.nlargest(10).reset_index()
data_plot.columns=('IP', 'Downloads')
data_plot.sort_values('Downloads', ascending = False)
plt.figure(figsize = (8,5))
sns.barplot(x = data_plot['IP'], y = data_plot['Downloads'])
plt.ylabel('Downloads', fontsize=16)
plt.xlabel('IP', fontsize=16)
plt.title('Top 10 bigget downloader', fontsize = 15)
```

The generated result is shown in Fig. 14.6. Note that there's a huge gap between the downloads at top 10 IP addresses.

In addition, simple statistics have been carried out, and the result is shown in Fig. 14.7. 70% of IP addresses have seen downloading only once (once), 18% of IP addresses have experienced downloading several times (multiple times), and 12% of IP addresses have no download happening (no), showing a very obvious long tail distribution. IP addresses can also be of great help to model prediction.

Extended Thinking
Although downloads vary significantly and can be used as a good feature, the downloads and download rate (predictive target) are not necessarily positively correlated to each other; for example, 100,000 clicks trigger 1000 downloads,

(continued)

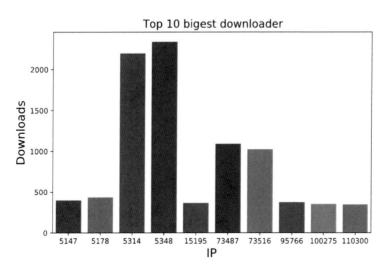

Fig. 14.6 Top 10 IP addresses for downloads

Fig. 14.7 Proportion of
download categories

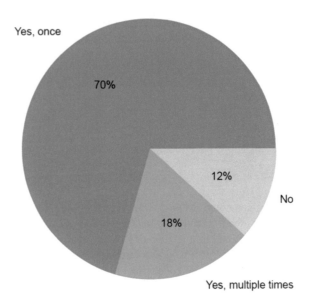

and 200 clicks cause 100 downloads. Although the former has a high quantity
of download, the download rate of it is far lower than that of the latter.
Therefore, there is a part of deviation if the download is used as a feature. A
more comprehensive approach is to construct features of times of clicks and
download rate to assist the model in better training.

Similarly, you can also use a similar method to analyze other distributions of
univariates and observe the relationship between quantities of clicks and downloads.
There may be some apps that are rarely clicked at ordinary times, but as long as users
click on these apps, they will probably download them.

14.2.2.2   The Relationship Between Univariates and Labels

Comparing the relationship between a single variable and a label is one of the
operations that can best find out the value of a variable. If there is a difference
between the attributes of the variable—that is, if the distribution of is_attributed is
inconsistent, then this variable can be regarded as a valuable feature; otherwise, it
needs to continue to dig deeper. Next, through a piece of code, the density distribution relationship between the feature variables app, device, os, and channel and the
label is_attributed is realized, and then visualized, as shown in Fig. 14.8.

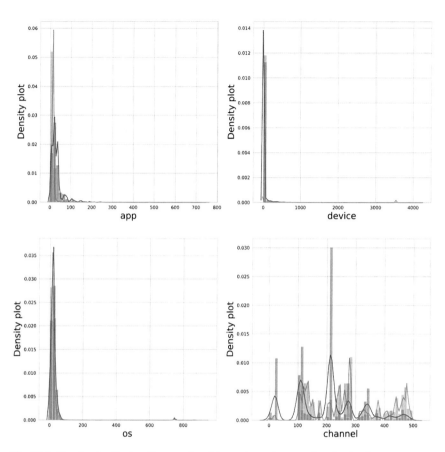

Fig. 14.8 Density distribution diagram of positive and negative samples of univariates

```
cols = ['app','device','os','channel']
train1 = train[train['is_attributed'] == 1][train['day'] == 8]
train0 = train[train['is_attributed'] == 0][train['day'] == 8]

sns.set_style('whitegrid')
plt.figure()
fig, ax = plt.subplots(2, 2, figsize=(16,16))
i = 0
for col in cols:
    i += 1
    plt.subplot(2,2,i)
    sns.distplot(train1[col], label="is_attributed = 1")
    sns.distplot(train0[col], label="is_attributed = 0")
    plt.ylabel('Density plot', fontsize=12)
    plt.xlabel(col, fontsize=12)
plt.show()
```

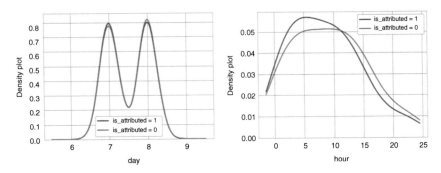

Fig. 14.9 Density distribution of positive and negative samples at different times

As can be noticed from Fig. 14.8, there are differences in the density distribution between the four variables and the labels, and the proportion of positive and negative samples among different attributes in the variables is different, which also shows that the feature variables have certain degree of distinction.

In addition to the above characteristics, the impact of time on tags is often very large. For example, the download rate of an app at night and the download rate of it at noon may vary. Not only in this competition question, similar recommendations and advertising scenarios are also affected by time, resulting in big differences in CTR and CVR. Therefore, time is also an important object which needs analyzing.

As shown in Fig. 14.9, first extract the features of day and hour, and then compare the density distribution differences of is_attributed labels. It is obvious that the density distribution of positive and negative samples in the unit of day is completely consistent, while the density distribution in the unit of hour shows variance to some extent.

Drawing a heat map using the following code can more clearly illustrate the change of download probability over time:

```
grouped_df = train.groupby(["day",
        "hour"])["is_attributed"].aggregate("mean").reset_index()
grouped_df = grouped_df.pivot('day', 'hour', 'is_attributed')

    plt.figure(figsize=(12,6))
    sns.heatmap(grouped_df)
    plt.show()
```

The result is shown in Fig. 14.10, where the abscissa is in hours and the ordinate is in days. It is explicit that there are differences in download rates at different times. For example, the download rate after 13 o'clock is relatively low (the darker the color, the lower the download rate). In addition, it can also be found that the data on the day 6 and day 9 are incomplete. In the case where data is not complete, the construction of count under the unit of day and other related features requires special processing, such as scaling the data amount.

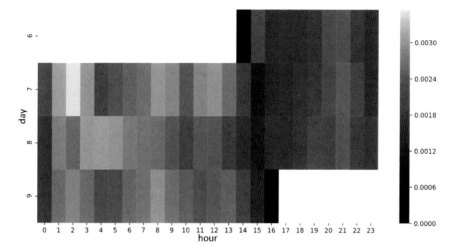

Fig. 14.10 Display of download rates at different times

14.2.3 Multivariate Analysis

This section will carry out more complex analysis work, combining multiple variables to explore more valuable information in the data, such as combining ip and app, which contributes to understanding the distribution of the apps under different ips. Statistical features constructed after combining multiple variables, such as count-related features, can reflect the preference level of the same IP over different apps. It is assumed that the higher the frequency of the same combination, the closer the connection between the combinations. If ip, app, and os are combined, the information reflected will be more detailed, and the data granularity at this time is also very fine.

Multiple variables are combined, and the count of the feature of the multivariate after combination is computed; the result is shown in Fig. 14.11, where the abscissa is the frequency (count), and the ordinate is the density distribution (distribution). It can be clearly seen that the multivariate combination features related to frequency obey the long tail distribution, and there are certain differences in the density distribution between positive and negative samples.

14.2.4 Data Distribution

Accurate understanding of the distribution of data is of great help to structural features and data modeling. In competitions, it is often encountered that the online and offline evaluation scores are inconsistent. Under such circumstance, the first

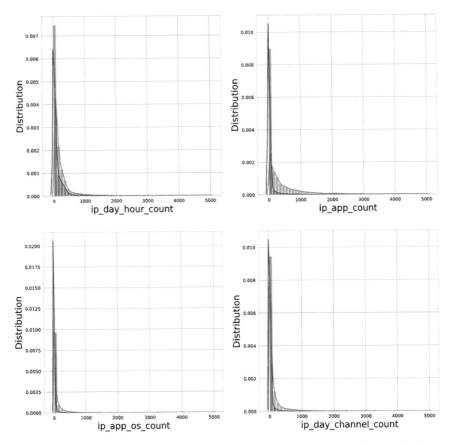

Fig. 14.11 Density distribution of positive and negative samples after multivariate combination

thing to observe is whether the data distribution is consistent. The data here specifically refers to that in the training set (verification set) and testing set.

First, let's introduce a concept—Covariate Offset. Covariate refers to the input variables (features) of the model, and covariate offset means that the input variables of the training set and testing set have different data distributions—that is, the data variables have shifted. However, what we expect is that the input variables of the training set and the testing set are equally distributed, which is also conducive to the prediction of the model. However, it is difficult to achieve this goal in real scenarios.

For example, the training set contains 30% app1, 40% app2, and 30% app3, while the testing set contains 10% app1, 20% app2, and 70% app3. Obviously, the two data sets have different proportions of app categories, that is, the distribution of input variables is different, which is the covariate offset. Next, there is a specific analysis of this through a visual way, as shown in Fig. 14.12.

As shown in Fig. 14.12, there are certain variance in the proportions of each variable on the training set and on the testing set, especially the variable channel. In addition, observing the density distribution map of variables, it can be found that the

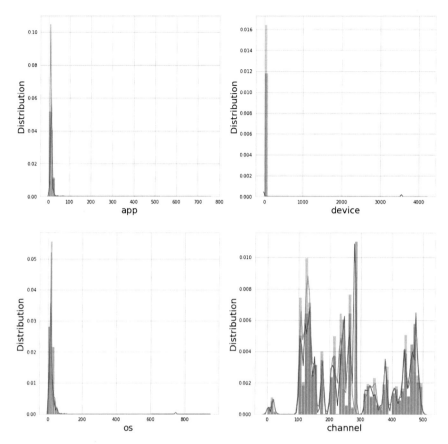

Fig. 14.12 Density distribution of variables in training set and testing set

numbers of unique values (unique count) of variables corresponding to the training set and the testing set also differ from each other, which indicates that many attributes in variables only exist in the training set, but not in the testing set.

As shown in Fig. 14.13, the distribution of features of different days and different hours on the training set and the testing set also varies, because the time intervals of the two data sets are inconsistent, which is the main reason for the difference. Therefore, the influence of time units should be considered when constructing features.

14.3 Feature Engineering

For problems related to CTR and CVR, there are many associated features, such as statistical features of different granularities (count, nunique, ratio, rank, lag, etc.), target coding, embedding features, etc., and different granularities are combinations

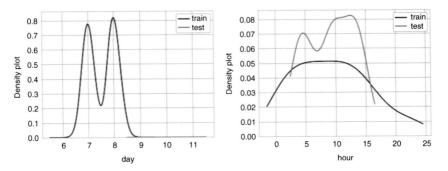

Fig. 14.13 Density distribution by days and hours of training set and testing set

of features in first-order, second-order or multi-order. Although these rich features all have their own significance and value, the amount of data in this question is too large to meet all the feature requirements. Therefore, how to select the most important features from a large number of features is particularly important. This section will introduce the four core features, namely statistical features, time difference features, ranking features, and target coding features. These features contribute positively to the final score largely and are also features contestants often use in competitions.

14.3.1   Statistical Features

Here, the core feature of this question, namely statistical feature groupby, will be constructed. Specifically, the first-order and second-order combinations of the five original category features (ip, os, app, channel, device) will be built, and then aggregated to obtain features of count, nunique, and ratio. The following is the first-order and second-order feature construction code, where the extracted first-order features consist count and ratio, and the second-order features contain count, nunique, and ratio.

```
# first-order features count, ratio
for f in tqdm(['ip','app','device','os','channel']):
    data[f+'_cnts'] = data.groupby([f])['click_time'].transform
('count')
    data[f+'_cnts'] = data[f+'_cnts'].astype('uint32')
    data[f+'_ratio'] = (data[f].map(data[f].value_counts())) / len
(data) *
    100).astype('uint8')
# second-order feature count
cols = ['ip','app','device','os','channel']
for i in tqdm(range(0, len(cols)-1)):
    for j in range(i+1, len(cols)):
        f1, f2 = cols[i], cols[j]
```

```
            data[f1+'_'+f2+'_cnts'] =
               data.groupby([f1,f2])['click_time'].transform('count')
            data[f1+'_'+f2+'_cnts'] = data[f1+'_'+f2+'_cnts'].astype
('uint32')
   # second-order features nunique, ratio
   for f1 in tqdm(['ip','app','device','os','channel']):
       for f2 in ['ip','app','device','os','channel']:
           if f1 != f2:
               data[f1+'_'+f2+'_nuni'] = data.groupby([f1])[f2].
transform('nunique')
               data[f1+'_'+f2+'_nuni'] = data[f1+'_'+f2+'_nuni'].astype
('uint32')

               data[f1+'_'+f2+'_ratio'] = (data.groupby([f1,f2])
['click_time'].
               transform('count') / data.groupby([f1])
               ['click_time'].transform('count') * 100).astype('uint8')
```

Of course, multi-level feature combinations can also be carried out to obtain information of different fine grains, such as third-order combinations of ip, app, and device. However, special attention should be paid at this time. The granularity of the constructed features cannot be too fine. Generally, features that are too fine cannot be directly put into the model. Such features can be called high-dimensional sparse features. The confidence level of the weights corresponding to such features is very low (many such feature combinations only appear once), which generally requires conversion or compression processing.

14.3.2 Time Difference Features

The time difference (time-delta) feature is also one of the core features of this competition. Specifically, the temporal difference between the first n clicks and the next n clicks of each click event can be extracted as a feature. The specific implementation code of the time-delta feature is as follows:

```
    for cols in [['ip','os','device','app'],['ip','os','device',
'app','day']]:
    for i in range(1,6):
        data['ct'] = (data['click_time'].astype(np.int64)//10**9).
astype(np.int32)
        name = '{}_next_{}_click'.format('_'.join(cols), str(i))
        data[name] = (data.groupby(cols).ct.shift(-i)-data.ct).astype
(np.float32)
        data[name] = data[name].fillna(data[name].mean())
        data[name] = data[name].astype('uint32')

        name = '{}_lag_{}_click'.format('_'.join(cols), str(i))
        data[name] = (data.groupby(cols).ct.shift(i)-data.ct).astype
(np.float32)
        data[name] = data[name].fillna(data[name].mean())
```

```
data[name] = data[name].astype('uint32')

data.drop(['ct'],axis=1,inplace=True)
```

So why does the time difference feature work? First of all, this feature can reflect the frequency of the user's activities; secondly, from the business perspective, the time difference between the users' clicking on an app can be used to reflect the possibility of downloading the app by the users. If you click on the same app many times in a short period of time, then the possibility of downloading will be greater. The extraction of such features is very skillful and can be regarded as a trick feature, which can play a role in many competitions at the same time.

According to this characteristic of time difference, it is also possible to construct a marking feature of successively clicking. There are no more than three types of clicking on one app by users: the first click, intermediate clicks, and the last click. As you can imagine, when we browse the app store, the last click is often the time when users are most likely to download. The specific implementation code is as follows:

```
subset = ['ip', 'os', 'device', 'app']
data['click_user_lab'] = 0
pos = data.duplicated(subset=subset, keep=False)
data.loc[pos, 'click_user_lab'] = 1
pos = (~data.duplicated(subset=subset, keep='first')) & data.
duplicated(subset=subset,
     keep=False)
data.loc[pos, 'click_user_lab'] = 2
pos = (~data.duplicated(subset=subset, keep='last')) & data.
duplicated(subset=subset,
     keep=False)
data.loc[pos, 'click_user_lab'] = 3
```

14.3.3 Ranking Features

Literally, the ranking feature has a strong traversal nature, which is mainly based on the number of interactions between the user and the app. The code for constructing the ranking feature is as follows:

```
for cols in tqdm([['ip','os','device','app'],['ip','os','device',
'app','day']]):
    name = '{}_click_asc_rank'.format('_'.join(cols))
    data[name] = data.groupby(cols)['click_time'].rank
(ascending=True)

    name = '{}_click_dec_rank'.format('_'.join(cols))
    data[name] = data.groupby(cols)['click_time'].rank
(ascending=True)
```

14.3.4   Target Coding Features

The target coding feature is a feature that directly presents entity information, that is, the target-based probability distribution feature. Because of its direct relationship with labels, it is necessary to avoid data traversal problems when constructing target coding features. In data sets containing direct time series information, statistical historical information is used as the current feature; in data sets without time series information, it is sufficient to use the common *K*-fold cross statistics. The code for constructing target coding features is as follows:

```
   for cols in tqdm([['ip'], ['app'], ['ip','app'], ['ip','hour'],
['ip','os','device'],
      ['ip','app','os','device'], ['app','os','channel']]):
      name = '_'.join(cols)
      res = pd.DataFrame()
      temp = data[cols + ['day', 'is_attributed']]
      for period in [7,8,9,10]:
      mean_ = temp[temp['day']<period].groupby(cols)
['is_attributed'].mean().
         reset_index(name=name + '_mean_is_attributed')
      mean_['day'] = period
      res = res.append(mean_, ignore_index=True)
  data = pd.merge(data, res, how='left', on=['day']+cols)
```

14.4   Model Training

14.4.1   LR

The LR model has always been a benchmark model for click-through rate prediction problems, and is widely used by virtue of its simplicity, easy parallel implementation, and strong interpretability. However, due to the limitations of the linear model itself, it cannot handle the nonlinear relationship between features and targets, so the prediction effect of the model relies heavily on the feature engineering experience of algorithm engineers. The code for constructing LR model is as follows:

```
from sklearn.linear_model import LogisticRegression
model = LogisticRegression(C=5, solver='sag')
model.fit(trn_x, trn_y)
val_preds = model.predict_proba(val_x)[:,1]
preds = model.predict_proba(test_x)[:,1]
```

In the end, the online private score is 0.82260, and the online public score is 0.79545. It can be found that the effect of the LR model is difficult to meet the expected requirements under the same characteristics. However, the LR model is often used as a second-level learner in stacking integration, which can reduce the risk of overfitting and learn the linear relationship between features and targets.

14.4.2   CatBoost

At first, the CatBoost model was proposed mainly to solve the problem of category features. It was the high-quality category coding and feature crossover ability that made it popular. Since the CatBoost model appeared relatively late, it could be called a young algorithm, which slowly made a name for itself in the second half of 2017. Therefore, CatBoost cannot be seen in this competition. Next, CatBoost modelling will be performed, and the construction code is as follows:

```
from catboost import CatBoostClassifier
params = {'learning_rate':0.02, 'depth':13, 'l2_leaf_reg':10,
    'bootstrap_type':'Bernoulli',
    'od_type': 'Iter','od_wait': 50,'random_seed':
11,'allow_writing_files': False}
clf = CatBoostClassifier(iterations=20000, eval_metric='AUC',
**params)
clf.fit(trn_x,trn_y, eval_set=(trn_x, trn_y),
    cat_features=categorical_features,
    use_best_model=True,
    early_stopping_rounds=20,
    verbose=10)
```

In the end, the offline verification set score is 0.9850, the online private score is 0.97538, and the online public score is 0.97655.

14.4.3   LightGBM

This section chooses a very stable LightGBM model. In addition, the percentage of positive samples of this competition question is 0.247%, and the label distribution is extremely unbalanced, so this section will try to carry out negative sampling processing, and judge from practice whether the overall training performance can be improved under the condition that the score is basically unchanged.

14.4.3.1   Full Data

In order to be able to obtain higher scores, two trainings will be conducted: the first training uses the data records of the seventh day and the eighth day as the training set and the data records of the ninth day as the verification set; the second time of

training first obtains the optimal number of iterations from the first training, and then
uses the data records of the seventh, eighth, and ninth days as the training set to train
and make the final prediction. The specific implementation code is as follows:

```
trn_x = data[data['day']<9][features]
trn_y = data[data['day']<9]['is_attributed']
val_x = data[data['day']==9][features]
val_y = data[data['day']==9]['is_attributed']

params = { 'min_child_weight': 25,
   'subsample': 0.7,
   'subsample_freq': 1,
   'colsample_bytree': 0.6,
   'learning_rate': 0.1,
   'max_depth': -1,
   'seed': 48,
   'min_split_gain': 0.001,
   'reg_alpha': 0.0001,
   'max_bin': 2047,
   'num_leaves': 127,
   'objective': 'binary',
   'metric': 'auc',
   'scale_pos_weight': 1,
   'n_jobs': 24,
   'verbose': -1,
   }
train_data = lgb.Dataset(trn_x.values.astype(np.float32),
label=trn_y,
   categorical_feature=categorical_features,
feature_name=features)
valid_data = lgb.Dataset(val_x.values.astype(np.float32),
label=val_y,
   categorical_feature=categorical_features,
feature_name=features)

clf = lgb.train(params, train_data, 10000,
   early_stopping_rounds=30,
   valid_sets=[test_data],
   verbose_eval=10
)
```

Then, the training set and the testing set are merged. Because the training set is
expanded, the optimal number of iterations is expanded 1.1 times, and finally the
trained model is used to predict the results of the testing set.

```
trn_x = pd.concat([trn_x, val_x], axis=0, ignore_index=True)
trn_y = np.r_[trn_y, val_y]
# Here the two matrices are joined by columns, that is, adding the two
matrices up and down, requiring the same number of columns, similar to
concat() in pandas
del val_x
```

```
del val_y
gc.collect()
train_data = lgb.Dataset(trn_x.values.astype(np.float32),
label=trn_y,
      categorical_feature=categorical_features,
feature_name=features)
trees = 400
clf = lgb.train(params,
      train_data,
      int(trees * 1.1),
      valid_sets=[train_data],
      verbose_eval=10
      )
```

14.4.3.2   Negative Sampling Data

When optimizing negative sampling, it should be noted that negative sampling needs to be performed after feature extraction is completed for all data. If negative sampling is performed before feature extraction, it will affect the true description of the original data source distribution, and the features constructed in this way do not have real meaning.

The following gives a random negative sampling code; after sampling of negative samples, the model can be trained directly.

```
# negative sampling is performed on the training set
df_train_neg = data[(data['is_attributed'] == 0)&(data['day'] < 9)]
df_train_neg = df_train_neg.sample(n=1000000)

# merge into new data sets
df_rest = data[(data['is_attributed'] == 1)|(data['day'] >= 9)]
data = pd.concat([df_train_neg, df_rest]).sample(frac=1)
del df_train_neg
del df_rest
gc.collect()
```

14.4.4   DeepFM

DeepFM is also a classic model used in CTR, CVR problems, and its structure is a combination of FM and deep neural networks, as shown in Fig. 14.14. Therefore, DeepFM not only has the ability of FM to automatically learn cross features, but also introduces implicit high-order cross information of neural networks. In the specific implementation part of DeepFM, it is mainly divided into three parts: FM layer, DNN layer, and Liner layer. Finally, the results of the three parts are spliced together and input to the output layer to obtain the final results.

The following is the specific implementation code of DeepFM:

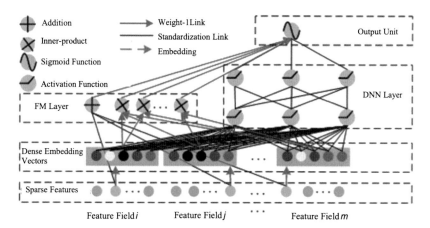

Fig. 14.14 Network structure diagram of DeepFM

```
from tensorflow.keras.layers import *
import tensorflow.keras.backend as K
import tensorflow as tf
from tensorflow.keras.models import Model
from keras.callbacks import *

def deepfm_model(sparse_columns, dense_columns, train, test):
    # the processing part of sparse features
    sparse_input = []
    lr_embedding = []
    fm_embedding = []
    for col in sparse_columns:
      _input = Input(shape=(1,))
      sparse_input.append(_input)
      nums = pd.concat((train[col], test[col])).nunique()
      embed = Embedding(nums, 1, embeddings_regularizer=tf.
        keras.regularizers.l2(0.5))(_input)
        embed = Flatten()(embed)

        lr_embedding.append(embed)
      embed = Embedding(nums,10,embeddings_regularizer=tf.
        keras.regularizers.l2(0.5))(_input)
        reshape = Reshape((10,))(embed)
        fm_embedding.append(reshape)

    # FM processing layer
    fm_square = Lambda(lambda x: K.square(x))(Add()(fm_embedding))
    square_fm = Add()([Lambda(lambda x:K.square(x))(embed) for embed
in fm_embedding])
    snd_order_sparse_layer = subtract([fm_square, square_fm])
    snd_order_sparse_layer = Lambda(lambda x: x * 0.5)
(snd_order_sparse_layer)
```

```
    # processing of numerical features
    dense_input = []
    for col in dense_columns:
        _input = Input(shape=(1,))
        dense_input.append(_input)
    concat_dense_input = concatenate(dense_input)
    fst_order_dense_layer = Activation(activation="relu")
        (BatchNormalization()(Dense(4)(concat_dense_input)))

    # splicing of linear parts
    fst_order_sparse_layer = concatenate(lr_embedding)
    linear_part = concatenate([fst_order_dense_layer,
fst_order_sparse_layer])

    # splicing the FM embedding vectors with the numerical features and then "feeding" into the
FC part after together
        concat_fm_embedding = concatenate(fm_embedding)
    concat_fm_embedding_dense = concatenate([concat_fm_embedding,
        fst_order_dense_layer])
    fc_layer = Dropout(0.2)(Activation(activation="relu"
                      )(BatchNormalization()(Dense(128)(concat_fm_
embedding_dense))))
  fc_layer = Dropout(0.2)(Activation(activation="relu")
          (BatchNormalization()(Dense(64)(fc_layer))))
   fc_layer = Dropout(0.2)(Activation(activation="relu")
  (BatchNormalization()(Dense(32)(fc_layer))))

    # output layer
    output_layer = concatenate([linear_part,
snd_order_sparse_layer, fc_layer])
    output_layer = Dense(1, activation='sigmoid')(output_layer)
          model = Model(inputs=sparse_input+dense_input,
outputs=output_layer)

    return model
```

The following is the code of the final training phase, which is divided into data conversion, compilation, callback function, and training. The overall structure is very common.

```
    train_sparse_x = [trn_x[f].values for f in categorical_features]
    train_dense_x = [trn_x[f].values for f in numerical_features]
    train_label = [trn_y]
    valid_sparse_x = [val_x[f].values for f in categorical_features]
    valid_dense_x = [val_x[f].values for f in numerical_features]
    valid_label = [val_y]
    # compilation part
    model = deepfm_model(categorical_features, numerical_features,
trn_x, val_x)
    model.compile(optimizer="adam",
        loss="binary_crossentropy",
        metrics=["binary_crossentropy", tf.keras.metrics.AUC
```

```
(name='auc')])
    # callback function
    file_path = "deepfm_model.h5"
    checkpoint = ModelCheckpoint(
    file_path, monitor='val_auc', verbose=1, save_best_only=True,
        mode='max', save_weights_only=True)
    earlystopping = EarlyStopping(
        monitor='val_auc', min_delta=0.0001, patience=5, verbose=1,
mode='max')

    callbacks = [checkpoint, earlystopping]

    hist = model.fit(train_sparse_x+train_dense_x,
        train_label,
        batch_size=8192,
        epochs=50,
        validation_data=(valid_sparse_x+valid_dense_x, valid_label),
        callbacks=callbacks,
        shuffle=True)
```

The above is the complete and runnable code of the DeepFM model. The overall implementation is still very simple, and you can get results that are very different from the tree model in the first two sections.

14.5   A Summary of the Competition Question

Throughout the competition process, we tried many feature extraction methods and different models. In addition to them, the more important part is the summary of the competition question. This section will introduce the solutions of the top ranked players, sort out the key knowledge points, and take you to learn and know more about similar competitions together.

14.5.1   More Schemes

14.5.1.1   Top 1 Scheme

The champion used multiple LightGBM models and neural networks models in terms of models and performed weighted integration. Different from most players, the champion player carried out negative sampling processing in the model training phase, that was, selecting all positive samples (is_attributed == 1), and negative samples of the same sample size, which meant that 99.8% of the negative samples were discarded. It could be noticed that the performance of the model was not greatly affected (the feature engineering part was to extract features on all data, not only on the data after sampling). In addition, different proportions of negative samples were sampled first to train the models respectively, and then the models were integrated. The final result would be improved (i.e. 5 models were trained and each model sampled different random seeds), and the training time of the models could be greatly reduced.

In the verification phase, the data of November seventh and November eighth were selected for training, and the data of November ninth was used for verification. After obtaining parameters such as the number of iterations, the model was re-trained on the data of November seventh to November 9th. There were 646 features in the data set. The score after fusing the five LightGBM models was 0.9833 on the online public leaderboard and 0.9842 on the online private leaderboard. The final submission scheme was a weighted average based on ranking, which combined 7 LightGBM models and 1 neural networks model, and the online public leaderboard score was 0.9834.

In the aspect of feature engineering, it mainly included aggregation statistics, the number of clicks (count feature) for the groupby feature construction in the next 1 h and 6 h, the calculation of time difference between forward and backward clicks for groupby feature construction, and average download rate of historical click time for groupby feature construction. In addition, for the combination of classification variables (20 in total), LDA, NMF, LSA were used to get embedding features (embedding), and n_component was set to 5, so that each method could get 100 features, and finally a total of 300 features (LDA, NMF, PCA).

14.5.1.2 Top 2 Scheme

The runners-up also chose the LightGBM model and the neural network model in terms of models. The best single model was LightGBM, with an online private leaderboard score of 0.9837, and the private lb. score of the best neural network model was 0.9834 (for classification variables, a Dot_Product layer was used to feed continuous features into the FC layer). In terms of model integration, everyone in the runner-up team trained a LightGBM and neural network model. There were a total of 6 models. The results of these 6 models could be directly weighted and averaged.

As for the model training and verification, the runner-up also performed negative sampling processing, and the specific proportion of negative sampling was not informed. Feature extraction was performed on all data, and then the features were merged into the sampled samples; that was, the features were constructed first and then sampled, and finally five-fold cross-validation was used for offline model evaluation.

Regarding feature engineering, statistical features were mainly constructed through aggregation, such as count, cumcount, nunique, and time difference features. There was also a special feature that combined IP addresses in apps, os, and channels respectively and calculated the number of clicks of each attribute, and then attributes of high-frequency were directly selected as category features.

Table 14.1 Description of different feature types

Feature types	Specific features	Descriptions
User Information	Age, gender, occupation, user level, interest preference	Extract the basic information features on the user side
Ad Information	Advertising types, advertising materials, advertisers, advertising industry	Extract the basic information features on the user side
Context	Advertising space type, operator, advertising location	Extract the basic characteristics of context information
First-order Features	count、nunique、rank、target encoding	The most basic statistical features all have structural significance
Second-order Features	Cross-combine count, ratio, groupby, target encoding	Perform cross-combination for a section of features to obtain more fine-grained feature description, and of course, cross-combination of third or higher orders can also be carried out
Time Correlation	Time feature (year, month, day, hour), time series features (historical statistics), time difference feature	Time information appears in many log data sets; time related statistics often can find out a lot of useful information
Embedding Features	Word2Vec, Graph embedding, TF-IDF, combined with PCA, LDA and other text to find out algorithms	Extract the entity representation vectors of the user or AD

14.5.1.3   Top 3 Scheme

In terms of models, the players who took the third place also chose the LightGBM model and the neural network model. The difference was that the team constructed many neural network models with different structures to increase the diversity of the models, such as modeling the time series information of clicks through circular neural networks, adding res-links to the FC layer, etc. Stacking integration was used in the model integration stage, and the output results of the model were used as features, combined with more new features to continue to participate in training.

With regard to feature engineering, 23 features were mainly used, of which the most important was the time difference feature, which extracted the time difference between the first 5 clicks and the last 5 clicks of each click event as a feature. In addition to the app, device, os, channel features in the original data source, there were also hour features and statistical features.

14.5.2   Sorting Out Knowledge Points

This section will sort out the feature extraction methods of CTR/CVR related competition questions in computational advertising in detail. When extracting features, you not only need to face very rich and high-dimensional data, but also have extremely high accuracy requirements for the results. Table 14.1 will describe different feature types.

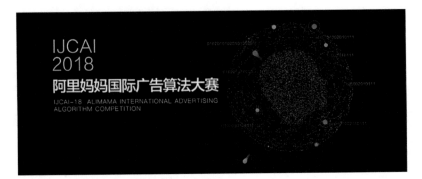

Fig. 14.15 Home page of competition question of IJCAI 2018 Alimama search ad conversion forecast

14.5.3   Extended Learning

This competition can be regarded as a competition related to CTR/CVR. The focus of this type of competition is on feature engineering and models. Feature engineering is mainly feature extraction, such as aggregate statistical features, target coding features, and embedding-related features. There are a variety of choices in models, mainly models related to ad click-through rate prediction, such as FM, FFM, DeepFM, Wide & Deep, etc. Next, similar classic competition questions will be given to help further understand such competition questions and achieve the effect of solving problems at ease.

14.5.3.1   IJCAI 2018 Alimama Search Ad Conversion Forecast

The first competition topic is "Alimama Search Ad Conversion Forecast" (the home page of the competition topic is shown in Fig. 14.15). The competition topic askes participants to build a forecast model through artificial intelligence technology, so as to predict the user's purchase intention, that is, to provide five types of information related to historical advertisement click events: users (user), advertising products (ad), search terms (query), context information (context), and shops (shop), and to predict the probability of purchase behavior (pCVR) of the next advertisement.

Combined with the business scenarios of Taobao platform and different traffic characteristics, the official definitions of the following two types of challenges are: daily conversion rate forecast (in the preliminary round) and conversion rate forecast for special dates (in the final round).

Basic ideas: The difficulty of this question lies in the conversion rate estimation for special dates during the quarter final, which not only requires the accuracy of the forecast results, but also asks for effective offline verification. In the preliminary round, the data of the last day can be used as the verification set, and the rest as the training set. Due to the need to estimate the conversion rate of the special date, and the difference between the special date—day 7- and days before is relatively large, so the data of the first half day of the seventh day is used for training, and the data of the last 2 h of it is used as the verification set.

The specific modeling scheme can be roughly divided into three categories:

1. Only the data of the first 7 days are used to predict, and the data of the last day is used for offline verification. Because the data distribution of the training set and the testing set is different, the effect of this scheme is relatively poor.
2. Only the data of the last half day is used to predict; thus, the data information of the first 7 days will be lost.
3. The champion uses the transfer learning method, first using the data of the first 7 days to train and predict, merging the obtained results into the last half day, and then using only the data of the last half day to complete the final training and prediction. In this way, not only can the information of the first 7 days be retained, but also the predicted result is closer to what in the last day.

The features constructed in this question can be roughly divided into original features, statistical features (count, nunique, mean, etc.), time difference features (between the last n clicks and the next n clicks), segmented features (discrete continuous features such as hour, score, rate, etc.), probability features (conversion rate and proportion calculation), etc.

14.5.3.2   2018 iFLYTEK AI Marketing Algorithms Competition

This competition provides a large amount of advertising data from iFLYTEK AI marketing cloud. Participants are required to construct a prediction model by taking advantage of artificial intelligence technology and estimate the probability of users clicking on advertisements, that is, to provide advertising, media, users, contextual information related to advertising click events, and to predict the probability of user's clicking advertisements.

This is also a question about CTR forecast. For such questions, the dominant factor for whether an advertisement is clicked is the users, followed by advertising information. So, what we need to do is to fully dig the information of users and user behavior, and then the information of advertisers, advertisements, etc. The evaluation index of this question is logarithmic loss.

Basic ideas: The difficulty of the question is the processing of user_tags multi-valued features. Because it contains the attribute information of the users, it is very important to be able to perfectly express user_tags (extract effective attributes and reduce redundancy). For the processing of multi-valued features, the most basic thing is to use CountVectorizer for expansion, and then use chi-square test for feature selection. Another more efficient way is to use LightGBM feature importance analysis to extract top tags, which is a certain degree of improvement compared to chi-square test. There is also the click-through rate feature, because this feature contains time information, so extracting historical click-through rates as features can largely avoid data overfitting and leakage.

Moreover, the competition data lacks the key information of user ID (uid), which makes it difficult to clearly establish a user profile. Therefore, how to fully find out the information contained in the user tag is crucial. There are still some anonymous data in the data. At this time, it is necessary to fully understand and analyze the data, and even try to perform reverse encoding according to business understanding, in order to point out the direction for feature engineering. In the modeling process, the interaction between user tags and other information is fully considered, the use of dimension and memory is reduced by using stacking to extract feature information, and the advertisement and user interaction information are fully discovered, so that the model can remain relatively stable in tests both in A and B list.

Part V
Listen to What You Say and Understand What You Write

Chapter 15
Natural Language Processing

With the continuous development of social platforms and content platforms, it has become increasingly common to use the Internet as a carrier for content dissemination. The application of recommendation systems, computational advertising, user profile analysis, and other related technologies make it possible to screen massive information streams for personalized information. In all data, the use of multimodal data (text, images, audio, video, and structured data, etc.) makes it more likely to predict more accurately. Text data, as the main information carrier of the content platform, has also become an indispensable data type for the above technologies, and how to use text data is the core problem to be solved in natural language processing.

15.1 Development of Natural Language Processing

The development of natural language processing can be divided into three stages:

1. From 1950 to 1970, a stage based on experience and rules;
2. From 1970 to 2008, a stage based on statistical methodology;
3. From 2008 till now, a stage based on deep learning technology.

At stage (1) (the early stage), Turing test was regarded as a test standard to judge the level of artificial intelligence. As part of Turing test, the recognition of natural language input started the research on natural language processing. At this stage, template construction and grammar analysis based on empiricism and artificial rules became the mainstream of growth and progress. However, due to timeliness and variability of languages and the fact that setting rules depended extremely on linguists and relevant knowledge in the field, fixed rules often could not cover language recognition in most common scenarios.

At stage (2), with the popularization of computers and the development of the Internet, statistical methods appeared as a new scheme under the background of the accumulation of a large amount of data. The traditional scheme required a large number of manual operations to summarize knowledge; the statistical method replaced this scheme and was applied to all kinds of industrial natural language processing scenarios, obtaining relatively good effect. However, the statistical methods that could be adopted at this stage, such as Bayesian model, bag of words model/TF-IDF, N-Gram language model, etc., were only able to handle some tasks that were not particularly complex. For tasks that contained rich language information, complex language structures, and contextual scenarios, this method was still very mediocre.

At stage (3), many deep learning algorithms have been applying to natural language processing. The early word embedding model and the subsequent development of convolutional neural networks and recurrent neural networks have played a very important role for this time period, greatly improving the accuracy baseline of the original statistical method and achieving more generalized effects in different fields (such as translation, voice recognition, and other tasks). In the current latest environment, the self-attention mechanism models like transformer structures are applicable to a sea of data, which could then generate training models. This ability further helps natural language processing to develop, which has even achieved baseline scores that exceed what human can get on some tasks.

15.2   Common Scenarios of Natural Language Processing

The goal of natural language processing technology is to recognize human language through various electronic machines such as computers, so as to understand human intentions. Natural language processing can better free labor from complicated tasks in some specific fields. Take Taobao as an example; dialog systems can analyze and identify customer questions, locate customer needs, and provide answers to corresponding questions or specific operations of the purchase process, thus saving merchants a lot of repetitive labor and time cost. In the vehicle-mounted voice system, the voice recognition system is combined with the natural language processing system to free up the hands of drivers while driving and provide corresponding services (such as wayfinding, playing music, etc.).

According to different task scenarios, the technologies used in the development of natural language will vary significantly. This section will introduce several common natural language processing tasks.

15.2.1   Classification and Regression Tasks

This kind of task is also the most common task in traditional machine learning. How to vectorize natural language features and use traditional or deep learning models to train predictions are the main concerns of this type of task. Its typical tasks include semantic analysis, sentiment analysis, intention recognition, etc., usually involving text representation and model selection.

15.2.2   Information Retrieval, Text Matching, and Other Tasks

Information retrieval, text matching, Q & A, and other tasks need to predict a large number of Q-Q, Q-A pairings. The focus of this type of problem is how to use data feature construction as well as how to select appropriate models to realize rapid retrieval and matching between text and text. The common work of searching based on keywords is actually a special subcategory of this category of task. And more complex matching based on semantics and even logical judgment is still a problem that has aroused great concern within the academic and industrial circles.

15.2.3   Sequence to Sequence, Sequence Labeling

This kind of problem pays more attention to the generation and annotation of sequences. Its common tasks consist of voice and text interconversion, machine translation, name entity recognition, etc. The methods used are usually deep learning CNN, RNN, transformer structure, etc. In addition, in appropriate scenarios, it will also be used in combination with other traditional models such as CRF, MRF, HMM.

15.2.4   Machine Reading

Machine reading is a method of giving questions and text, and then finding out the answers that meet the requirements from the text according to the questions. In the past, traditional methods often had limitations in data and technical means and did not achieve the desired results. With the continuous development of deep learning and pre-training language models, an increasingly number of latest technologies are applied to the field of machine reading. By mining the semantics of the context, the attention mechanism is used to identify the answers to questions in specific scenarios.

Natural language processing technology is usually not applied independently, but integrates other feature data, and cooperates with other media, structured data, etc. to fulfill multimodal data prediction.

15.3 Common Technologies of Natural Language Processing

For different tasks, the feature generation scheme of natural language processing also differs greatly. According to the characteristics of the task to be solved, how to choose the data processing method and model should be considered first. Although deep learning has a very high prediction upper limit, it does not mean that traditional natural language feature processing will be abandoned in normal applications. What is the opposite is in some tasks with high requirements on response time, model complexity and size, and model interpretability, traditional statistical-based feature extraction methods and machine learning models will play an extremely important role. Next, we will list some common text feature extraction methods.

15.3.1 Feature Extraction Based on Bag-of-Words and TF-IDF

The bag-of-words model is the simplest and most direct feature extraction method. This model is often applied to the field of information retrieval. The bag-of-words model usually ignores the context relationship of words in the text and assumes that words are context-independent. Such assumptions can well characterize the information of the entire sentence by the frequency of word occurrence without losing a certain prediction accuracy.

By constructing a dictionary of corpus, the originally discrete word set can be characterized as a sparse vector with dictionary size.

For example, when we have the following two statements:

```
We have noticed a new sign in to your Zoho account.
We have sent back permission.
```

Then the dictionary structure for these two sentences is:

```
{'We': 2, 'have': 2, 'noticed': 1, 'a': 1, 'new': 1, 'sign': 1, 'in': 1,
'to': 1, 'your': 1, 'Zoho': 1, 'account.': 1, 'sent': 1, 'back': 1,
'permission.': 1}
```

The BOW features generated by the two statements are: [1, 0, 1, 1, 1, 1, 0, 0, 1, 1, 1, 1, 1] and [0, 1, 1, 0, 0, 0, 1, 1, 0, 0, 1, 0, 0].

By vectorizing the discrete features in the above way, traditional machine learning models (such as logistic regression models, neural network models, tree models, SVM, etc.) can be used for training.

The bag-of-words model only considers whether a word appears in a sentence or not but does not take into account the importance of the word itself in the sentence.

Therefore, the TF-IDF method is proposed. The value of TF × IDF is used to weight each word that appears. The word has better representation ability in the text.

TF is calculated as: the number of occurrences of words in a sentence/the total number of words in a doc.

IDF is calculated as: log (total number of docs/total number of docs containing words).

After calculating TF and IDF based on the above two formulas, multiply the two to obtain TF-IDF. Among the algorithms that characterize the meaning of sentences by constructing sparse matrices, the bag of words model BOW and TF-IDF methods have the advantages like simplicity, ease of use, and fast speed. At the same time, the disadvantages are also obvious; that is, when the text corpus is scarce and the dictionary size is greater than the text corpus, due to the lack of sufficient corpus for feature construction, the statistical information basis for the representation of words will be insufficient and therefore will lead to overfitting of the model in the training process.

15.3.2   N-Gram Models

In natural language processing, the representation of sentences is an important issue. Early statistics-based methods proposed such a scheme: there is a sentence $S(w_1, w_2, w_3, \ldots w_n)$, where w_i represents the word in the sentence. It is required to calculate the occurrence probability $p(S)$ of the sentence, and its expression function is $p(S) = p(w_1) \times p(w_2) \times p(w_3) \times \ldots \times p(w_n)$. The Markov hypothesis is added to the prefix of this public formula. Assuming that the occurrence probability of the current word is only related to the first n words, the N-Gram model can be modified to $p(S) = p(w_1) \times p(w_2| w_1) \times \ldots \times p(w_n \mid w_{n-1})$. Combining the concepts of N-Gram models and bag-of-words models can further improve the forecasting ability of text features.

BOW and TF-IDF models can be combined with N-Gram models to generate additional sparse feature vectors by constructing Bi-Gram, Tri-Gram, etc. The constructed features are better than those BOW and TF-IDF features using Uni-Gram, having more representational capabilities and being able to obtain certain contextual information, but they still cannot handle long sequence dependence well.

15.3.3   Word Embedding Models

There is an unsolved problem in the bag-of-words model: if synonyms appear in different texts, then when calculating the similarity of this type of text or making predictions, there will be occasions when it is not be possible to identify the semantic words with similar contexts if the training data does not contain a large number of annotations.

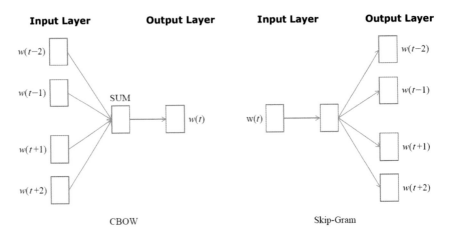

Fig. 15.1 Structure of CBOW and skip-gram algorithms

The later appeared word embedding model solves this kind of problem well, and the currently commonly used word vector algorithms include Word2Vec, glove, fastText, etc. In addition, for the pre-training of Chinese word vectors, there is also the AI Lab word vector disclosed by Tencent.

A prior assumption of word vectors is that the information of the current word can be inferred from the context, so it has good generalization ability for some rare words, polysemous words, and even common misspelled words, etc. Taking Word2Vev as an example, the common model training methods are divided into two algorithms: CBOW and Skip-Gram, as shown in Fig. 15.1.

The word vector matrix generated by training will record the vectors generated by each word training in the form of a query table. These vectors correspond to different tasks and can play different feature extraction roles as follows.

- In the traditional feature extraction, a weighted summation method can be used to sum the vectors of all words in the sentence, and finally generate a sentence vector that can be used to characterize the sentence. Sentence vectors can be used to calculate the cosine similarity between text and text, etc.
- As the initialization parameters of the word embedding layer of the deep learning natural language processing model, higher accuracy can be obtained than what is obtained by the model trained by the end-to-end method.
- Construct the feature of aggregate class similarity of the matching task, which can be used for word-level similarity calculation and calculate the statistical numerical construction features such as average value, median value, maximum value, and minimum value based on different dimensions.

15.3.4   Context-Sensitive Pre-training Models

Although the word embedding model can solve the problem of different words having the same or nearly the same meaning, in actual natural language processing task scenarios, even same words often have different meanings in a variety of context scenarios. The result of the word embedding model is the static vector of the word. We often need to use in-depth learning models to construct context relations on the basis of this vector, which leads to the research on context-related pre-training models. The development of this type of model is highly related to the word vector model. From the early ELMo model built based on bi-directional Bi-LSTM to the GPT and BERT structural models constructed by introducing Multi-Head Attention mechanism, all of them use the Seq2Seq language model, apply massive text data, adopt language models or autoencoder modes to train the context semantics of words, thereby encoding and compressing a sea of text semantic information in the sequence model.

Existing context-sensitive pre-training models include: ELMo, GPT, BERT, BERT-wwm, ERNIE_1.0, XLNet, ERNIE_2.0, RoBERTa, ALBERT, ELECTRA. Next, we will list some common sequence models.

- **ELMo models**. ELMo is a language model trained in the form of an autoregressive language model. The essence of an autoregressive language model is to predict the next word through the input text sequence. By continuously optimizing the accuracy of the prediction, the model gradually learns the semantic relationship of the context. The structure of the ELMo model includes the forward direction LSTM layer and the reverse LSTM layer, and could achieve better prediction results by optimizing the next word of forward direction and the reverse next word respectively.
- **GPT models**. After Google announced the Multi-Head Attention mechanism and the transformer structure, GPT applied them to the pre-training of language models. GPT models adopt the forward transformer structure, remove the decoder, and use the autoregressive language model for training just like ELMo models. Compared with the ELMo pre-training model, the GPT structure based on massive data training greatly exceeds the original benchmark at that time.
- **BERT/RoBERTa models**. Unlike ELMo models and GPT models using autoregressive language model, BERT models use autoencoder mode for training, and the model structure contains forward and reverse transformer structures. In order to reduce the information overflow caused by bidirectional Transformer structure and autoencoder mode, BERT introduces MLM (Masked Language Model) in the training process, and 15% of the entries (tokens) in the pre-training will be obscured. For these 15% tokens, there is an 80% probability of using [MASK] replacement, a 10% probability of random replacement, and a 10% probability of remaining the same. This replacement strategy plays a regularization role in the model training process and can prevent the BERT model from learning the word in the "future" during training due to its two-way Self-Attention structure, and thus from causing overfitting.

- **RoBERTa models**. It was proposed by Facebook. It has removed NSP (Next Sentence Prediction) mechanism from the BERT model, modified the MLM mechanism, adjusted its parameters, and finally obtained the accuracy beyond original BERT models.
- **ERNIE models**. Baidu optimized the pre-training of Chinese data on the basis of BERT models and added three levels of pre-training on the basis of BERT: Basic-Level Masking (the first layer), Phrase-Level Masking (the second layer), and Entity-Level Masking (the third layer). They add prior knowledge from the three levels of words, phrases, and entities respectively to improve the prediction ability of the model on Chinese corpus.
- **BERT/RoBERTa-wwm**. This kind of model was released by the Joint Laboratory of Harbin Institute of Technology and iFLYTEK. It is not a new model in the strict sense; wwm (whole word mask) is a training strategy. In the MLM strategy mentioned in the original BERT paper, it has a certain randomness and will mask out the word pieces of the original word. By comparison, in the wwm strategy, after the word pieces to which the same word belongs are hidden, other parts of the same word will also be masked out together, which ensures the integrity of the word without affecting its independence.

As for the application of context-sensitive pre-training models, massive data can be used to pre-train the language model first, and then carry out finetune on its downstream tasks. Compared with end-to-end training methods, choosing a suitable pre-training model can usually achieve better results, but this does not mean that the pre-training model can achieve the same effect on all tasks. The final effect is closely related to the training corpus, training strategies, and the fields to which the downstream tasks belong in the pre-training process of the model. For upstream models and downstream tasks with huge differences in their fields, the results of fine tuning do not achieve the expected results, and in some cases are even worse than the results obtained by using word vector + Bi-LSTM.

15.3.5 Common Deep Learning Model Structures

Pre-training models can usually bring high accuracy benefits, but their complex model structure also brings additional time cost for training and prediction. Compared with complex pre-training models such as BERT, the use of general convolutional neural networks and recurrent neural networks to train models has a wider range of applications in practice. These model structures are simpler, the number of parameters is less, and the training and inference time is shorter.

The following part describes common deep learning models and corresponding structures.

- **TextCNN**. This model is characterized by its simple structure, fast training and prediction speed, and higher accuracy than traditional models. Its design concept is derived from the N-Gram model, which uses multi-scale convolution kernels to

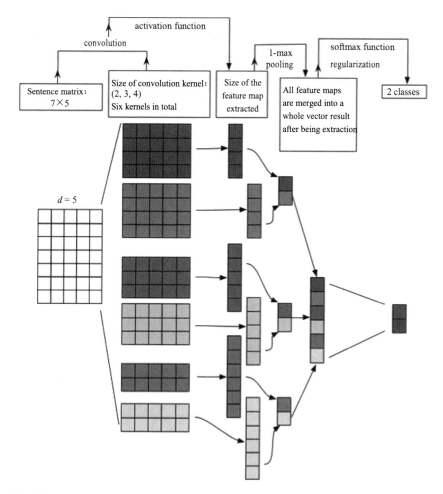

Fig. 15.2 TextCNN model structure

simulate the operation of the N-Gram model on the text, and finally merges before making prediction, suitable for short texts and corpus with obvious phrase structures. Its structure is shown in Fig. 15.2.

- **LSTM, Bi-LSTM/Bi-GRU + Attention**. These are very typical bidirectional recurrent neural network structures. In recurrent neural networks, both the LSTM layer and the GRU layer have very good time series fitting ability, and the model after the Attention mechanism is added to weight the state values at different times, which can further improve the prediction ability of the model and is relatively suitable for texts with complex semantic contexts.

The LSTM module is a recurrent neural networks structure module, and its structure diagram is shown in Fig. 15.3. Compared with the ordinary cyclic neural networks module, LSTM alleviates the gradient disappearance in the training

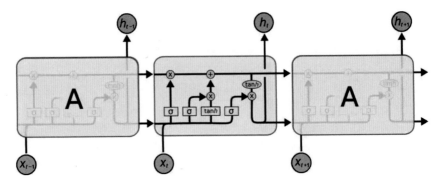

Fig. 15.3 LSTM structure diagram

process, and its added "gate" structure is used to control the memory, output, forgetting state, which is conducive for the model to obtain better results in long sequences.

By arranging the sequence forward and backward, and using Bi-LSTM (bidirectional LSTM), the reverse coding information that cannot be obtained by LSTM can be obtained, which increases the fitting ability of the model. The Attention mechanism is analyzed in principles. It is an operation that weights the different states of words in a sentence. From the most primitive weighted average, it has gradually developed into the most popular Self-Attention, etc. Its core idea has always been to use the similarity matrix of the word to carry out calculation, so as to adjust the weight corresponding to the word in the sentence, thereby allowing the weighted sum of the word to be used as the output or the input of the next layer. The model constructed by combining Bi-LSTM and the Attention layer can better train the input text.

- **DPCNN.** TextCNN structure can simulate the N-Gram models and has the ability to extract phrases from words. However, when faced with texts with complex text structures, rich semantics, or strong context dependence (such as words or phrases with long sequence head-to-tail dependence), TextCNN often fails to achieve good results, which is caused by the capacity of its shallow structure itself. Therefore, it is necessary to adopt a deeper convolutional neural network structure—DPCNN, which draws on the concept of residual block from the ResNet structure to simulate the layer-by-layer extraction of image features in the CV task, which is relatively suitable for long text and text with complex grammatical structures. Its structure is shown in Fig. 15.4.

15.4   Thinking Exercises

1. If you want to design a dialog system, what should you design and what model should you apply?

Fig. 15.4 DPCNN
structure

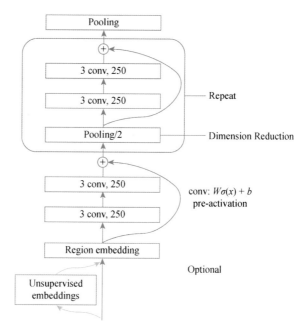

2. What task does the core of natural language processing technology solve? Which can be used in applications, and which cannot?
3. If the timeliness, storage medium, and inference time of a task are all restricted, how should models be selected and designed?

Chapter 16
Case Study: Quora Question Pairs

This chapter will take the Quora Question Pairs competition on the Kaggle competition platform (as shown in Fig. 16.1) as an example to explain the text matching case of natural language processing. Quora is a question-and-answer SNS website, similar to Zhihu (a Chinese question-and-answer website), which provides social question-and-answer services. There are many questions created by users on the Quora platform, and these questions often contain a lot of repetitive content; there will also be related questions, or the same type of questions being asked many times. In the scenario of real business, the above phenomenon often disperses the flow of high-quality answers, so it is necessary to match the same problems in business, so as to normalize repeated problems.

16.1 Understanding the Competition Question

Although the type of competition of text matching in natural language processing has similarities with the traditional ML problems, it also has its own unique characteristics. The feature construction and model selection of this competition are often different from those of other tasks, requiring additional reserved knowledge of feature engineering.

16.1.1 Competition Background

For question-and-answer platforms such as Zhihu in China and Quora, whether they can integrate high-quality answers determines whether their traffic is centralized. For repeated questions, it is difficult to integrate the answers manually or simply by extracting and matching keywords. For problems containing complex semantics and grammar, unexpected situations often occur according to their features. For example,

Fig. 16.1 Home page of quora question pairs competition

if the method of matching sentence segmentations or phrases is adopted, even if more than 90% of the contents in the sentence are the same, matching cannot be completed as long as there are opposite semantics or different keywords.

Considering these situations, further manual processing of text features should be carried out, or a deep learning model that can capture such features should be adopted to build the whole scheme.

16.1.2   Competition Data

After downloading the data compression file from the official website and decompressing it, the specific fi les obtained are train.csv (training set), test.csv (testing set), sample_submission.csv (correct and standardized example submission file, containing all card_id that needs to be predicted by the contestant), Train.csv. zip, and Test.csv.zip.

16.1.3   Competition Tasks

Predict whether the competition questions on the Quora platform repeat each other, that is, predict the probability of repetition of question1 and question2 in the test. csv file.

16.1.4   Evaluation Indicators

This competition uses logarithmic loss for computation, and the specific computation code is as follows:

```
import numpy as np
def log_loss(y,pred):
      return -np.mean(y * np.log(pred) + (1 - y) * np.log(1 - pred))
```

16.1.5   Competition FAQ

🅀How can we better grasp the tasks related to natural language processing?

🅰For problems related to natural language processing, we can start from two aspects. One is to perform model construction and training by using traditional ML methods, namely artificial extraction features, based on statistics and related technologies in the field of natural language processing; the other is to consider using the deep learning model for model training and optimization, automatically extracting features, which can save a lot of physical work of preprocessing.

🅀For natural language processing, deep learning is currently popular. Is it necessary to master traditional methods?

🅰It is necessary to use in-depth learning to solve problems related to natural language processing, often because its high computer performance depends on the response speed (the more complex the model, the slower the speed of training and inference), and cannot meet some instantaneous response requirements. Outside the competition, you often encounter scenes with strict requirements for response efficiency, as well as scenes with rough sorting and other requirements for efficiency that are greater than the accuracy requirements when traditional ML plus manual feature extraction methods tend to have higher priority.

16.2   Data Exploration

For data exploration of text types, the following aspects can be considered:

- The number of samples. The number of samples for each training set and verification set;
- The maximum and minimum length of the text, and the distribution map of the length. This aspect determines the efficiency of model training. In the process of training deep learning models, it is not always necessary to use the longest sample length as the text truncation length. Appropriate selection of the truncation length according to the text length distribution helps to improve the speed of model training, and sometimes can also improve accuracy (reduce overfitting);
- The number of words in the text dictionary, analysis of stop words: to determine the word vector or deep learning pre-training model used
- Key words, word clouds, etc.: to visually understand the hot content of the text

16.2.1   Field Category Meaning

As shown in Fig. 16.2, the meaning of the each field in the data set is described.

16.2.2   Basic Quantity of Data Sets

First, analyze the total amount of data and judge whether the labels are evenly distributed. This part will decide what model will be selected as well as whether there will be a potential overfitting situation in training. If the sample size is too small, you need to add more prior knowledge or use a better pre-training model to make up for lack of model training samples and choose better data enhancement methods as well. The specific code is as follows:

```
print('Total number of question pairs for training: {}'.format(len
(df_train)))
print('Total number of question pairs for tes 的 ting: {}'.format(len
(df_test)))
print('Duplicate pairs: {}%'.format(round(df_train
['is_duplicate'].mean()*100, 2)))
qids = pd.Series(df_train['qid1'].tolist() + df_train['qid2'].
tolist())
print('Total number of questions in the training data: {}'.format(len
(np.unique(qids))))
print('Number of questions that appear multiple times:
      {}'.format(np.sum(qids.value_counts() > 1)))
```

The result is:

```
Total number of question pairs for training: 404290
Total number of question pairs for testing: 2345796
Duplicate pairs: 36.92%
```

id	training set id
qid1, qid2	the unique key of the questions, existing only in the training set
question1,	the complete text content of the questions
is_duplicate	whether the two questions repeat each other: 1 indicates yes, and 0 indicates no

Fig. 16.2 Field meaning description

```
Total number of questions in the training data: 537933
Number of questions that appear multiple times: 111680
```

From the operation result above, it can be seen that the quantity degree of the training set and verification set in the data are above the 100,000 level, which is not very large for a natural language processing task. Some of the potential prior knowledge may not be automatically identified from the sample set, and therefore needs to be compensated by using pre-training models or artificial structural features. Through analysis, we have found that there is a certain imbalance in the labels in the training set, with repeated data accounting for 36.92% of the total samples. Whether this finding will impact the verification of logarithmic loss functions needs to be further verified by the training model.

16.2.3   Distribution of Text

This section first needs to determine whether the text length maintains a consistent distribution in the training set and testing set, and at the same time has the need to find out the longest and shortest text. If there is empty text, some processing is required. By observing the distribution of text length, we will try to find an effective and reasonable text truncation length.

For English text, there are character level and word level distributions. Let's look at the two types of distributions.

The distribution code at the character level is as follows:

```
train_qs = pd.Series(df_train['question1'].tolist() +
    df_train['question2'].tolist()).astype(str)
  test_qs = pd.Series(df_test['question1'].tolist() +
    df_test['question2'].tolist()).astype(str)

dist_train = train_qs.apply(len)
dist_test = test_qs.apply(len)
plt.figure(figsize=(15, 10))
plt.hist(dist_train, bins=200, range=[0, 200], color=pal[2],
normed=True,
    label='train')
plt.hist(dist_test, bins=200, range=[0, 200], color=pal[1],
normed=True, alpha=0.5,
    label='test')
plt.title('Normalised histogram of character count in questions',
fontsize=15)
plt.legend()
plt.xlabel('Number of characters', fontsize=15)
plt.ylabel('Probability', fontsize=15)
```

The length distribution of character text in the training set and testing set is shown in Fig. 16.3.

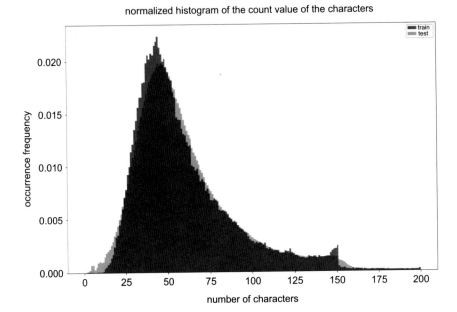

normalized histogram of the count value of the characters

Fig. 16.3 Length distribution of character text

The distribution code at word level is as follows:

```
dist_train = train_qs.apply(lambda x: len(x.split(' ')))
dist_test = test_qs.apply(lambda x: len(x.split(' ')))

plt.figure(figsize=(15, 10))
plt.hist(dist_train, bins=50, range=[0, 50], color=pal[2],
normed=True, label='train')
plt.hist(dist_test, bins=50, range=[0, 50], color=pal[1],
normed=True, alpha=0.5,
      label='test')
plt.title('Normalised histogram of word count in questions',
fontsize=15)
plt.legend()
plt.xlabel('Number of words', fontsize=15)
plt.ylabel('Probability', fontsize=15)
```

The length distribution of word text in the training set and testing set is shown in Fig. 16.4.

From Figs. 16.3 and 16.4, it can be generally concluded that the length of the text basically maintains the same distribution on the training set and the testing set. In the above analysis process, we have carried out the maximum length truncation, using 200 as the threshold at the character level and 50 as the threshold at the word level. It is explicit that in the process of analyzing the length distribution of the character text, the longest length of the text exceeds 200, but the number of samples in this part is

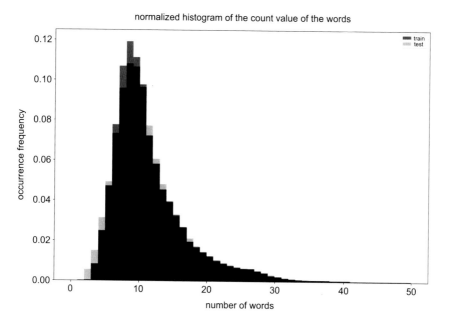

Fig. 16.4 Length distribution of word text

very small, and the same is true for the word text. At the same time, there are samples with text lengths close to 0 or being 0. Readers who have interest can further analyze these samples, print them out, and visualize them, and observe what these particularly short samples are. Is it noise, or is there a special sample? In addition, it is necessary to consider whether the appearance of particularly short text will influence the subsequent feature construction or model training, and the missing value interpolation or filling should be done in advance to avoid errors in code.

16.2.4 Number of Words and Word Cloud Analysis

Calculating the number of words helps to better understand the text and can preliminarily judge things like the proper nouns in the text. In Sect. 16.2.3, we have analyzed the distribution of word text, and next we will analyze word text from the aspects of size and content. The code is as follows:

```
txt_tmp = ' '.join(train_qs.values.tolist())+' '.join(test_qs.
values.tolist())
    words = set(txt_tmp.lower().split(' '))
    print('max number of words is %s'%len(words))
    # get the result
    # max number of words is 327537
```

Fig. 16.5 The word cloud
generated

As can be seen from the code result, the total number of words on the training set and testing set of this sample is about 320,000, which is relatively small in the quantity degree. It is almost one-tenth of that of the commonly used Word2Vec, glove, fastText, and other pre-training word vectors. Therefore, it may be necessary to use more external data or pre-training word vectors to enrich the semantics of text content.

The code for generating the word cloud is as follows:

```
import matplotlib.pyplot as plt
from sklearn.datasets import fetch_20newsgroups # Import the data set that
comes with SkLearn
import jieba
from wordcloud import WordCloud

newsgroups_train = fetch_20newsgroups(subset='train')
text = newsgroups_train.data
text = ' '.join(text)
wc = WordCloud(background_color='white',scale=32)
wc.generate(text)
plt.axis('off')
plt.imshow(wc)
plt.show()
```

With the help of the word cloud, we can simply draw a set of hot words to understand which words are the most popular in the corpus. Many samples of these words may dominate the evaluation indicators of the model. The word cloud generated is shown in Fig. 16.5.

We will find an interesting phenomenon in Fig. 16.5. Words such as "best", "difference", and "will" appear more frequently. Whether these words are accurately classified in subsequent model predictions may influence the final score.

16.2.5   Preprocessing Text Data Based on Traditional Means

For text data, the preprocessing methods will differ due to different ways of using traditional methods and deep learning models. Here, some text preprocessing methods are listed for different scenarios.

For the feature construction of text, first of all, stop words, such as "can", "is", "are", should be removed. Since such words appear in almost all sentences, they often affect the feature construction. For example, it will have a certain impact on the construction of TF-IDF features or other features based on TF-IDF features. You can try to remove stop words to get better results. For the problem of English stop words, you can use the corpus that comes with the nltk package. The code is as follows:

```
from nltk.corpus import stopwords
stops = set(stopwords.words("english"))
```

The following is an optional operation. Since there are still various tense problems in the English corpus, you can also try (not a must) to use the stem extractor (stemer) that comes with the nltk package. The nltk package provides two different stemmers. Which specific stemmer to choose is determined according to how well you use it. No matter which one is applied, the goal is to convert English words in different tenses into the same stem. The specific code is as follows:

```
from nltk.stem.porter import PorterStemmer
from nltk.stem.snowball import SnowballStemmer
def stem_str(x,stemmer=SnowballStemmer('english')):
    x = text.re.sub("[^a-zA-Z0-9]"," ", x)
    x = (" ").join([stemmer.stem(z) for z in x.split(" ")])
    x = " ".join(x.split())
    return x
```

Meanwhile, the errors of some words can also be corrected through manual effort. For cases where there are misspellings and synonyms, the editing distance can be used to make a simple recall of English words, and then manually re-label or replace the words. For some important words, when using traditional feature engineering construction methods such as TF-IDF, it is often impossible to recognize character-level misspellings, and these words have a great contribution to the forecast accuracy and need special processing.

16.2.6 Preprocessing Text Data Based on Deep Learning Models

Text data preprocessing based on the deep learning model is also divided into two parts, which are not the same as those considered in Sect. 16.2.5; it is often not necessary to take into account too many ways to deal with stop words or stems here.

The first is the training method of deep learning model which uses word vector convolution neural networks or recurrent neural networks. Such method has made great progress in the past few years. Adjusting the model structure based on the model and selecting better pre-trained word vectors are the core concerns for such methods to obtain better results. Therefore, pre-processing needs to reduce the proportion of oov (out of vocabulary) in the words for the selected pre-training model, and make all words appear in the pre-trained word vectors as much as possible. For those words that do not appear in the pre-trained word vectors, you can consider using Word2Vec or glove word vector trained by yourself to find similar words that exist in the dictionary by comparing the similarity between words and then replace them. The replacement of such oov enables the deep learning model scheme based on the pre-trained word vector to get a very large room for improvement.

The second is the training method based on pre-training sequence models. In recent years, from the development of transformer to GPT-3, the pre-training and migration of the deep learning model based on transformer on massive data sets have made considerable progress in natural language processing. Models such as BERT, RoBERTa, and GPT rarely have scenes containing oov due to the internal Word Segment mechanism, so they do not need to perform a large number of word replacement operations like the scheme of depth learning model does on the basis of pre-trained word vectors. This kind of model is more often deployed in the re-pre-training, fine-tuning and formulation of training strategies for the model.

In addition, for the Chinese BERT model, its Tokenizer (an analyzer) does not use the Word Segment mechanism in English words; rather, it uses every single Chinese character as a token for training.

16.3   Feature Engineering

The Quora Question Pairs mentioned in this chapter is a text matching task under the natural language processing category. In the scene of text matching, participants will not only need to consider extracting the features of the text itself, but also think about constructing the relationship between text. How to express "text and text are semantically similar" is the biggest challenge for feature engineering regarding this proposition.

For the feature engineering that traditional machine learning models need to utilize, in addition to using TF-IDF\ word embedding to build original features such as sentence embedding, it is also necessary to figure out how to build text similarity at word level and sentence level and other ways to achieve semantic matching expressions. The following part will introduce some common feature construction schemes for text matching tasks.

16.3.1   Common Text Features

The simplest way to build features is to use bag of word (bow) or TF-IDF to build sparse features for model training and forecast.

The method of setting up sparse features using TF-IDF can also refer to the sklearn package mentioned in the previous chapters of this book. The code is as follows:

```
from sklearn.feature_extraction.text import TfidfVectorizer
len_train = df_train.shape[0]

data_all =pd.concat([df_train,df_test])

max_features = 200000
ngram_range = (1,2)
min_df = 3
print('Generate tfidf')
feats= ['question1','question2']
vect_orig = TfidfVectorizer(max_features=max_features,
ngram_range=ngram_range,
    min_df=min_df)

corpus = []
for f in feats:
    data_all[f] = data_all[f].astype(str)
    corpus+=data_all[f].values.tolist()

vect_orig.fit(corpus)

for f in feats:
    train_tfidf = vect_orig.transform(df_train[f].astype(str).
values.tolist())
    test_tfidf = vect_orig.transform(df_test[f].astype(str).values.
tolist())
    pd.to_pickle(train_tfidf,path+'train_%s_tfidf_v2.pkl'%f)
    pd.to_pickle(test_tfidf,path+'test_%s_tfidf_v2.pkl'%f)
```

In the above code, the maximum number of words that can be used (equivalent to the dimension of the sparse vector) can be limited by adjusting the size of the max_features. In some scenarios, limiting the dimension plays a role in reducing the dimension and decreasing the occurrence of overfitting. The exact number of words that need to be limited requires debugging of super parameters, where the default is 200,000; use pickle to cache the generated features to obtain the TF-IDF sparse features of question1 and question2.

At the same time, ngram_range = (1,2) is a more critical parameter setting. The settings of ngram_range parameters affect the choice of N-Gram by the vectorizer, that is, the maximum and minimum N-Gram values that can be used by the text TF-IDF features it constructs. When the upper limit of ngram_range is set to 2, unigram and bigram term features will be created; when set to 3, unigram, bigram, and trigram features will be generated. The larger the upper limit value of the ngram_range parameter, the higher the dimension of backup features generated will be.

For text classification in natural language processing, a sparse TF-IDF feature plus a linear model (LR, Linear SVM) might be enough as a baseline scheme. However, for text matching, simply splicing the sparse features of question1 and question2 together, and then using linear model training cannot guarantee good results, because the linear model cannot capture the nonlinear relationship between text features in question1 and question2.

For tree models such as XGBoost and LightGBM, if the training set and testing set are sparse, then the efficiency and accuracy of model training will be a challenge, because the training process of the tree model will involve the computation of the division and gain of leaf nodes. When the dimension of the sparse matrix is too high, even models such as XGBoost and LightGBM that use parallel training methods often suffer from low training efficiency due to dimension explosion. Therefore, a better method is to generate relevant features by constructing the relationship between text and text, and then carry out subsequent model training.

In short, here, on the basis of constructing TF-IDF features, the sparse matrix can be reduced to a dense matrix, which is also very suitable for feeding tree models for training. Specific dimensionality reduction methods can be LSI (latent semantic index, equivalent to Truncated SVD), or other topic models. Readers can consider using the decomposition module provided in the sklearn package to try to reduce the dimension of the TF-IDF and try to do it by yourself to compare the verification accuracy of the TF-IDF features and the data model after dimension reduction. Common decomposition modules include the use of Truncated SVD, NFM, and LDA methods, where LDA can be used as a topic model to extract topic features or as a dimensionality reduction method.

In addition to text representation methods such as topic models, another text representation method uses pre-trained word vector model to construct sentence-level features and obtains a dense feature matrix with fixed dimensions through weighted average or summation of words. This type of feature matrix can be directly trained and used by the tree model due to its fixed dimensions (usually around 200 to 300 dimensions). The specific scheme will be discussed in Sect. 16.3.2.

16.3.2   Similarity Features

Neither the original TF-IDF features nor the dimension-reduced LSI features nor the Topic Model features can solve the above-mentioned correlation problems or the phenomenon that the same type of problems have been raised many times. Therefore, it is necessary to adopt one or more text similarity matching methods to construct the similarity of texts at different levels. This kind of similarity can be at the syntactic level, the keyword set level or the semantic level.

The following will list some methods of similarity computation to construct similarity features from different angles.

16.3.2.1   The First Method of Constructing Similarity Features

The simplest way to build text similarity features is to compute them based on edit distance or set similarity. The common practice is to calculate jaccard distance and dice distance. The following is the calculation code:

```
def get_jaccard(seq1, seq2):
    """Compute the Jaccard distance between the two sequences `seq1` and
`seq2`.
    They should contain hashable items.

    The return value is a float between 0 and 1, where 0 means equal, and 1
totally
    different.
    """

    set1, set2 = set(seq1), set(seq2)
    return 1 - len(set1 & set2) / float(len(set1 | set2))

def get_dice(A,B):
    A, B = set(A), set(B)
    intersect = len(A.intersection(B))
    union = len(A) + len(B)
    d = try_divide(2*intersect, union)
    return d
```

Of course, there are many other ways.

The biggest advantage of calculation based on set similarity is that it is efficient and can well express the similarity between two texts from the sentence level. For jaccard distance and dice distance, sentence patterns and stop words need to be taken into account when constructing similarity features. When there are a lot of same stop words in the two sentence patterns, features constructed using set similarity will be highly similar, resulting in decline in resolution of features. Therefore, in the process of using set similarity, it is suggested to first compare the characteristics of the original text with the similar text after removing the stop words, and in some cases, both characteristics can be used at the same time to improve the accuracy.

16.3.2.2 The Second Method of Constructing Similarity Features

The second method is to construct the angle of two vectors or Euclidean distances based on TF-IDF features. The similarity obtained by directly calculating the cosine included angles or Euclidean distances using the sparse matrix of TF-IDF features can characterize the similarity of vectors to a certain extent. Because TF-IDF features directly express the importance of the word level, it contains more semantic information and has a higher degree of distinction between important words and unimportant words than set similarity. You can use the TF-IDF file we have generated to carry out the construction. The code is as follows:

```
def calc_cosine_dist(text_a,text_b,metric='cosine'):
    return pairwise_distances(text_a, text_b, metric=metric)[0][0]
```

The above code can be used to calculate the similarity between vectors. The similarity type can be cosine included angle or Euclidean distance. After getting the function to calculate the similarity, we can implement the following operations:

```
train_question1_tfidf = pd.read_pickle(path+'train_question1_tfidf.
pkl')
    test_question1_tfidf = pd.read_pickle(path+'test_question1_tfidf.pkl')
    train_question2_tfidf = pd.read_pickle(path+'train_question2_tfidf.
pkl')
    test_question2_tfidf = pd.read_pickle(path+'test_question2_tfidf.pkl')

    train_tfidf_sim = []
    for r1,r2 in zip(train_question1_tfidf,train_question2_tfidf):
        train_tfidf_sim.append(calc_cosine_dist(r1,r2))
    test_tfidf_sim = []
    for r1,r2 in zip(test_question1_tfidf,test_question2_tfidf):
        test_tfidf_sim.append(calc_cosine_dist(r1,r2))
    train_tfidf_sim = np.array(train_tfidf_sim)
    test_tfidf_sim = np.array(test_tfidf_sim)
    pd.to_pickle(train_tfidf_sim,path+"train_tfidf_sim.pkl")
    pd.to_pickle(test_tfidf_sim,path+"test_tfidf_sim.pkl")
```

This is based on the pairwise_distances function of the sklearn package to build features, which is actually a slower feature building function. A better way is to directly use the sparse function of the scipy library to perform dot product operations, which can get results faster and avoid the pressure caused by the for loop. Readers can think about how to use the scipy library to build their own sparse matrix cosine similarity calculation method.

16.3.2.3 The Third Method of Constructing Similarity Features

The third method depends on word vectors, which will appear in vector space to some extent. The word_a + word_b is similar to word_c + word_d. In other words, the vector at sentence level is generated by summing or averaging the word vectors, and then the sum of angles is measured by using the sentence vector, which could also characterize the similarity degree of the text.

The advantage of this method is that the semantic information based on word vectors is richer than TF-IDF features. Due to the pre-training information of external data, for polysemy cases that cannot be handled by TF-IDF features, the similarity based on embedding vectors has stronger representation capabilities.

When using word vectors, there are many alternatives. First, you can use pre-training models, such as Word2Vec, glove, fastText, and other commonly used and informative pre-training models; secondly, you can use the existing corpus on Quora platform to which you have access to self-training, in order to capture some context information that may be missed in the pre-training scene; thirdly, you can also consider the way to calculate the similarity after the weighted summation of multiple embedding vectors, so that multiple embedding vectors can be integrated together, and the obtained similarity information is relatively more precise.

Convert the embedding matrix built into a dictionary and maintain it through the mapping relationship between words and embedding vectors. Before calculating the similarity, you can try to weight words with IDF values, or if there are predefined weighting coefficients, you can also weight words that need to be emphasized. In some scenarios, it is weighted based on IDF values. Similarity has stronger ability to forecast, but it does not ensure 100% correct prediction. You still need to have a try.

The corresponding computing code is as follows:

```
def calc_w2v_sim(row,embedder,idf_dict=None,dim=300):
    '''
    Calc w2v similarities and diff of centers of query\title
    '''
    a2 = [x for x in row['question1'].lower().split() if x in embedder.
vocab]
    b2 = [x for x in row['question2'].lower().split() if x in embedder.
vocab]

    vectorA = np.zeros(dim)
    for w in a2:
        if w in idf_dict:
            coef = idf_dict[w]
        else:
            coef = idf_dict['default_idf']
        vectorA += coef*embedder[w]
    if len(a2)!=0:
        vectorA /= len(a2)

    vectorB = np.zeros(dim)
    for w in b2:
        if w in idf_dict:
            coef = idf_dict[w]
        else:
            coef = idf_dict['default_idf']
        vectorB += coef*embedder[w]
    if len(b2)!=0:
        vectorB /= len(b2)

    return (vectorA,vectorB)
```

Suppose we use a pre-trained word vector with a dimension of 300. Here we construct a function calc_w2v_sim, whose parameter row is the original text containing question1 and question2; the parameter idf_dict is optional, representing a dictionary, which can calculate the item in advance, use it to store the IDF coefficient of each word, and then output two sentence vectors after weighting.

The vectorA and vectorB returned by the function can be directly used as sentence-level semantic features. After splicing, they participate in the training. By calculating the cosine angle or Euclidean distance between the two sentence vectors of A and B, features with the ability to characterize sentence similarity are obtained. According to experience, such features usually have relatively strong representational ability. Whether both distance measures should be calculated depends on the contribution of the feature to the model after it is constructed, and whether it helps to improve the evaluation index. In addition, there is a certain risk that there will be over-fitting after both distance measures are used.

16.3.3   Further Application of Word Vectors: Unique Word Matching

In addition to the above-mentioned text sentence matching based on word vectors, another usage of word vectors is to measure the difference in the semantics of text details. In Sect. 16.3.2, we have introduced various methods of vectorizing sentences and calculating similarity in order to determine the degree of difference between sentences. However, under normal circumstances, the part that computes the difference between text sentences is exactly where there are no repeated words in the sentence.

For example:

Do you know apple?

Do you know banana?

Do you know machine learning?

In these three sentences, the questions of the first two sentences mainly focus on a type of fruit, while the last sentence focuses on a subject. When the sentence pattern of a text paragraph is the same, how to judge the similarity of the different parts of the remaining words and sentences is another factor that needs to be considered for text matching.

The code to deduplicate is as follows:

```
def distinct_terms(lst1, lst2):
    lst1 = lst1.split(" ")
    lst2 = lst2.split(" ")
    common = set(lst1).intersection(set(lst2))
    new_lst1 = ' '.join([w for w in lst1 if w not in common])
    new_lst2 = ' '.join([w for w in lst2 if w not in common])

    return (new_lst1, new_lst2)
```

With the use of the distinct_terms function, the repeated parts of the input text string can be removed, only the unique words are retained, and their sequence is maintained. Then you can use the feature calculation method of sentence vectors in Sect. 16.3.2 to calculate and generate the same feature again.

16.3.4   Further Application of Word Vectors: Pairwise Matching of Words

Another advanced method of using word vectors for matching is do the pairwise matching of word-to-word. There is a possibility that the similarity between sentences does not depend on the sentence vector similarity after the weighting and summing of all the words in the whole sentence; rather, it can be derived from the similarity between the words of the two sentences. In other words, we can first calculate the Cartesian product of words between two sentences to obtain the word similarity between the two sentences, and then use statistical methods to generate a series of feature values such as the maximum, minimum, mean, and median of word similarity. This set of features can usually better characterize the similarity of the sentence as a whole. Compared with the sentence vector similarity, it contains more complementary information.

16.3.5   Other Similarity Calculation Methods

In addition to the common word vector algorithms such as Word2Vec, you can also use Doc2Vec to directly generate sentence text features (there is Doc2Vec's interface in the gensim package), or you can use the simhash method to compute the similarity of the text. For more details about the simhash method, you can refer to a paper "Detecting Near-duplicates for web crawling" published by Google in 2007.

In short, any method that can be deployed to assess the similarity between sentences can be used as a feature engineering scheme for training text matching scenes via a traditional machine learning model.

16.4   Model Training

The processes for training and parameter adjustment of common machine learning models have been described in the previous sections, and this section will introduce more deep learning models. In the development of natural language processing, there have been a variety of deep learning models based on convolutional neural networks, recurrent neural networks, and attention (attention mechanism). Today, BERT-like models are constantly refreshing various benchmarks in the field of natural language processing with their large pre-training data sets and weight parameters.

Taking text matching as an example, the deep learning model used can be divided into two types: representation-based and interaction-based. Both models can use the deep learning model structure related to the natural language processing mentioned in the previous chapter as the backbone models. Generally speaking, the training efficiency of representation-based model is higher, but the final result is worse than that founded on the interaction-based model.

Before explaining the differences between the two, let's list some common natural language processing models, and then explore some structural differences between representation-based and interaction-based models.

16.4.1   TextCNN Model

The first is the shallow TextCNN model, which is the simplest (and more frequently used in the engineering scenario) model in natural language processing scenarios.

A shallow TextCNN model consists of three components, namely, the embedding layer, the shallow cnn layer, and the output layer. In common natural language processing scenarios, the convolution layer usually used is one-dimensional convolution without considering paragraphs. Here we use PyTorch to build our TextCNN model. The code is as follows:

```python
import torch
import torch.nn as nn
import torch.nn.functional as F
from torch.autograd import Variable

class TextCNN(nn.Module):
    def __init__(self, args):
        super(TextCNN, self).__init__()
        self.args = args

        self.embed = nn.Embedding(args.sequence_length, args.embed_dim)
        self.convs1 = nn.ModuleList([nn.Conv2d(1, args.kernel_num,
            (ks, args.embed_dim)) for ks in args.kernel_sizes])
        self.dropout = nn.Dropout(args.dropout)
        self.fc1 = nn.Linear(len(args.kernel_sizes) * args.kernel_num,
args.class_num)

    def forward(self, x):
        x = self.embed(x) # (batch_size, sequence_length, embedding_dim)
        if self.args.static:
            x = torch.tensor(x)
        x = x.unsqueeze(1) # (batch_size, 1, sequence_length,
embedding_dim)
        ## input size (N,Cin,H,W) output size (N,Cout,Hout,1)
        x = [F.relu(conv(x)).squeeze(3) for conv in self.convs1]
        x = [F.max_pool1d(i, i.size(2)).squeeze(2) for i in x]
        x = torch.cat(x, 1) # (batch_size, len(kernel_sizes)*kernel_num)
        x = self.dropout(x)
        logit = self.fc1(x)
        return logit
```

The structure of the model is very simple. The output of the embedding layer is used to connect multiple convolution layers with convolution kernels of different sizes. The output of the convolution layer passes through the maximum pooling layer and then splices to connect the output layer. Such a model structure integrates the operation of the convolution layer in deep learning and simulates an artificial extraction of N-Gram through convolution calculation. The size of the convolution kernel here represents the range of the convolution calculation sliding window, which is somewhat equivalent to taking N-Gram trem to create a new phrase. The model will automatically learn such a feature extraction behavior by updating parameters.

Because only one layer of convolutional neural networks is used in parallel, even if multiple convolutional kernels of different sizes are used, the efficiency of the model will not be affected. At the same time, convolutional calculation operations like N-Gram can effectively extract local phrase features of text. Therefore, TextCNN is very suitable for scenarios that require fast response and have certain requirements for prediction accuracy (better than traditional TF-IDF + linear models).

16.4.2   TextLSTM Model

This is also the most widely used and simplest recurrent neural network model in natural language processing scenarios. Because recurrent neural networks such as LSTM and GRU have the ability to fit sequences, and natural language processing is a very typical sequence data scenario, TextLSTM (or TextGRU) and TextCNN are often deployed together as a baseline reference for a deep learning model.

The code used to build the TextLSTM model is as follows:

```
class TextLSTM(nn.Module):

    def __init__(self, args):
        super(TextLSTM, self).__init__()
        self.hidden_dim = args.hidden_dim
        self.batch_size = args.batch_size

        self.embeds = nn.Embedding(args.vocab_size, args.
embedding_dim)
        self.lstm = nn.LSTM(input_size=args.embedding_dim,
            hidden_size=args.hidden_dim,
            num_layers=args.num_layers,
            batch_first=True, bidirectional=True)

        self.hidden2label = nn.Linear(args.hidden_dim, args.
num_classes)
        self.hidden = self.init_hidden()

    def init_hidden(self):
        h0 = Variable(torch.zeros(1, self.batch_size, self.
hidden_dim))
        c0 = Variable(torch.zeros(1, self.batch_size, self.
hidden_dim))
        return h0, c0

    def forward(self, sentence):
        embeds = self.embeds(sentence)
        # x = embeds.view(len(sentence), self.batch_size, -1)
        lstm_out, self.hidden = self.lstm(embeds, self.hidden)
        y = self.hidden2label(lstm_out[-1])
        return y
```

In this TextLSTM, the embedding layer is connected to the LSTM layer. The LSTM layer we use here is a bidirectional LSTM. The specific code of this layer is as follows:

```
nn.LSTM(input_size=args.embedding_dim,
    hidden_size=args.hidden_dim,
    num_layers=args.num_layers,
        batch_first=True, bidirectional=True)
```

Here bidirectional = True means that LSTM includes both forward and reverse directions. Bidirectional LSTM has stronger sequence fitting ability than unidirectional LSTM, and usually the backward sequence contains additional information that cannot be captured in the forward sequence. The decision-making layer of the TextLSTM model only uses the last state output of the bidirectional LSTM as input. In the model discussed in the following sections, there will be other solutions to optimize TextLSTM for how to make the decision-making layer obtain better input representation.

16.4.3   TextLSTM with Attention Model

Neither taking the last layer of LSTM output nor performing pooling for all state outputs of LSTM can well capture the importance of features in different temporal states, so Attention (Attention Mechanism) comes into being. The goal of Attention is that the model can give different degree of attention to each word of the sentence during the training process, and then increase or decrease the importance of each word by weighting, in order to improve the overall accuracy of the model. Compared with pooling operations or taking the last state output of the LSTM, Attention can highlight the role of important words in the model training and prediction process without losing long-term information (using the last state will result in loss). Here will take the simplest weighted sum Attention as an example to introduce a simple Attention layer and its modification for the TextLSTM model. The code is as follows:

```
class SimpleAttention(nn.Module):
  def __init__(self,input_size):
    super(SimpleAttention,self).__init__()
    self.input_size = input_size
    self.word_weight = nn.Parameter(torch.Tensor(self.input_size))
    self.word_bias = nn.Parameter(torch.Tensor(1))
    self._create_weights()

  def _create_weights(self, mean=0.0, std=0.05):
    self.word_weight.data.normal_(mean, std)
    self.word_bias.data.normal_(mean, std)

  def forward(self,inputs):
    att = torch.einsum('abc,c->ab',(inputs,self.word_weight))
    att = att+self.word_bias
    att = torch.tanh(att)

    att = torch.exp(att)
    s = torch.sum(att,1,keepdim=True)+1e-6
    att = att / s
    att = torch.einsum('abc,ab->ac',(inputs,att))

    return att
```

In the above code, a weight matrix is constructed, and each state dimension of the output is normalized to the $(0,1)$ interval, so that each time state (the position of the word) has a corresponding weight. In the PyTorch program, we can use Einstein summation to calculate the tensor of any dimension, simplifying the complexity of our code.

The corresponding TextLSTM model can be modified to:

```
class TextLSTMAtt(nn.Module):

    def __init__(self, args):
        super(TextLSTMAtt, self).__init__()
        self.hidden_dim = args.hidden_dim
        self.batch_size = args.batch_size

        self.embeds = nn.Embedding(args.vocab_size, args.embedding_dim)
        self.lstm = nn.LSTM(input_size=args.embedding_dim,
            hidden_size=args.hidden_dim, num_layers=1,
            batch_first=True, bidirectional=True)

        self.simple_att = SimpleAttention(args.hidden_dim*2)

        self.hidden2label = nn.Linear(args.hidden_dim*2, args.
num_classes)
        self.hidden = self.init_hidden()
        def init_hidden(self):
            h0 = Variable(torch.zeros(1, self.batch_size, self.
hidden_dim))
            c0 = Variable(torch.zeros(1, self.batch_size, self.
hidden_dim))
            return h0, c0
        def forward(self, sentence):
            embeds = self.embeds(sentence)
            # x = embeds.view(len(sentence), self.batch_size, -1)
            lstm_out, self.hidden = self.lstm(embeds, self.hidden)
            x = self.simple_att(lstm_out)
            y = self.hidden2label(x)
            return y
```

Under normal circumstances, the model works better with Attention.

16.4.4   Self-Attention Layer

Transformer structures or models that use multi-head-attention mostly employ the Self-Attention layer and gain the ability to fit longer text through the block stack built by the Self-Attention layer. The Self-Attention layer itself can also be taken out separately to build weighting operations in deep CNN/RNN models. In essence, the Self-Attention layer first calculates the word similarity matrix of the input query, and then uses this matrix to generate weighting coefficients to re-weight the input query to improve its ability to capture word-level interaction information. Without using any convolutional neural networks or recurrent neural networks, a layer of Self-

Attention can usually only capture the interaction information between any two words, with a computational complexity of $O(n^2)$, which has a lower computational efficiency than the weighted sum Attention in Sect. 16.4.3. Models such as Transformers obtain the similarity matrix between phrases by continuously stacking the blocks of the Self-Attention layer to increase complexity.

Here are some more complex Self-Attention layer structures:

```python
class MatchTensor(torch.nn.Module):
  def __init__(self,size_a,size_b,channel_size=8,max_len=10):
      super(MatchTensor,self).__init__()
      self.size_a=size_a
      self.size_b=size_b
      self.channel_size=channel_size
      self.max_len = max_len

      self.M = nn.Parameter(torch.Tensor(channel_size,size_a,
size_b).to(device))

      self.W = nn.Parameter(

      torch.Tensor(channel_size,size_a+size_b,max_len).to(device))
      self.b = nn.Parameter(torch.Tensor(channel_size).to(device))

      self._create_weights()

  def _create_weights(self, mean=0.0, std=0.05):
    self.M.data.normal_(mean, std)
      self.W.data.normal_(mean, std)
      self.b.data.normal_(mean, std)

  def forward(self, x1,x2):

    matching_matrix = torch.einsum('abd,fde,ace->afbc', [x1, self.M,
x2])
      tmp = torch.cat([x1,x2],2)
      linear_part = torch.einsum('abc,fcd->afbd', [tmp,self.W])
      matching_matrix = matching_matrix+linear_part+self.b.view(
          self.channel_size,1,1)
      matching_matrix = F.relu(matching_matrix)

      return matching_matrix

class SelfMMAttention(nn.Module):
  def __init__(self,input_size,max_len=100,channel_size=3):
      super(SelfMMAttention,self).__init__()
      self.input_size = input_size
      self.channel_size = channel_size
      self.max_len = max_len
      self.match_tensor = MatchTensor(input_size,input_size,
channel_size,max_len)
      self.V = nn.Parameter(torch.Tensor(max_len,channel_size).to
(device))
```

```
      self._create_weights()

   def _create_weights(self, mean=0.0, std=0.05):
      self.V.data.normal_(mean, std)

   def forward(self,inputs,mask=None,output_score=False):

      batch_size,len_seq,embedding_dim = inputs.size()
         x = self.match_tensor(inputs,inputs)

      att_softmax = torch.tanh(x)

      att_softmax = torch.softmax(att_softmax,dim=1)
      x = torch.einsum('abc,adbe->adc', [inputs,att_softmax])

      if output_score:
         return x,att_softmax
         return x
```

In the above code, the MatchTensor structure constructs a calculation layer to calculate the similarity matrix between any two input sequences and uses the calculation layer to add weight in the subsequent Self-Attention layer. For the calculation of the similarity matrix, there are usually many complex methods, such as matrix dot product, concat, Euclidean distance, and cosine included angle.

Using Self-Attention can usually achieve better results than using weighted summation Attention, but it is also more time-consuming.

16.4.5   Transformer and BERT Family of Models

Before the release of Transformer structures, sequence pre-training models such as ELMo were regarded as the SOTA structure at that time. The LSTM hierarchical structure was pre-trained in a semi-supervised way instead of only pre-training word vectors, so that the obtained pre-training model can capture the context of different inputs. This is an effect that cannot be got only by pre-training word vectors. However, models such as BERT, GPT, and RoBERTa are large-scale mass corpus pre-training models realized by stacking Transformer block layers as encoders. By using a large amount of pre-training data, the semantic and syntax information is compressed and represented in the super-large-scale parameters of the model.

The use of the BERT models is relatively convenient. PyTorch's third-party package provides a complete BERT model implementation, whose code is as follows:

```
# coding: UTF-8
import torch
import torch.nn as nn
from pytorch_pretrained import BertModel, BertTokenizer
class Config(object):

   """ configuration parameters """
   def __init__(self, dataset):
```

```
        self.model_name = 'bert'
        self.train_path = dataset + '/data/train.txt' # training set
        self.dev_path = dataset + '/data/dev.txt' # verification set
        self.test_path = dataset + '/data/test.txt' # testing set
        self.class_list = [x.strip() for x in open(
            dataset + '/data/class.txt').readlines()] # class list
        self.save_path = dataset + '/saved_dict/' + self.model_name + '.
ckpt'
        # result of model training

        self.device = torch.device('cuda' if torch.cuda.is_available()
else 'cpu')
        # device

        print('device',self.device)
        self.require_improvement = 1000 # If the effect is not improved after 1000batch,
the training will be finished in advance
        self.num_classes = len(self.class_list)
        print('num_classes',self.num_classes) # category number
        self.num_epochs = 10 # epoch number
        self.batch_size = 32 # size of mini-batch
        self.pad_size = 100 # length of each sentence after being processed (fill in short and
cut long)
        self.learning_rate = 5e-5 # learning rate
        self.bert_path = './bert_pretrain'
        self.tokenizer = BertTokenizer.from_pretrained(self.bert_path)
        self.hidden_size = 768

    class Model(nn.Module):
        def __init__(self, config):
            super(Model, self).__init__()
            self.bert = BertModel.from_pretrained(config.bert_path)
            for param in self.bert.parameters():
                param.requires_grad = True
            self.fc = nn.Linear(config.hidden_size, config.num_classes)

        def forward(self, x):
            context = x[0] # the sentence input
            mask = x[2]
            # carrying out mask for padding, with the same size as the sentence, the padding being
represented by 0
            # for example: [1, 1, 1, 1, 0, 0]
            _, pooled = self.bert(context, attention_mask=mask,
              output_all_encoded_layers=False)
            out = self.fc(pooled)
            return out
```

The above code is the simplest model to complete classification or regression tasks based on BERT. Load the pre-trained BERT model parameters by calling the BertModel package in the pytorch_pretrained library. BertModel can not only load the native Google BERT model, but also use RoBERTa, ERNIE, RoBERTa-wwm-ext, and other pre-training models which employ the BERT structure but have different training schemes.

16.4.6 Differences Between Representation-Based and Interaction-Based Deep Learning Models

The models introduced in the previous sections can be used as the backbone models for training text matching models. The differences between the two models introduced in this section can be explained by the two small graphs in Fig. 16.6.

The representation-based deep learning model usually connects the input text to the recurrent neural network layer or the convolutional neural network layer, then obtains a one-dimensional text representation through pooling or flatten operations, and at last calculates the final output result through the representation matching layer. Usually, the weight parameters of the convolutional neural networks layer or the recurrent neural network layer used for one-dimensional feature vector extraction can be reused, and the input text 1 and input text 2 can simultaneously obtain the representation vector result through the representation layer.

The following is a simple Siamese Network (twin network, one of the deep learning models based on representation) in accordance with the combination of LSTM and Attention. The code is as follows:

```
class SpatialDropout1D(nn.Module):
def __init__(self,p=0.5):
    super(SpatialDropout1D,self).__init__()
    self.p = p
    self.dropout2d = nn.Dropout2d(p=p)

  def forward(self,x):
  # b,h,c
  x = x.permute(0,2,1)
  # b,c,h
  x = torch.unsqueeze(x,3)
  # b,c,h,w
  x = self.dropout2d(x)
  x = torch.squeeze(x,3)
```

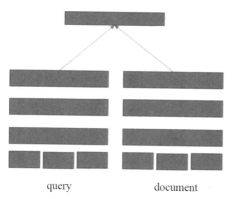

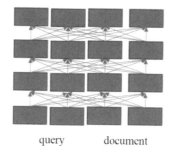

Representation-Based   Interaction-Based

Fig. 16.6 Models comparison

```python
   x = x.permute(0,2,1)

    return x
class LSTM_ATT(torch.nn.Module):
  def __init__(self,embedding_dim=200,hidden_dim=128,
              voacb_size=10000,target_size=66,embedding_matrix=None,
          layer_num = 1,num_heads=40):
    super(LSTM_ATT,self).__init__()
    self.hidden_dim=hidden_dim
    self.voacb_size=voacb_size
    self.target_size=target_size
    self.embedding_dim = embedding_dim
    self.layer_num = layer_num
    self.num_heads = num_heads
    self.embed_scale = np.sqrt(embedding_dim)

    if embedding_matrix is not None:
    self.emb = nn.Embedding.from_pretrained(
        torch.FloatTensor(embedding_matrix),freeze=True)
    print('use pretrained embedding')
    else:
        self.emb = nn.Embedding(voacb_size,embedding_dim)

    self.dropout = SpatialDropout1D(0.15)
      embedding_dim = int(embedding_dim/2)
      self.embedding_dim = embedding_dim
     self.lstm=nn.LSTM(input_size=embedding_dim,
       hidden_size=hidden_dim,batch_first=True,
       bidirectional=True,num_layers=1)
     self.simple_att = SimpleAttention(hidden_dim*2)

      self.out = nn.Sequential(
       nn.LayerNorm(hidden_dim*2),
       nn.Linear(hidden_dim*2, target_size),
       )
         self.log_softmax=torch.nn.LogSoftmax(dim=1)

  def init_hidden(self,batch_size=None):

   h0 = torch.zeros((2*self.layer_num,batch_size,
   self.hidden_dim),dtype=torch.float32).to(device)
   h0 = Variable(h0)
   c0 = torch.zeros((2*self.layer_num,batch_size,
   self.hidden_dim),dtype=torch.float32).to(device)
   c0 = Variable(c0)

   return (h0, c0)
  def forward(self,x,mask=None):

    x = self.emb(x)
       x = self.dropout(x)
       x, _ = self.lstm(x)
       x = self.simple_att(x)
       x = self.out(x)

    return x
```

What we have done at the beginning is building a LSTM_ATT model, which is a standard single-text input model with an input and output that can be used as regression or classification tasks. In order to transform it and adapt it to a twin neural network (Siamese Network), we need to inherit it and reuse some of its attributes. The code is as follows:

```
class LSTM_ATT_SIA(LSTM_ATT_AM):

  def forward(self,x,x1,mask=None):

    x = self.emb(x)
       x = self.dropout(x)
       x, _ = self.lstm(x)
         x = self.simple_att(x)

    x1 = self.emb(x1)
       x1 = self.dropout(x1)
       x1, _ = self.lstm(x1)
       x1 = self.simple_att(x1)

    x = F.cosine_similarity(x, x1)

    return x
```

By reusing the LSTM_ATT class, inputting x and x1 at the same time is accomplished, then the same LSTM_ATT model is used as the backbone model (framework) and is mapped into a one-dimensional text vector, and finally cosine similarity is calculated and output.

The use of twin neural networks has its unique strengths, such as simple model construction, reuse of text categorization models, and high training efficiency. Compared with the deep learning model based on interaction, twin neural networks do not need to interact at the sequence level, so it is very simple for GPUs to carry out parallel training, which is suitable for some scenarios where the text is relatively simple, the matching is time-sensitive, and the accuracy is not too high.

The deep learning model based on interaction has more complex prior assumptions. For the deep learning model based on representation, because the operation of calculating similarity occurs at the final output layer, the text will be compressed into vectors before calculation. In this process, the word-level sequences of input text 1 and input text 2 have no perception of each other. For some texts, the interaction between word and word pairs may be useful, and the deep learning model based on representation will not be able to construct the characteristics of such information. The capture of word-level similarity or interactive information is the focus of the deep learning model based on interaction. Its core idea is to make input text 1 and input text 2 interact with each other in advance through the Attention mechanism. The generated similarity matrix can be used for subsequent classification tasks to determine whether the text matches or not.

Take the deep correlation matching model DRMM as an example. DRMM is a typical deep learning model based on interaction. The following carefully decomposes the composition logic of the Attention mechanism by observing the code. The code is as follows:

```python
"""An implementation of DRMM Model."""
import typing

import keras
import keras.backend as K
import tensorflow as tf

from matchzoo.engine.base_model import BaseModel
from matchzoo.engine.param import Param
from matchzoo.engine.param_table import ParamTable

class DRMM(BaseModel):
    """
    DRMM Model.
    Examples:
        >>> model = DRMM()
        >>> model.params['mlp_num_layers'] = 1
        >>> model.params['mlp_num_units'] = 5
        >>> model.params['mlp_num_fan_out'] = 1
        >>> model.params['mlp_activation_func'] = 'tanh'
        >>> model.guess_and_fill_missing_params(verbose=0)
        >>> model.build()
        >>> model.compile()
    """

    @classmethod
    def get_default_params(cls) -> ParamTable:
        """:return: model default parameters."""
        params = super().get_default_params(with_embedding=True,
            with_multi_layer_perceptron=True)
        params.add(Param(name='mask_value', value=-1,
            desc="The value to be masked from inputs."))
        params['optimizer'] = 'adam'
        params['input_shapes'] = [(5,), (5, 30,)]
        return params

    def build(self):
    """"Build model structure."""

            # Scalar dimensions referenced here:
            #   B = batch size (number of sequences)
            #     D = embedding size
            #     L = `input_left` sequence length
            #     R = `input_right` sequence length
            #     H = histogram size
            #     K = size of top-k

        # Left input and right input.
        # query: shape = [B, L]
        # doc: shape = [B, L, H]
        # Note here, the doc is the matching histogram between original
query
        # and original document.
        query = keras.layers.Input(
            name='text_left',
```

```
            shape=self._params['input_shapes'][0]
        )
        match_hist = keras.layers.Input(
            name='match_histogram',
            shape=self._params['input_shapes'][1]
        )
        embedding = self._make_embedding_layer()
        # Process left input.
        # shape = [B, L, D]
        embed_query = embedding(query)
        # shape = [B, L]
        atten_mask = tf.not_equal(query, self._params['mask_value'])
        # shape = [B, L]
        atten_mask = tf.cast(atten_mask, K.floatx())
        # shape = [B, L, D]
        atten_mask = tf.expand_dims(atten_mask, axis=2)
        # shape = [B, L, D]
        attention_probs = self.attention_layer(embed_query,
atten_mask)

        # Process right input.
        # shape = [B, L, 1]
        dense_output = self._make_multi_layer_perceptron_layer()
(match_hist)

        # shape = [B, 1, 1]
        dot_score = keras.layers.Dot(axes=[1, 1])(
            [attention_probs, dense_output])

        flatten_score = keras.layers.Flatten()(dot_score)

        x_out = self._make_output_layer()(flatten_score)
        self._backend = keras.Model(inputs=[query, match_hist],
outputs=x_out)

    @classmethod
    def attention_layer(cls, attention_input: typing.Any,
        attention_mask: typing.Any = None
        ) -> keras.layers.Layer:
        """
    generate the input of Attention
    :param attention_input: input tensor .
    :param attention_mask: enter the mask of the tensor .
    : returns: the result of the tensors masked out .
    """
    # shape = [B, L, 1]
    dense_input = keras.layers.Dense(1, use_bias=False)
(attention_input)
    if attention_mask is not None:
     # Since attention_mask is 1.0 for positions we want to attend and
     # 0.0 for masked positions, this operation will create a tensor
     # which is 0.0 for positions we want to attend and -10000.0 for
     # masked positions.

     # shape = [B, L, 1]
     dense_input = keras.layers.Lambda(
```

```
        lambda x: x + (1.0 - attention_mask) * -10000.0,
        name="attention_mask"
        ) (dense_input)
    # shape = [B, L, 1]
    attention_probs = keras.layers.Lambda(
        lambda x: tf.nn.softmax(x, axis=1),
        output_shape=lambda s: (s[0], s[1], s[2]),
        name="attention_probs"
        ) (dense_input)
    return attention_probs
```

The deep learning model based on interaction is usually more accurate than the deep learning model based on representation, but its efficiency is lower. How to choose between them in the using process still needs to be judged according to the actual situation. However, for model integration, both are backup models that can be used.

16.4.7 A Special Deep Learning Model Based on Interaction

When training BERT family models or all models with Self-Attention as the core layer, there is a special operation that can greatly simplify the model building method. Specifically, during the training process, you can simply splice input text 1 and input text 2 into a new long text 3, and directly put long text 3 into the model as input text, so that the entire task is converted from text matching to text binary classification. Only fine-tuning of a mode of BERT category is needed to achieve the effect of text matching.

The reason why the above-mentioned special operation can be carried out is that the deep learning model based on interaction is essentially a mutual Attention operation on the input text. The BERT model uses multi-head attention as the basis in its own block layer and calculates the similarity matrix of the text itself on world levels. Therefore, this special using skills of the model can be approximately regarded as an interaction-based deep learning model operation. And making use of this operation training model usually contributes to better accuracy and higher efficiency than what is achieved by twin neural networks that adopt BERT model as the backbone model.

16.4.8 Translation Enhancement of Deep Learning Text Data

Broadly speaking, for image tasks, images can be enhanced after various rotations, offsets, and scaling, and data can also be enhanced during training through mix-up techniques, etc. The meaning behind the enhancement is to increase the robustness of the model, and to make the model better generalized by adding more specious images.

For text data, operations such as truncation, translation, and extraction can usually be performed, but due to the uncertainty of the length of the text content, the thresholds for truncation, translation, and extraction are often difficult to set, and it is likely that the expected enhancement effect cannot be achieved. In this regard, a trickier method is to translate and flip back the corpus that needs training by using an open-source machine translation model or a public interface. The text thus obtained can increase the diversity of the text while keeping certain semantic information unchanged. Attention should be paid to using popular languages and commonly used foreign languages for translation as much as possible in this process. The enhancement effect of using unpopular languages is poor and even counterproductive.

16.4.9   Preprocessing of Deep Learning Text Data

In order to use depth learning models, text data needs to be converted into corresponding word coding values, so as to obtain the word vector of its mapping relationship in the embedding layer. Here we take the training of BERT model as an example to show the code for preprocessing its text:

```python
def build_dataset(config):

    def load_dataset(path, pad_size=32):
        contents = []
        with open(path, 'r', encoding='UTF-8') as f:
            for line in tqdm(f):
                lin = line.strip()
                if not lin:
                    continue
                try:
                    content, label = lin.split('\t')
                    label = int(label)
                    label = config.le.transform([label])[0]
                except Exception as e:
                    continue
                token = config.tokenizer.tokenize(content)
                token = [CLS] + token
                seq_len = len(token)
                mask = []
                token_ids = config.tokenizer.convert_tokens_to_ids(token)

                if pad_size:
                    if len(token) < pad_size:
                        mask = [1] * len(token_ids) + [0] * (pad_size - len(token))
                        token_ids += ([0] * (pad_size - len(token)))
                    else:
                        mask = [1] * pad_size
                        token_ids = token_ids[:pad_size]
                        seq_len = pad_size
                contents.append((token_ids, int(label), seq_len, mask))
        return contents
```

```python
train = load_dataset(config.train_path, config.pad_size)
dev = load_dataset(config.dev_path, config.pad_size)
test = load_dataset(config.test_path, config.pad_size)
return train, dev, test
        class DatasetIterater(object):
            def __init__(self, batches, batch_size, device, shuffle=False):
                self.batch_size = batch_size
                self.batches = batches
                self.n_batches = len(batches) // batch_size
                self.residue = False # record if the number of batch is an
integer
                self.shuffle=shuffle
                if len(batches) % self.n_batches != 0:
                    self.residue = True
                self.index = 0
                self.device = device

            def _to_tensor(self, datas):
                x = torch.LongTensor([_[0] for _ in datas]).to(self.device)
                y = torch.LongTensor([_[1] for _ in datas]).to(self.device)

                # Pad length (if the length exceeds pad_size, set this
parameter to pad_size)
                seq_len = torch.LongTensor([_[2] for _ in datas]).to(self.
device)
                mask = torch.LongTensor([_[3] for _ in datas]).to(self.
device)
                return (x, seq_len, mask), y

            def __next__(self):
                if self.residue and self.index == self.n_batches:
                    batches = self.batches[self.index * self.batch_size:
len(self.batches)]
                    self.index += 1
                    batches = self._to_tensor(batches)
                    return batches

                elif self.index > self.n_batches:
                    self.index = 0
                    raise StopIteration
                else:
                    batches = self.batches[self.index * self.
batch_size: (self.index + 1) *
                        self.batch_size]
                    self.index += 1
                    batches = self._to_tensor(batches)
                    return batches
            def __iter__(self):
                if self.shuffle:
                    np.random.shuffle(self.batches)
                return self
            def __len__(self):
                if self.residue:
                    return self.n_batches + 1
```

```
        else:
              return self.n_batches
    def build_iterator(dataset,config,shuffle=False):
        iter = DatasetIterater(dataset,config.batch_size,config.
device,shuffle)
        return iter
```

In the above code, the data iteration class DataIterator is constructed, and its function is to carry out iterative loading of data. In the actual training process, we first splice text 1 and text 2, and the splicing relationship between the two needs to be separated using a special symbol [sep]. The amount of data may be very large, and it cannot be fully loaded into memory at one time. At this time, the data iterator DataIterator can be constructed to carry out iterative training of small batches of data loading.

16.4.10   Training of BERT Model

In model training, there is also a skill about data enhancement. We can employ the pre-trained word vector model to set a random ratio in the training process of the model by setting a threshold, randomly use the most similar word calculated by taking the word vector similarity as a synonym word on the premise of satisfying the similarity threshold, and perform sentence-level text replacement to increase the variability of the text. The code is as follows:

```
def synonyms_augmentation(texts,tokenizer,aug_rate=0.2,
sample_size=3):
    candidate_texts = texts[:int(len(texts)*aug_rate)]
    raw_texts = texts[int(len(texts)*aug_rate):]
    new_texts = []
    for idx,text in enumerate(candidate_texts):

        words = text.split(' ')
        indices = np.random.choice(len(words),size=sample_size)
        flag=0
        for idx in indices:
            word = words[idx]

          res = word_syn_dict.get(word,[])
            if len(res)>0:
                res_idx = np.random.choice(len(res),size=1)[0]
                syn_word,syn_score = res[res_idx]

            if syn_score>0.75 and syn_word in tokenizer.word_index:
                words[idx] = syn_word
                flag+=1

          text = ' '.join(jieba.cut(''.join(words)))
```

```
        new_texts.append(text)
    texts = list(new_texts)+list(raw_texts)
  return texts
```

However, this method will lead to certain errors, which may cause the model to learn the wrong relationship because the selected synonyms have opposite semantics (the text that should not be matched after replacement is trained as a matching label), so it is necessary to strictly control the enhancement rate and the threshold value. In addition, increasing the weight in the training process can be used as further solutions for optimization.

Since we cannot directly enhance the semantics of the model at the word token level, we can carry out adversarial learning on the parameters of the BERT pre-training model, and further improve the generalization ability of the model by adding certain disturbances and noise errors during the training process. As early as 2016, Goodfellow proposed FGM, which increased the perturbations as follows:

$$r_{adv} = \varepsilon \cdot / \|g\|_2$$
$$g = \nabla_x L(\theta, x, y)$$

The newly added adversarial samples are:

$$x_{adv} = x + r_{adv}$$

By adding adversarial samples, it can be compared with the image transformation operation in the category of CV task training to achieve the effect of data enhancement. The corresponding code is:

```
class FGM():
  def __init__(self, model):
    self.model = model
    self.backup = {}

  def attack(self, epsilon=0.9, emb_name=["word_embeddings"]):
    # the paremater emb_name should be changed into the parameter name
of embedding in you model
    for name, param in self.model.named_parameters():
      if param.requires_grad and any([p in name for p in emb_name]):
        self.backup[name] = param.data.clone()
        norm = torch.norm(param.grad)
        if norm != 0:
          r_at = epsilon * param.grad / norm
          param.data.add_(r_at)

  def restore(self, emb_name=["word_embeddings"]):
    # the paremater emb_name should be changed into the parameter name
of embedding in you model
    for name, param in self.model.named_parameters():
      if param.requires_grad and any([p in name for p in emb_name]):
        assert name in self.backup
```

```
            param.data = self.backup[name]
        self.backup = {}
```

The training function code of the model is as follows:

```
    def train(config, model, train_iter, dev_iter, test_iter):
        if hasattr(config, 'loss_type'):
            loss_type = config.loss_type
    else:
            loss_type='cce'

    if loss_type=='bce':
      loss_function = nn.BCEWithLogitsLoss()
    else:
      loss_function = nn.CrossEntropyLoss()
     start_time = time.time()
     model.train()
    param_optimizer = list(model.named_parameters())
    no_decay = ['bias', 'LayerNorm.bias', 'LayerNorm.weight']
    optimizer_grouped_parameters = [
   {'params': [p for n, p in param_optimizer if not any(nd in n for nd in
no_decay)],
             'weight_decay': 0.05},
        {'params': [p for n, p in param_optimizer if any(nd in n for nd in
no_decay)],
        'weight_decay': 0.0}]

  optimizer = BertAdam(optimizer_grouped_parameters,
  lr=config.learning_rate,
  warmup=0.05,
  t_total=len(train_iter) * config.num_epochs)
  total_batch = 0 # record how many batches are carried out
  dev_best_loss = float('inf')
  last_improve = 0 # record the batch number of the last verification set loss reduction
  flag = False # record whether the effect has not been improved for a long time
  model.train()
  fgm = FGM(model)
  for epoch in range(config.num_epochs):
  print('Epoch [{}/{}]'.format(epoch + 1, config.num_epochs))
  for i, (trains, labels) in enumerate(train_iter):
        outputs = model(trains)

        model.zero_grad()
        if loss_type=='bce':
          bs = labels.size()[0]
          labels_one_hot = torch.zeros(bs, config.num_classes).
          to(config.device).scatter_(1, labels.view(-1,1), 1)
          loss = loss_function(outputs,labels_one_hot)
          else:
        loss = loss_function(outputs,labels)

    loss.backward()
    fgm.attack()
```

```
    outputs_adv = model(trains)
    loss_adv = loss_function(outputs_adv,labels)
    loss_adv.backward()
    fgm.restore()

    optimizer.step()
    if total_batch % 100 == 0:
        # output the effect on the training set and validation set after a certain number of rounds,
which is used to display the current effect and cache it
        # model weight
        true = labels.detach().cpu()
        predic = torch.max(outputs.data, 1)[1].cpu()
        train_acc = metrics.accuracy_score(true, predic)
        dev_acc, dev_loss = evaluate(config, model, dev_iter)
        if dev_loss < dev_best_loss:
            dev_best_loss = dev_loss
            torch.save(model.state_dict(), config.save_path)
            improve = '*'
            last_improve = total_batch
        else:
            improve = ''
        time_dif = get_time_dif(start_time)
        msg = 'Iter: {0:>6}, Train Loss: {1:>5.2}, Train Acc: {2:>6.2%}, Val
            Loss: {3:>5.2}, Val Acc: {4:>6.2%}, Time: {5} {6}'
        print(msg.format(total_batch, loss.item(), train_acc, dev_loss,
            dev_acc, time_dif, improve))
        model.train()
    total_batch += 1
    if total_batch - last_improve > config.require_improvement:
        # if the number exceeds 1000Batch but the loss of validation sets does not decrease, the
training is terminated
        print("No optimization for a long time, auto-stopping...")
        flag = True
        break
    if flag:
        break
test(config, model, test_iter)
```

During the training process, the code for adversarial attack of the word vector part by using FGM is as follows:

```
gm.attack()
outputs_adv = model(trains)
loss_adv = loss_function(outputs_adv,labels)
loss_adv.backward()
fgm.restore()
```

In the whole training process, we should pay attention to the following contents as much as possible:

```
no_decay = ['bias', 'LayerNorm.bias', 'LayerNorm.weight']
optimizer_grouped_parameters = [
    {'params': [p for n, p in param_optimizer if not any(nd in n for nd in
no_decay)],
        'weight_decay': 0.05},
    {'params': [p for n, p in param_optimizer if any(nd in n for nd in
no_decay)],
        'weight_decay': 0.0}]
optimizer = BertAdam(optimizer_grouped_parameters,
    lr=config.learning_rate,
    warmup=0.05,
    t_total=len(train_iter) * config.num_epochs)
```

This part of the code is mainly to adjust the finetune learning rate of the model, the percentage of warmup, and the number of training rounds.

The parameters of the BERT model will affect the convergence effect of the model to a certain extent, so it is necessary to adjust the parameters to find out the appropriate configuration before model training. The main parameters involved include weight_decay, learning_rate, and warmup. When the percentage setting of warmup is not available, it may even cause the BERT model to fail to converge during the training process.

In addition, in the training process of the BERT model, attention must be paid to the size of the batch_size used for training. According to different scenes and text lengths, the value of the batch_size should be appropriately reduced or increased to maximize the use of video memory capacity.

Generally speaking, for the Medium model of BERT class, the use of single card at 2080ti level can meet most of the training scenarios, but if you need to use large sales of BERT model, you need to consider increasing the number of graphics cards, through multi-card training, to meet its dependence on video memory.

16.5   Model Integration

Under normal circumstances, the traditional model training scheme of natural language processing can adopt the integration strategy mentioned in the previous chapters, use weighted average, bagging, or stacking schemes for model integration, and train multiple models to construct meta-features, and two-layer models to output results.

For deep learning models, due to their own model complexity and time consumption, stacking scheme is not recommended for multi-fold cross-training of them. Under this condition, weighted average is a common scheme to merge the results of deep learning models.

16.6   A Summary of the Competition Question

16.6.1   More Schemes

16.6.1.1   Other Solutions to Pre-training Models

For deep learning models to be adopted, BERT family models are not the only selection. Rather, you can consider the most recent models including but not limited to RoBERTawmm-ext, GPT, GPT3, etc.

16.6.1.2   Other Schemes for Artificial Feature Extraction

For traditional text features extracted manually, schemes such as simhash and wordnet can also be adopted to calculate the degree of similarity between sentences or words, and at the same time judge the length gap of the text, count the punctuation marks in the text, and construct N-Gram terms for the text in advance. Schemes such as TF-IDF feature similarity calculation can also increase the generalization ability of the model.

16.6.1.3   The Pattern Mining of Problems

You can use Python code for sentence mining of problems and construct one-hot features to join the model for training.

16.6.1.4   Other Models of Traditional Machine Learning

Since tree models such as XGBoost and LightGBM are not suitable for training sparse features, and feature interactions cannot be captured with traditional linear models, FM models (Factorization Machine) can be employed for training, which not only effectively combines the feature interactions of polynomial degree 2, but also guarantees the linear time complexity $O(kn)$ calculated by itself to a certain extent, where k is the hidden_dim of the FM model.

16.6.2   Sorting Out Knowledge Points

16.6.2.1   Feature Engineering

There are three main types of features used in this chapter, namely, the original features of the text, the statistical features of the text, and the similarity features of the text.

16.6.2.2   In-Depth Learning

This chapter lists the construction of a variety of deep learning models for common natural language processing, the model construction based on PyTorch framework, and pre-training BERT model, etc.

16.6.2.3   Modeling Ideas

On the whole, the competition topic mentioned in this chapter is a relatively standard data mining and machine learning modeling problem. The distribution of its training set and testing set is highly coupled, so that participants only need to focus on depicting the user's own consumption behavior, and then train and predict the model through machine learning algorithm.

16.6.3   Extended Learning

For natural language processing models, in addition to the problems encountered in the Quora Question Pairs competition on the Kaggle platform, there are many other issues in practice. One of the most common problems data engineers in the industry encounter is how to build a model that meets both the online timeliness requirements and the accuracy requirements. Facing these issues, large-scale pre-training models like BERT with large parameters often cannot complete millisecond-level responses under given hardware conditions, and the TextCnn model that can achieve this timeliness requirement cannot capture the deeper semantic relationships and information behind the text during the training process due to its own complexity.

How to choose a suitable middle value between the predictive power of the model and the operating efficiency is a problem that often needs to be considered when building a model. Without thinking about enhancing hardware, the industry has proposed many common solutions to improve the efficiency of the model, such as tailoring the model or using a relatively small model for model distillation.

Here, additional problems related to model distillation are expanded. Common model distillation methods include output-based distillation and layer-by-layer distillation. For some tasks that require online efficiency and need to improve accuracy as much as possible, you can consider first using the BERT model for training to get a teacher model, and then building a convolutional neural network model or a recurrent neural network model as a student model for distillation. For the distillation of labels, it is often impossible to choose the appropriate evaluation index—what kind of loss function to use to assess that the teacher model and the student model are "similar"—resulting in unsatisfactory distillation results.

Therefore, there is another GAN-based distillation training scheme, in which
there is no need to define a loss function to evaluate the degree of similarity, but let
the discriminator learn it by itself. The following is a code based on the LSTM model
to distill the BERT pre-training results, which uses DRAGAN (Deep Regret Ana-
lytic GAN) to transform the loss function of GAN and obtain better distillation
results:

```python
if config.do_train:
    early_stop_rounds = 20
    best_loss = 999
    patient_count = 0
    for epoch in trange(int(200), desc="Epoch"):
        tr_loss = 0
        tr_loss_D = 0
    nb_tr_examples, nb_tr_steps = 0, 0
    lr_scheduler.step()

    model.train()
    tr_gen = test_batch_generator(data_tr,batch_size=batch_size,
        shuffle=True,maxlen=config.MAX_SEQUENCE_LENGTH,
        tokenizer=tokenizer,bert_pred = bert_prediction_tr,
        use_mlm=config.use_mlm,
        use_synonym_aug=config.use_synonym_aug)
    step = 0
    fgm = FGM(model)
    for batch in tr_gen:
        input_ids,label_ids,bert_pred_tr = batch[0],batch[1],batch[2]
        input_ids = torch.LongTensor(input_ids).to(device)
        label_ids = torch.LongTensor(label_ids).to(device)
        if bert_pred_tr is not None:
            bert_pred_tr = torch.FloatTensor(bert_pred_tr).to(device)

        bs = input_ids.size()[0]

    if config.use_distill:
    def train_discriminator(model,model_discriminator,
        input_ids,label_ids,bert_pred_tr,
          lambda_=0.25,K=5.0):
            logits = model(input_ids)

            real_teacher_label = np.ones(bert_pred_tr.size()[0])
            fake_student_label = np.zeros(logits.size()[0])

            bin_label = np.concatenate([real_teacher_label,
            fake_student_label])
            bin_label = torch.FloatTensor(bin_label).to(device)

            label_ids_class = torch.cat([label_ids,label_ids])

            x = torch.cat([bert_pred_tr,logits])
```

```
            D_real,real_classification = model_discriminator
 (bert_pred_tr)
            D_real_loss = loss_function_discriminator(D_real,
            torch.FloatTensor(real_teacher_label).to(device))

            D_fake,fake_classification = model_discriminator(logits)
            D_fake_loss = loss_function_discriminator(D_fake,
            torch.FloatTensor(fake_student_label).to(device))

            out_classification = torch.cat([real_classification,
            fake_classification])
            loss_clf = loss_function_classification(out_classification,
            label_ids_class)

            """ DRAGAN Loss ( gradient penalty term ) """
            alpha = torch.rand((bs, 1))
            alpha = alpha.to(device)
            x_p = bert_pred_tr + 0.5 * bert_pred_tr.std() *
            torch.rand(bert_pred_tr.size()).to(device)

            differences = x_p - bert_pred_tr
            interpolates = bert_pred_tr + (alpha * differences)
            interpolates.requires_grad = True
            pred_hat,_ = model_discriminator(interpolates)

            gradients = grad(outputs=pred_hat,
               inputs=interpolates,
               grad_outputs=torch.ones(pred_hat.size()).to(device),
                    create_graph=True, retain_graph=True,
 only_inputs=True)[0]

         gradient_penalty = lambda_ * ((gradients.view(gradients.size()[0],
               -1).norm(2, 1) - 1) ** 2).mean()

            loss_bin = (D_real_loss + D_fake_loss) + gradient_penalty

            return loss_bin,loss_clf

         loss_bin,loss_clf = train_discriminator(model,
 model_discriminator,
            input_ids,label_ids,bert_pred_tr)
         loss_discriminator = loss_bin*0.5+loss_clf*0.5
         loss_discriminator.backward()
         optimizer_D.step()

         tr_loss_D += loss_discriminator.item()

         def train_student(model,model_discriminator,
            input_ids,label_ids,bert_pred_tr,K=5.0):
```

```
          logits = model(input_ids)
          out_bin,out_classification = model_discriminator(logits)

          real_student_label = Variable(torch.ones(
            bert_pred_tr.size()[0]).to(device))
          if config.multi_bin:
            label_ids_one_hot = torch.zeros(bs, target_size).to
(device).
              scatter_(1, label_ids.view(-1,1), 1)
            loss = loss_function(logits,label_ids_one_hot)
            else:
            loss = loss_function(logits,label_ids)

        loss2 = loss_function_distil(logits,bert_pred_tr)

        loss_clf = loss_function_classification(
            out_classification,label_ids)
        loss_bin = loss_function_discriminator(out_bin,
real_student_label)

        return loss,loss2,loss_bin,loss_clf

        loss,loss2,loss_bin,loss_clf = train_student(model,
          model_discriminator,input_ids,label_ids,bert_pred_tr,)

        loss = loss+loss2+loss_bin*0.5-loss_clf*0.5

        loss.backward()

        if config.use_adv:
          fgm.attack()
          loss,loss2,loss_bin,loss_clf = train_student(model,
            model_discriminator,input_ids,label_ids,bert_pred_tr)
          loss_adv = loss+loss2+loss_bin*0.5-loss_clf*0.5
          loss_adv.backward()
          fgm.restore()
      else:
        logits = model(input_ids)

        if config.multi_bin:
          label_ids_one_hot = torch.zeros(bs, target_size).to(device).
            scatter_(1, label_ids.view(-1,1), 1)
          loss = loss_function(logits,label_ids_one_hot)
          loss.backward()
        else:
              loss = loss_function(logits,label_ids)
              loss.backward()

        if config.use_adv:
          fgm.attack()
          logits_adv = model(input_ids)
```

```
                    loss_adv = loss_function(logits_adv,label_ids)
                    loss_adv.backward()

                    fgm.restore()

            tr_loss += loss.item()

            torch.nn.utils.clip_grad_norm(model.parameters(),2)
            optimizer.step()

            nb_tr_examples += input_ids.size(0)
            nb_tr_steps += 1

            model.zero_grad()
            if config.use_distill:
                model_discriminator.zero_grad()

            step+=1

    tr_loss /= nb_tr_steps
    tr_loss_D /= nb_tr_steps
    model.eval()
    eval_loss,eval_acc = evaluation(data_te,model)
    if best_loss>eval_loss:
        best_loss = eval_loss
        torch.save(model.state_dict(), dump_path+generator_file_name)
        if use_distill:
          torch.save(model_discriminator.state_dict(),
            dump_path+discriminator_file_name )
        patient_count = 0
        else:
          if patient_count>=early_stop_rounds:
                break
          else:
                patient_count+=1
```

Printed in the United States
by Baker & Taylor Publisher Services